Sperm Quality: Past, Present and the Future Knowledge We Need

Sperm Quality: Past, Present and the Future Knowledge We Need

Editors

Pilar Santolaria Blasco
Rosaura Pérez-Pe
Jessica Rickard

MDPI • Basel • Beijing • Wuhan • Barcelona • Belgrade • Manchester • Tokyo • Cluj • Tianjin

Editors

Pilar Santolaria Blasco
Department of Animal Production and Food Sciences
Universidad de Zaragoza
Huesca
Spain

Rosaura Pérez-Pe
Dp. Biochemistry and Molecular and Cell Biology
Universidad de Zaragoza
Zaragoza
Spain

Jessica Rickard
School of Life and Environmental Sciences
The University of Sydney
Sydney
Australia

Editorial Office
MDPI
St. Alban-Anlage 66
4052 Basel, Switzerland

This is a reprint of articles from the Special Issue published online in the open access journal *Biology* (ISSN 2079-7737) (available at: www.mdpi.com/journal/biology/special_issues/sperm_quality).

For citation purposes, cite each article independently as indicated on the article page online and as indicated below:

LastName, A.A.; LastName, B.B.; LastName, C.C. Article Title. *Journal Name* **Year**, *Volume Number*, Page Range.

ISBN 978-3-0365-8437-9 (Hbk)
ISBN 978-3-0365-8436-2 (PDF)

Contents

About the Editors

Pilar Santolaria Blasco

Dr. Santolaria Blasco has an extensive research career in the field of animal reproduction. Her journey began in 1990 at the Department of Animal Production within the Faculty of Veterinary Medicine at the University of Zaragoza. Over the years, she has pursued further training and conducted research visits to various prestigious research centers around the world. These include the University of Bristol (UK), Institute de Recherches Agroalimentaires (France), Politechique Virginia University (USA) and the University of Liverpool (UK).

Her contributions have primarily been through her association with the research group BIOFITER (Biology, Physiology, and Technologies of Reproduction). This group has been dedicated to studying the reasons behind reduced reproductive performance after artificial insemination, focusing mainly on cows and sheep. The research conducted by the group spans more than three decades and covers various aspects of reproductive biology, including female infertility, pregnancy loss and male-associated factors.

Dr. Santolaria Blasco's specific area of expertise lies in the study of factors affecting male fertility. She has been involved in numerous projects aimed at understanding and improving male reproductive health. These projects encompass a wide range of topics, such as conducting epidemiologic surveys to identify different causes affecting reproductive efficacy, developing new techniques for assessing sperm quality, predicting male fertility using unconventional sperm analysis methods, enhancing male fertility through improved management practices and creating new semen diluents to optimize semen preservation.

Her dedication to the field has led her to achieve the position of Full Professor at the University of Zaragoza, where she continues to contribute to the advancement of knowledge in animal reproduction.

Rosaura Pérez-Pe

Dr Pérez-Pe began her research career in 1996 in the Department of Biochemistry, Molecular and Cellular Biology within the Faculty of Veterinary Medicine at the University of Zaragoza. She defended her doctoral thesis in 2001 which focused on the effect of seminal plasma on ram sperm functionality and was granted an extraordinary doctorate award. She completed further training and visits at the University College of Dublin (Ireland), Universidad Nacional de La Pampa (Argentina) and the University of Agricultural Sciences (Uppsala, Sweden). Her research has included investigating the ability of ram seminal plasma proteins to protect sperm from cryodamage, the isolation and cloning of specific proteins of interest, the development of new methods for analysing sperm functionality, including a free and open-source software (Open-CASA), predicting the fertility of seminal doses; examining molecular pathways involved in sperm capacitation, apoptosis and cold-shock, the effect of female hormones on ram sperm functionality, the role of melatonin and melatonin receptors on sheep reproduction and, finally, effects of dietary supplementation with plant by-products rich in phytomelatonin on the ram reproductive behavior and semen quality.

She has also developed several research contracts with companies within the sheep industry and is part of the research group BIOFITER (Biology, Physiology and Technologies of Reproduction) which is linked to the Environmental Sciences Institute of Aragon (IUCA). She has combined her research activity with teaching from 2001 and has been a full Professor at the University of Zaragoza since 2010.

Jessica Rickard

Following the completion of her PhD in 2014, Dr Rickard undertook a number of post-doctoral research positions at the University of Sydney. She has worked on a wide variety of projects related to applied artificial reproductive technologies in sheep such as the commercialization of sexed ram semen, the effect of heat stress on the Merino ewe, the use of transrectal ultrasound to assess early embryo loss, the role of Melatonin in advancing the ram breeding season, the use of on-animal sensors to assess ram-mating behaviour and factors affecting the survival of liquid stored ram spermatozoa. She is currently the McCaughey Senior Research Fellow at the University of Sydney and is leading a research program investigating factors affecting the success of artificial insemination within the Australian sheep industry. She is also continuing the work of her PhD, which focused on the role of ram seminal plasma, using a combined transcriptomic and proteomic approach to identify molecular sperm traits responsible for improved sperm survival and fertility in the ovine cervix. She is currently serving as a Vice chair and board director of Merinolink Pty Ltd and is a standing committee member of the International Congress of Animal Reproduction.

biology

Editorial

Understanding Sperm Quality for Improved Reproductive Performance

Pilar Santolaria [1,*], Jessica P. Rickard [2] and Rosaura Pérez-Pe [3]

1 Grupo BIOFITER, Departamento Producción Animal y Ciencia de los Alimentos, Instituto Universitario en Ciencias Ambientales de Aragón (IUCA), Escuela Politécnica Superior, Universidad de Zaragoza, 22071 Huesca, Spain
2 RMC Gunn Building, School of Life and Environmental Sciences, Faculty of Science, The University of Sydney, Sydney, NSW 2006, Australia; jessica.rickard@sydney.edu.au
3 Grupo BIOFITER, Departamento Bioquímica y Biología Molecular y Celular, Instituto Universitario en Ciencias Ambientales de Aragón (IUCA), Facultad de Veterinaria, Universidad de Zaragoza Miguel Servet, 50013 Zaragoza, Spain; rosperez@unizar.es
* Correspondence: psantola@unizar.es

The assessment of semen quality is used to identify factors that influence sperm performance and diagnose male infertility. From the earliest works to the present day, a considerable number of analytical techniques have been developed. These tests study many aspects of the morphology and physiology of the spermatozoan; however, their ability to predict male fertility remains low. Some of these techniques have been automated, which can make the test results more objective, but others still require subjective evaluation. The introduction of computer-assisted sperm analysis (CASA) systems and flow cytometry has revolutionized sperm quality analysis in recent decades. However, the use of these advanced techniques remains experimental, with only a few of them having successfully led to practical applications in routine commercial semen evaluation. Future sperm biology research should focus on developing analytical techniques that have a greater capacity to predict male fertility and can be used in both the laboratory and the field. Field adaptation will require progress in automation and simplification, to produce precise, economical and efficient techniques. Nevertheless, as technology advances in these analytical tests and research continues, our greater understanding of male fertility will aid the development of new methods of sperm evaluation.

This Special Issue of Biology entitled "Sperm quality: Past, present and the future knowledge we need" focuses on understanding the quality of spermatozoa in three subsections: (1) the function of the sperm cell, (2) its ability to withstand cryopreservation and (3) its performance both in vitro and in vivo. Combined, this Special Issue contains 11 published peer-reviewed papers.

Citation: Santolaria, P.; Rickard, J.P.; Pérez-Pe, R. Understanding Sperm Quality for Improved Reproductive Performance. *Biology* **2023**, *12*, 980. https://doi.org/10.3390/biology12070980

Received: 30 June 2023
Accepted: 3 July 2023
Published: 10 July 2023

1. Function of the Sperm Cell

An important endeavor in the creation of this Special Issue was to delve into aspects related to the biology and physiology of spermatozoa. In this regard, sperm capacitation is essential for the acquisition of fertilizing capacity, and, although it has been extensively studied in many species, there are still aspects that remain unclear. Many studies have been conducted to establish a suitable capacitation medium for different species, because effective in vitro sperm capacitation is required for successful in vitro fertilization (IVF). Most of the media examined include bovine serum albumin (BSA) and bicarbonate, but these compounds may not be necessary to capacitate sperm from all species. Chaves et al. [1] demonstrated that BSA is crucial for pig sperm to elicit in vitro capacitation and trigger subsequent progesterone-induced acrosome exocytosis. In contrast, although exogenous bicarbonate does not appear to be indispensable, it shortens the time needed to reach that capacitated status. The authors of this work concluded that media containing BSA and low levels of or no bicarbonate were the most suitable for inducing the capacitation of pig sperm, maintaining higher motility and plasma membrane integrity than those with high bicarbonate levels. These results do highlight the existence of species-specific mechanisms that are regulating the successful performance of sperm capacitation in vitro.

The development and maturation of spermatozoa, as well the acquisition of motility and capacitation, are strongly coordinated by sperm protein phosphorylation, among other protein post-translational modifications. Serrano et al. [2] reviewed and summarized the current knowledge of protein phosphorylation in human spermatozoa as the mechanism responsible for the regulation of processes necessary to achieve successful fertilization. The application of global phosphoproteomic profiling technology in evaluating sperm quality would enable the identification of male infertility biomarkers and could reveal insights into cases of idiopathic infertility in humans. In this regard, many couples attending infertility clinics are eventually enrolled in an intracytoplasmic sperm injection (ICSI) cycle after several failed attempts of pregnancy. In this procedure, the selection of the most adequate sperm to be injected inside the egg is crucial to the cycle's success. Magnetic activated cell sorting (MACS) is a technique that removes sperm that have begun undergoing an apoptotic process from a semen sample. This technique is not routinely performed prior to ICSI, but is usually offered as an option to increase the chances of success. In the retrospective, multicenter, observational study conducted by Gil-Juliá et al. [3], the researchers assessed the impact of MACS on reproductive outcomes. Their results suggested that non-apoptotic sperm selection via MACS prior to ICSI with autologous oocytes reduces the number of embryos required to be transferred in order to obtain a live birth, but the method has limited clinical impacts. Consequently, MACS should not be recommended to all infertile couples before performing an ICSI cycle. Interestingly, this technique in combination with the use of antibodies has been shown to be effective for sperm sexing. In the dairy industry, the use of sexed sperm to increase the percentage of female calves in offspring has been long considered. The technique currently used for this purpose is flow cytometry combined with a cell sorter which separates X-chromosome-bearing sperm (X-sperm) and Y-chromosome-bearing sperm (Y-sperm) based on the difference in their DNA content. Immunoseparation is presented as a cheaper and less stressful alternative for spermatozoa sexing by Sringarm et al. [4], who showed that the use of magnetic microbeads coupled with scFv antibodies against Y-sperm (PY-microbeads) produced 82.65% of X-chromosome sperm in the selected fraction with acceptable sperm quality.

2. Impact of Cryopreservation on Sperm Quality

The use of sperm preservation has long been considered a vital reproductive technology in human medicine, livestock artificial breeding and wildlife conservation programs. Cryopreservation, in particular, not only facilitates biobanks and the storage of precious genetic material but also maximizes long-term fertility and genetic gain. Ever since glycerol was discovered to be an effective cryoprotectant in 1947, the species-specific protocols and media used have undergone extensive research to optimize sperm survival and quality post-thaw. The basic principles involved in the freeze–thaw process, including the cooling curve, freezing rate and thawing, will always remain; however, as technology advances and procedures are automated, there is always the opportunity to update and improve the quality of spermatozoa post-thaw. Within this Special Issue, knowledge on the impact of cryopreservation on the metabolome of turkey spermatozoa was reported, recommendations were made regarding the use of the sperm-rich fraction or entire ejaculate for improved boar sperm storage and fertility, and a novel enzyme was suggested to reduce the viscosity associated with wildlife ejaculates, improving the cryosurvival and availability of rhinoceros spermatozoa for wildlife conservation programs. These scientific papers represent some of the compelling research being conducted to advance our knowledge and understanding of sperm quality post-cryopreservation.

Of all the animal industries, research on the impact of cryopreservation on avian sperm quality is likely to be the most limited in comparison to that of cattle, pigs or sheep. Paventi et al. [5] used nuclear magnetic resonance (NMR) to create a profile of turkey spermatozoa both prior to and after freezing. They observed a positive correlation between amino acid levels with physiological changes in sperm parameters, with the exception of glycine, which was revealed to have a negative association. Interestingly, glycine was also seen to increase following cryopreservation. From these results, not only was a metabolomic profile of turkey spermatozoa created, but authors also identified biomarkers for sperm freezability and quality. Furthermore, they identified potential strategies through the supplementation of metabolites which could be used to improve the success of turkey sperm cryopreservation.

Boar sperm storage protocols (both liquid and frozen) have undergone extensive development and optimization over time. A recent objective was to maximize the number of doses produced for artificial insemination from one ejaculate per male. Luongo et al. [6] investigated the impact of different boar ejaculate fractions on the quality of sperm liquid stored from the sperm-rich fraction. There was no impact of incubating the entire ejaculate or mixing the sperm-rich and intermediate fraction on sperm quality, pregnancy following artificial insemination or offspring survival. This suggests the possibility that boar ejaculates could be used more effectively for both sperm storage and insemination protocols, increasing the reproductive and economic efficiency of artificial reproductive technologies for the pig industry.

Ejaculates collected from wildlife species via electro-ejaculation are often highly viscous. It is suggested that this method of semen collection, artificially stimulates the accessory sex glands, altering the contribution of seminal plasma and physiological parameters of the ejaculate. Like that of the boar, the ejaculate of the rhinoceros is also fractionated, with a sperm-rich fraction predominantly being used for cryopreservation and artificial insemination. However, this renders the very expensive and labor-intensive procedure of collecting semen very inefficient, especially when considering the risk posed to valuable endangered animals. Any semen obtained from these viscous fractions are often of poor quality and do not survive the freezing process. Rickard et al. [7] investigated the use of the enzyme papain, which has successfully been used in alpaca freezing protocols, to reduce the viscosity of rhinoceros ejaculates. Here, they reported that papain not only increased the quantity of spermatozoa available for use from one ejaculate but also improved the motility and kinematics of spermatozoa post-thaw compared to those of the sperm-rich fraction, without detriments to viability, DNA integrity or acrosome integrity. This work substantially increases our knowledge of rhinoceros sperm quality during cryopreservation, the reduced viscosity of samples also enables the use of advanced semen assessment technology such as CASA and flow cytometry. It also raises the possibility of improving the ejaculate quality collected of other wildlife species for conservation and breeding purposes.

3. Relationship between Sperm Quality and Fertility

There is a direct relationship between sperm quality and fertility. Sperm quality refers to the health and characteristics of sperm cells, including their count, motility, morphology, and viability. These factors are crucial for successful the fertilization of an egg and the establishment of a viable pregnancy. Evaluating and addressing any issues with sperm quality through proper diagnosis and appropriate treatments can significantly improve the chance of inducing pregnancy. However, achieving satisfactory predictions based on the in vitro evaluation of sperm quality remains a challenge. Overall, the combination of scientific discoveries, technological innovations, advancements in optical physics, and computing techniques has significantly advanced our understanding of semen quality and its impact on fertility. These advancements could improve, significantly, the accuracy of semen analysis.

In this sense, testicular ultrasound, a non-invasive diagnostic procedure, could be a useful tool. Recent ultrasound-video analysis and software developments have allowed the visualization of tissue at the microscopic level. Carvajal-Serna et al. [8] revealed that echotexture analysis via ultrasound video could be a valuable tool for assessing the breeding soundness of rams. An increase in the number of white and grey pixels could indicate a decrease in seminal quality, and tubular density and lumen area could be predictors of good seminal quality. Therefore, ultrasound video analysis could be a useful tool for evaluating the fertility of rams, either for storage, artificial insemination or natural joining on the farm.

On the other hand, the analysis of sperm performance under in vitro conditions provides a good indication of fertilizing potential. Parameters such as motility, swimming kinetics, acrosome integrity, and ATP content are thus examined in efforts to characterize such potential. Hamster species are a good model for studying sperm parameters that are key determinants of fertilizing capacity because these species are at the higher end of the diversity of mammalian sperm morphology and performance. In vitro functional studies demand that sperm remain viable during a long period of time under conditions that resemble those in the female tract. Sperm from certain species require a supplementation of the incubation medium with factors that stimulate viability and motility, or that promote the acquisition of fertilizing capacity. Molecules important for sperm performance in hamsters have been identified, namely D-penicillamine, hypotaurine and epinephrine (PHE). Tourmente et al. [9] investigated the effect of PHE on spermatozoa from five hamster species incubated for up to 4 h, revealing that supplementation with a combination of D-penicillamine, hypotaurine and epinephrine maintains or improves the performance of spermatozoa from five hamster species in different manners, depending on the species.

Regarding the relationship between the quality and fertility parameters, Yániz et al. [10] investigated whether or not differences in bull fertility are associated with variations of sperm quality. Differences between high- and low-fertility bulls were found mainly in parameters related to sperm acrosome integrity when using a new fluorescence method that allowed the clear and precise detection of the sperm plasma membrane and acrosome: the ISAS3Fun method. It was concluded that the simultaneous assessment of sperm viability and acrosome integrity with the ISAS3Fun method is accurate and seems to have greater potential in discriminating between high- and low-fertility bulls than do more conventional in vitro sperm quality tests. These results may help to predict the breeding soundness of bulls used in artificial insemination, which is important for the dairy industry.

Finally, swine reproduction efficiency is determined by the fertility potential of the sow and sperm quality. The objective of Barquero et al.'s study [11] was to compare boar sperm motility and kinematic features to evaluate their relationship with reproductive success after artificial insemination. The movement patterns of boar ejaculates were analyzed using a computer-assisted semen analysis (CASA)-Mot system, and the kinematic values of ejaculate clusters were assessed. They showed that kinematic analysis of boar ejaculates reveals kinematically separate populations. There were also differences between the sperm kinematic variables in terms of sire line. However, there was no overall significant difference between dam lines assessed via multivariate procedures. The fertility variables characterized according to the sire genetic line did not show differences, except for the significantly fewer stillbirths in Pietrain boars. Sperm kinematic variables may provide the capacity to predict litter size variables, albeit a limited one. Nevertheless, the analysis of ejaculates, organizing them into clusters, did not provide the capacity to predict litter size variables.

Acknowledgments: The editors of this Special Issue would like to thank all the authors of the submitted works which have contributed to the excellent research contained within. The editors would also like to acknowledge the reviewers for donating their time and peer-reviewing.

Conflicts of Interest: The authors declare no conflict of interest.

References

1. Chaves, B.R.; Pinoti Pavaneli, A.P.; Blanco-Prieto, O.; Pinart, E.; Bonet, S.; Zangeronimo, M.G.; Rodríguez-Gil, J.E.; Yeste, M. Exogenous Albumin Is Crucial for Pig Sperm to Elicit In Vitro Capacitation Whereas Bicarbonate Only Modulates Its Efficiency. *Biology* **2021**, *10*, 1105. [CrossRef]
2. Serrano, R.; Garcia-Marin, L.J.; Bragado, M.J. Sperm Phosphoproteome: Unraveling Male Infertility. *Biology* **2022**, *11*, 659. [CrossRef] [PubMed]

3. Gil-Juliá, M.; Hervás, I.; Navarro-Gómez Lechón, A.; Quintana, F.; Amorós, D.; Pacheco, A.; González-Ravina, C.; Rivera-Egea, R.; Garrido, N. Sperm Selection by Magnetic-Activated Cell Sorting before Microinjection of Autologous Oocytes Increases Cumulative Live Birth Rates with Limited Clinical Impact: A Retrospective Study in Unselected Males. *Biology* **2021**, *10*, 430. [CrossRef] [PubMed]
4. Sringarm, K.; Thongkham, M.; Mekchay, S.; Lumsangkul, C.; Thaworn, W.; Pattanawong, W.; Rangabpit, E.; Rachtanapun, P.; Jantanasakulwong, K.; Sathanawongs, A.; et al. High-Efficiency Bovine Sperm Sexing Used Magnetic-Activated Cell Sorting by Coupling scFv Antibodies Specific to Y-Chromosome-Bearing Sperm on Magnetic Microbeads. *Biology* **2022**, *11*, 715. [CrossRef]
5. Paventi, G.; Di Iorio, M.; Rusco, G.; Sobolev, A.P.; Cerolini, S.; Antenucci, E.; Spano, M.; Mannina, L.; Iaffaldano, N. The Effect of Semen Cryopreservation Process on Metabolomic Profiles of Turkey Sperm as Assessed by NMR Analysis. *Biology* **2022**, *11*, 642. [CrossRef]
6. Luongo, C.; Llamas-López, P.J.; Hernández-Caravaca, I.; Matás, C.; García-Vázquez, F.A. Should All Fractions of the Boar Ejaculate Be Prepared for Insemination Rather Than Using the Sperm Rich Only? *Biology* **2022**, *11*, 210. [CrossRef] [PubMed]
7. Rickard, J.P.; Pool, K.; de Graaf, S.P.; Portas, T.; Rourke, N.; Wiesner, M.; Hildebrandt, T.B.; Göritz, F.; Hermes, R. Increasing the Yield and Cryosurvival of Spermatozoa from Rhinoceros Ejaculates Using the Enzyme Papain. *Biology* **2022**, *11*, 154. [CrossRef] [PubMed]
8. Carvajal-Serna, M.; Miguel-Jiménez, S.; Pérez-Pe, R.; Casao, A. Testicular Ultrasound Analysis as a Predictive Tool of Ram Sperm Quality. *Biology* **2022**, *11*, 261. [CrossRef] [PubMed]
9. Tourmente, M.; Sanchez-Rodriguez, A.; Roldan, E.R.S. Effect of Motility Factors D-Penicillamine, Hypotaurine and Epinephrine on the Performance of Spermatozoa from Five Hamster Species. *Biology* **2022**, *11*, 526. [CrossRef] [PubMed]
10. Yániz, J.L.; Palacín, I.; Silvestre, M.A.; Hidalgo, C.O.; Tamargo, C.; Santolaria, P. Ability of the ISAS3Fun Method to Detect Sperm Acrosome Integrity and Its Potential to Discriminate between High and Low Field Fertility Bulls. *Biology* **2021**, *10*, 1135. [CrossRef] [PubMed]
11. Barquero, V.; Roldan, E.R.S.; Soler, C.; Vargas-Leitón, B.; Sevilla, F.; Camacho, M.; Valverde, A. Relationship between Fertility Traits and Kinematics in Clusters of Boar Ejaculates. *Biology* **2021**, *10*, 595. [CrossRef] [PubMed]

biology MDPI

Article

Exogenous Albumin Is Crucial for Pig Sperm to Elicit In Vitro Capacitation Whereas Bicarbonate Only Modulates Its Efficiency

Bruna Resende Chaves [1,2,3], Ana Paula Pinoti Pavaneli [1,4], Olga Blanco-Prieto [5], Elisabeth Pinart [1,2], Sergi Bonet [1,2], Márcio Gilberto Zangeronimo [3], Joan E. Rodríguez-Gil [5,*,†] and Marc Yeste [1,2,*,†]

1 Biotechnology of Animal and Human Reproduction (TechnoSperm), Institute of Food and Agricultural Technology, University of Girona, ES-17003 Girona, Spain; brunarufla@gmail.com (B.R.C.); anap.pavaneli@gmail.com (A.P.P.P.); elisabeth.pinart@udg.edu (E.P.); sergi.bonet@udg.edu (S.B.)
2 Unit of Cell Biology, Department of Biology, Faculty of Sciences, University of Girona, ES-17003 Girona, Spain
3 Department of Veterinary Medicine, Federal University of Lavras, BR-37200-000 Lavras, MG, Brazil; zangeronimo@ufla.br
4 Laboratory of Andrology and Technology of Swine Embryos, Department of Animal Reproduction, School of Veterinary Medicine and Animal Science, University of São Paulo, BR-13635-900 Pirassununga, SP, Brazil
5 Unit of Animal Reproduction, Department of Animal Medicine and Surgery, Faculty of Veterinary Medicine, Autonomous University of Barcelona, ES-08193 Bellaterra (Cerdanyola del Vallès), Spain; obprieto@gmail.com
* Correspondence: juanenrique.rodriguez@uab.cat (J.E.R.-G.); marc.yeste@udg.edu (M.Y.); Tel.: +34-93-5811045 (J.E.R.-G.); +34-972-419514 (M.Y.)
† Joint senior authors.

Citation: Chaves, B.R.; Pinoti Pavaneli, A.P.; Blanco-Prieto, O.; Pinart, E.; Bonet, S.; Zangeronimo, M.G.; Rodríguez-Gil, J.E.; Yeste, M. Exogenous Albumin Is Crucial for Pig Sperm to Elicit In Vitro Capacitation Whereas Bicarbonate Only Modulates Its Efficiency. *Biology* **2021**, *10*, 1105. https://doi.org/10.3390/biology10111105

Academic Editors: Pilar Santolaria Blasco, Rosaura Pérez-Pé, Jessica Rickard and Paul S. Cooke

Received: 25 August 2021
Accepted: 22 October 2021
Published: 26 October 2021

Publisher's Note: MDPI stays neutral with regard to jurisdictional claims in published maps and institutional affiliations.

Simple Summary: In this work, we addressed if the presence of exogenous bicarbonate required for pig sperm capacitation, which is a necessary step to acquire fertilizing ability. While sperm incubated in media without BSA or BSA/bicarbonate did not achieve in vitro capacitation, those incubated with BSA reached that status under any bicarbonate concentration, even when bicarbonate was absent. Interestingly, there were differences related to the concentration of bicarbonate, since sperm incubated in media with BSA and with no bicarbonate or 5 mM bicarbonate showed lower overall efficiency in achieving in vitro capacitation than those incubated in the presence of BSA and higher concentration of bicarbonate. Additionally, at the end of the experiment, sperm incubated in the presence of BSA and 38 mM bicarbonate showed lower motility and plasma membrane integrity than those incubated in media with BSA and lower concentrations of bicarbonate. In conclusion, BSA is crucial in for pig sperm to elicit in vitro capacitation and trigger the subsequent progesterone-induced acrosome exocytosis. In contrast, although exogenous bicarbonate does not appear to be indispensable, it shortens the time needed to reach that capacitated status.

Abstract: This work sought to address whether the presence of exogenous bicarbonate is required for pig sperm to elicit in vitro capacitation and further progesterone-induced acrosome exocytosis. For this purpose, sperm were either incubated in a standard in vitro capacitation medium or a similar medium with different concentrations of bicarbonate (either 0 mM, 5 mM, 15 mM or 38 mM) and BSA (either 0 mg/mL or 5 mg/mL). The achievement of in vitro capacitation and progesterone-induced acrosomal exocytosis was tested through the analysis of sperm motility, plasma membrane integrity and lipid disorder, acrosome exocytosis, intracellular calcium levels, mitochondria membrane potential, O_2 consumption rate and the activities of both glycogen synthase kinase 3 alpha (GSK3α) and protein kinase A (PKA). While sperm incubated in media without BSA or BSA/bicarbonate, they did not achieve in vitro capacitation; those incubated in media with BSA achieved the capacitated status under any bicarbonate concentration, even when bicarbonate was absent. Moreover, there were differences related to the concentration of bicarbonate, since sperm incubated in media with BSA and with no bicarbonate or 5 mM bicarbonate showed lower overall efficiency in achieving in vitro capacitation than those incubated in the presence of BSA and 15 mM or 38 mM bicarbonate. Additionally, at the end of the experiment, sperm incubated in the presence of BSA and 38 mM bicarbonate showed significantly ($p < 0.05$) lower values of motility and plasma membrane integrity than those incubated in media with BSA and lower concentrations of bicarbonate. In conclusion, BSA is instrumental for pig sperm to elicit in vitro capacitation and trigger the subsequent progesterone-induced acrosome exocytosis. Furthermore, while exogenous bicarbonate does not seem to be essential to launch sperm capacitation, it does modulate its efficiency.

Keywords: sperm capacitation; pig; bicarbonate; bovine serum albumin (BSA); glycogen synthase kinase (GSK3); protein kinase A (PKA)

1. Introduction

Mammalian sperm are unable to fertilize oocytes upon ejaculation because, despite being mature and motile, they need to reside within the female reproductive tract and interact with that environment [1]. Within the female reproductive tract, sperm undergo capacitation, which was first described in the 1950s [2,3]. Following this, sperm are able to interact with oocyte vestments, trigger acrosome exocytosis and subsequently fuse with the oocyte plasma membrane (reviewed in [4]). The capacitated spermatozoon undergoes a series of functional changes that include, amongst others, modifications in the composition of the sperm plasma membrane, acrosome remodeling, an increase in mitochondrial activity and sperm motility, a noticeable rise in intracellular Ca^{2+} and ROS levels, and the tyrosine-phosphorylation of certain sperm proteins [5–10].

At present, sperm capacitation can be performed in vitro in several (including pigs), although not all, mammalian species. Because effective in vitro sperm capacitation is required for successful in vitro fertilization (IVF), the main attempts to establish a proper capacitation medium have been related to the development of this assisted reproductive technique. Most previous studies have aimed to set a medium mimicking the oviductal environment, where the capacitation status is fully achieved (reviewed in [11]). It is widely accepted that a capacitation medium should contain a protein source, which is usually bovine serum albumin (BSA), and a variety of ions, including bicarbonate (HCO_3^-) and Ca^{2+} [5,12,13]. During capacitation, removal of cholesterol from sperm plasma membrane is induced by BSA, which leads to the displacement and rearrangement of phospholipids, increases membrane fluidity, primes acrosome exocytosis and prepares sperm for the penetration of oocyte vestments (including zona pellucida) prior to gamete fusion [14–17]. Furthermore, high levels of bicarbonate destabilize the sperm plasma membrane and trigger a signaling cascade that involves a sperm-specific soluble adenylyl cyclase (sAC), the increase in cAMP levels and the activation of protein kinase A (PKA; [18]). The activation of PKA is one of the most important signals that elicits sperm capacitation and is also involved in triggering acrosome exocytosis [1,19,20]. Additionally, the concentration of bicarbonate changes across epididymal regions and in the oviductal fluid [19]. These variations play a crucial role in the modulation of sperm function, as they are involved in events such as sperm maturation during the epididymal transit, the acquisition of motility upon ejaculation and the modulation of capacitation timing within the female reproductive tract [19]. After ejaculation, sperm encounter a high bicarbonate concentration in the female tract, ranging between 35 mM and 90 mM [5,21]. In vitro, these variations in the concentration of bicarbonate do not exist, but rather sperm are exposed to a standard concentration throughout all the process [22], this scenario likely being one of the limiting factors in the efficiency of in vitro capacitation.

Previous research used different concentrations of bicarbonate and BSA to elicit in vitro capacitation in pig sperm. Whereas several studies used a medium containing 25 mM $NaHCO_3$ and 0.4% BSA [23], others used 36–38 mM $NaHCO_3$ and 0.5% of BSA [9,24–28]. In addition, other authors even found that a medium with 5 mg/mL BSA but without $NaHCO_3$ is able to induce in vitro capacitation in pig sperm [29,30]. These differences clearly indicate that while both BSA and $NaHCO_3$ are important in the achievement of in vitro capacitation in pig sperm, neither the precise role of these two effectors nor their particular relevance in the achievement of sperm capacitation are well known. Since, to the best of our knowledge, no previous study interrogated the specific importance of both exogenous bicarbonate and BSA during the achievement of in vitro capacitation, this work aimed to explain to which extent these two components are crucial for pig sperm to elicit in vitro capacitation and undergo progesterone-induced acrosome exocytosis.

2. Materials and Methods

2.1. Reagents

Unless otherwise declared, all the reagents were purchased from Sigma–Aldrich (Saint-Louis, MO, USA) and Merck (Darmstadt, Germany).

2.2. Semen Samples

Semen samples were obtained from 12 separate boars of proven fertility according to the farm records. Ejaculates were purchased from a local farm (Servicios Geneticos Porcinos, S.L., Roda de Ter, Spain), where boar semen was routinely collected twice a week through the gloved-hand method. Sperm-rich fractions were diluted in a commercial extender (Duragen®; Magapor, S.L.; Ejea de los Caballeros, Spain) to a final concentration of 3.3×10^7 sperm/mL. Samples were cooled to 17 °C, packed into 90 mL doses and finally transported in an insulated container at 17 °C for approximately 45 min, which was the time required to arrive at the laboratory. Upon arrival, sperm quality was evaluated to confirm that conventional spermiogram parameters were above quality standards.

All procedures involving animals were performed according to the EU Directive 2010/63/EU for animal experiments, the Animal Welfare Law issued by the Regional Government of Catalonia, and the current regulation on Health and Biosafety issued by the Department of Agriculture, Livestock, Food and Fisheries, Regional Government of Catalonia (Spain). In addition, the production of seminal doses by the farm followed the ISO certification (ISO-9001:2008). The authors of this study did not manipulate any animal but rather purchased the semen samples from a local farm, which operates under commercial standard conditions. For this reason, no approval from a local ethics committee was required.

2.3. In Vitro Capacitation and Progesterone-Induced Acrosome Exocytosis

All media were freshly prepared on the day of use and consisted of Tyrode's medium containing lactate and pyruvate, and different concentrations of bicarbonate and BSA. The composition of this basic medium was: 20 mM 4-(2-hydroxyethyl)-1-piperazineethanesulfonic acid (HEPES) buffer, 112 mM NaCl, 3.1 mM KCl, 5 mM glucose, 0.3 mM Na_2HPO_4, 0.4 mM $MgSO_4$, 4.5 mM $CaCl_2$, 21.7 mM sodium lactate and 1 mM sodium pyruvate. Eight different treatments were tested: (1) negative control (No BSA/No Bic), which was the basic medium without bicarbonate or BSA; (2) basic medium supplemented with 5 mg/mL BSA (BSA/No Bic); (3) basic medium supplemented with 5 mg/mL BSA and 5 mM $NaHCO_3$ (BSA + 5 mM Bic); (4) basic medium supplemented with 5 mM $NaHCO_3$ (No BSA + 5 mM Bic); (5) basic medium supplemented with 5 mg/mL BSA and 15 mM $NaHCO_3$ (BSA + 15 mM Bic); (6) basic medium supplemented with 15 mM $NaHCO_3$ (No BSA + 15 mM Bic); (7) basic medium supplemented with 5 mg/mL BSA and 38 mM $NaHCO_3$ (BSA + 38 mM Bic); and (8) basic medium supplemented with 38 mM $NaHCO_3$ (No BSA + 38 mM Bic). In all cases, pH was previously adjusted to 7.4 and media were equilibrated at 38.5 °C and 5% CO_2 for 60 min before use.

Seminal doses were separated from the diluent through centrifugation at $600\times g$ and 17 °C for 10 min, washed with PBS, and resuspended with the previously described media to a final concentration of 2×10^7 sperm/mL. Sperm were then incubated at 38.5 °C and 5% CO_2 in humidified air and evaluated after 0 min, 120 min and 240 min of incubation, since 240 min has been reported to be enough time for pig sperm to become in vitro capacitated [16,30]. After 240 min of incubation, progesterone was added to a final concentration of 10 μg/mL to induce acrosome exocytosis [31,32]. In order to evaluate the response of capacitated sperm to progesterone, especially with regard to acrosome exocytosis, samples were further incubated under the same conditions and evaluated after 5 min, 30 min and 60 min of progesterone addition. At each relevant time point, separate aliquots were taken for the assessment of sperm motility, plasma membrane integrity, membrane lipid disorder, intracellular calcium levels, mitochondrial membrane potential and O_2 consumption. Another aliquot was also taken for protein extraction and analysis. In this latter case, aliquots were centrifuged at 2400× *g* and 17 °C for 5 min and supernatants were discarded. Pellets were then stored at −80 °C until protein extraction.

2.4. Sperm Motility

Evaluation of sperm motility was performed using a CASA system (Integrated Sperm Analysis System V1.0; Proiser, Valencia, Spain). Briefly, 5 μL of each sample was placed onto a Makler counting chamber (Sefi-Medical Instruments). Samples were observed under a negative phase-contrast microscope (Olympus BX41; Olympus, Hamburg, Germany) at 100× magnification. Three replicates with a minimum of 1000 sperm per replicate were evaluated before calculating the corresponding mean ± SEM. The recorded sperm motility parameters were those described in Yeste et al. [33], namely: curvilinear velocity (VCL), which is the mean path velocity of the sperm head along its actual trajectory (μm/s); straight-line velocity (VSL), which is the mean path velocity of the sperm head along a straight line from its first to its last position (μm/s); average path velocity (VAP), which is the mean velocity of the sperm head along its average trajectory (μm/s); linearity coefficient (LIN), (VSL/VCL) × 100 (%); straightness coefficient (STR): (VSL/VAP) × 100 (%); wobble coefficient (WOB): (VAP/VCL) × 100 (%); mean amplitude of lateral head displacement (ALH), which is the mean value of the extreme side-to-side movement of the sperm head in each beat cycle (μm); and frequency of head displacement (BCF), which is the frequency with which the actual sperm trajectory crosses the average path trajectory (Hz). Total motility was defined as the percentage of sperm showing VAP > 10 μm/s, and progressive motility as the percentage of sperm exhibiting STR > 45%.

2.5. Flow Cytometry Analyses

Information about flow cytometry analyses is provided according to the recommendations of the International Society for Advancement of Cytometry (ISAC; [34]). Flow cytometry was used to evaluate plasma membrane integrity, membrane lipid disorder, acrosome integrity, mitochondrial membrane potential, and intracellular calcium levels in all samples and time-points. Prior to evaluation, sperm concentration was adjusted to 1×10^6 sperm/mL in a final volume of 0.5 mL [35]. Sperm were then stained with the appropriate combinations of fluorochromes, following the protocols described below. In all cases, a minimum of 10,000 events per replicate were evaluated through a Cell Lab QuantaSC Flow Cytometer (Beckman Coulter, Fullerton, CA, USA). Particles were excited through an argon ion laser (488 nm) set at a power of 22 mW and cell diameter/volume was directly measured employing the Coulter principle for volume assessment. The EV channel was periodically calibrated using 10-μm Flow-Check fluorospheres (Beckman Coulter) by positioning this size of the bead at channel 200 on the volume scale. To capture fluorochrome signals, three different optical filters were used: FL1 (green fluorescence)—Dichroic/Splitter, DRLP: 550 nm, BP filter: 525 nm, detection width 505–545 nm; FL2 (orange fluorescence)—DRLP: 600 nm, BP filter: 575 nm, detection width: 560–590 nm; FL3 (red fluorescence)—LP filter: 670 nm, detection width: 655–685 nm. Debris (particle diameter < 7 μm) and aggregates (particle diameter > 12 μm) were excluded from the analysis by gating out the particles based on EV/side scatter (SSC) plots. In all tests except SYBR14/PI, data were corrected following the procedure described by Petrunkina et al. [36]. Each assessment per sample and parameter was repeated three times in independent tubes and the corresponding mean ± SEM was calculated.

2.5.1. Evaluation of Sperm Membrane Integrity

Plasma membrane integrity was evaluated using the LIVE/DEAD Sperm Viability Kit (SYBR14/PI) following the protocol set by Garner and Johnson [37]. For this purpose, sperm samples were incubated with SYBR14 (final concentration: 100 nM) at 38 °C in darkness for 10 min and then with propidium iodide (PI; final concentration: 10 μM) at the same temperature for 5 min. Three sperm populations were identified in flow-cytometry dot plots: (i) membrane-intact sperm (green-stained), which were positive for SYBR14 and negative for PI ($SYBR14^+/PI^-$); (ii) non-viable sperm (red-stained), which were negative for SYBR14 and positive for PI ($SYBR14^-/PI^+$); and (iii) non-viable sperm (orange-stained), which were stained both in green and red ($SYBR14^+/PI^+$).

2.5.2. Evaluation of Sperm Membrane Lipid Disorder

Lipid disorder of plasma membrane was determined with Merocyanine 540 (M540) and YO-PRO-1, following the procedure set by Harrison et al. [38]. Briefly, sperm were stained with M540 (final concentration: 2.6 μM) and YO-PRO-1 (final concentration: 25 nM) and incubated at 38 °C for 10 min in the dark. A total of four sperm populations were identified: (i) non-viable sperm with low membrane lipid disorder ($M540^-/YO\text{-}PRO\text{-}1^+$); (ii) non-viable sperm with high membrane lipid disorder ($M540^+/YO\text{-}PRO\text{-}1^+$); (iii) viable sperm with low membrane lipid disorder ($M540^-/YO\text{-}PRO\text{-}1^-$); and (iv) viable sperm with high membrane lipid disorder ($M540^+/YO\text{-}PRO\text{-}1^-$).

2.5.3. Evaluation of Acrosome Integrity

Acrosome integrity was determined through staining with the lectin from *Arachis hypogaea* (peanut agglutinin, PNA) conjugated with fluorescein isothiocyanate (FITC) and ethidium homodimer (3,8-diamino-5-ethyl-6-phenylphenanthridinium bromide; EthD-1), following the protocol described by Rocco et al. [9]. Samples were incubated with EthD-1 (final concentration: 2.5 μg/mL) at 38 °C for 5 min in the dark. Following centrifugation at 2000× *g* and 17 °C for 30 s, sperm were resuspended with PBS containing 4 mg/mL BSA. Samples were again centrifuged under the same conditions and subsequently added to 100 μL of ice-cold methanol (100%) for 30 s. After centrifugation at 2000× *g* and 17 °C for 30 s, samples were resuspended with 250 μL PBS before adding 0.8 μL PNA-FITC (final concentration: 2.5 μM). Samples were incubated at 25 °C for 15 min in the dark and then centrifuged at 2000× *g* and 17 °C for 30 s. Pellets were resuspended with 0.6 μL PBS and centrifuged at 2000× *g* and 17 °C for 30 s. This step was repeated twice. Sperm were evaluated with the flow cytometer and four sperm populations were identified: (i) viable sperm with an intact acrosome ($PNA^+/EthD\text{-}1^-$); (ii) viable sperm with an exocytosed acrosome ($PNA^-/EthD\text{-}1^-$); (iii) non-viable sperm with an intact acrosome ($PNA^+/EthD\text{-}1^+$); and (iv) non-viable sperm with an exocytosed acrosome ($PNA^-/EthD\text{-}1^+$). Fluorescence of EthD-1 was detected through FL3, whereas that of PNA was detected through FL1.

2.5.4. Evaluation of Intracellular Calcium Levels

Intracellular calcium levels were evaluated through two different fluorochromes (Fluo3 and Rhod5; [7]). For Fluo3, sperm were incubated at 38 °C in the dark for 10 min with Fluo3-AM (final concentration: 1 μM) and PI (final concentration: 12 μM), which were detected through FL1 and FL3, respectively. Four sperm populations were identified: (i) viable sperm with low levels of intracellular calcium ($Fluo3^-/PI^-$), (ii) viable sperm with high levels of intracellular calcium ($Fluo3^+/PI^-$), (iii) non-viable sperm with low levels of intracellular calcium ($Fluo3^-/PI^+$), and (iv) non-viable sperm with high levels of intracellular calcium ($Fluo3^+/PI^+$). The geometric mean of Fluo3 intensity was recorded for all sperm populations.

With regard to Rhod5, sperm were incubated at 38 °C for 10 min in the dark with Rhod5-N (final concentration: 5 μM) and YO-PRO-1 (final concentration: 25 nM). Filters used to detect the fluorescence from Rhod5 and YO-PRO-1 were FL3 and FL1, respectively.

A total of four sperm populations were identified: (i) viable sperm with low levels of intracellular calcium (Rhod5$^-$/YO-PRO-1$^-$), (ii) viable sperm with high levels of intracellular calcium (Rhod5$^+$/YO-PRO-1$^-$), (iii) non-viable sperm with low levels of intracellular calcium (Rhod5$^-$/YO-PRO-1$^+$), and (iv) non-viable sperm with high levels of intracellular calcium (Rhod5$^+$/YO-PRO-1$^+$). The geometric mean of Rhod5 intensity was recorded for all sperm populations.

2.5.5. Evaluation of Mitochondrial Membrane Potential

Mitochondrial membrane potential (MMP) was determined through JC1 (5,5′,6,6′-tetrachloro-1,1′,3,3′-tetraethylbenzimidazolylcarbocyanine iodide) fluorochrome, following the protocol described by Guthrie and Welch [39]. For this purpose, sperm samples were incubated with JC1 (final concentration: 0.3 μM) at 38 °C in the dark for 30 min, and two sperm populations were distinguished: (i) sperm with high mitochondrial membrane potential, and (ii) sperm with low mitochondrial membrane potential. When MMP was high, JC1 inside mitochondria formed orange aggregates that were detected through FL2. Ratios between orange (FL2) and green (FL1) fluorescence in the sperm population with high MMP ($JC1_{agg}$) were also calculated.

2.6. Sodium Dodecyl Sulfate Polyacrylamide Gel Electrophoresis (SDS–PAGE)

Sperm pellets stored at −80 °C were thawed and resuspended in 400 μL ice-cold lysis buffer and maintained at 4 °C for 30 min under constant agitation. The lysis buffer was made up of 2% SDS, 1% Triton-X-100, 8 M urea, 2 mM dithiothreitol (DDT), 0.5% Tween 20 and 50 mM Tris-HCl; the pH was adjusted to 7.4. On the day of use, the lysis buffer was mixed with 1% commercial protease inhibitor cocktail (Sigma–Aldrich), 1% phenylmethanesulfonyl fluoride (PMSF), and 0.15% sodium orthovanadate. Following this, samples were homogenized by sonication (50% amplitude; 10 long-lasting pulses; Bandelin Sonopuls HD 2070; Bandelin Electronic GmbH and Co., Heinrichstrasse, Berlin), and then centrifuged at 10,000× *g* and 4 °C for 15 min. Supernatants were carefully collected and total protein was quantified in triplicate with a detergent-compatible protein assay (DC Protein Assay; BioRad, Hercules, CA, USA). Standard curves were produced with different concentrations of BSA (Quick Start Bovine Serum Albumin Standard; Bio-Rad).

For SDS–PAGE separation, 10 μg protein of each sample was diluted with 2× Laemmli reducing buffer containing 5% (*v:v*) β-mercaptoethanol (Bio-Rad Laboratories). Samples were boiled at 90 °C for 5 min and then loaded onto 1-mm SDS–PAGE gels, together with a molecular-weight marker (All Blue Precision Plus Protein Standards; Bio-Rad Laboratories). The separating gel contained 12% (*w:v*) acrylamide, whereas the stacking gel contained 5% (*w:v*) (Bio-Rad Laboratories). Subsequently, gel electrophoresis was run at constant voltage (80 V for stacking gels and 120 V for separating gels). Proteins were transferred onto polyvinylidene fluoride membranes (Immobilon-P; Millipore, Darmstadt, Germany) at 120 mA for 120 min. Thereafter, membranes were incubated with blocking solution, consisting of 5% BSA (*w:v*) in TBST at 4 °C overnight with agitation.

2.7. Immunoblotting of Tyr-P-GSK3

Membranes were incubated with a primary anti-phospo-GSK3 (Tyr279/Tyr216) antibody (Tyr-P-GSK3; clone 5G-2F, 05- 413, Millipore) diluted (1:1000; *v:v*) in blocking solution at room temperature and agitation for 60 min. After washing membranes four times (5 min each) in TBST, they were incubated with a secondary horseradish peroxidase (HRP)-conjugated anti-mouse antibody (P0260; Dako Denmark A/S, Glostrup, Denmark) diluted 1:5000 (*v:v*) in blocking solution at room temperature with agitation for 60 min. Membranes were subsequently washed eight times (5 min each) in TBST.

Reactive bands were visualized with a chemiluminescent substrate (Immobilon Western Chemiluniniscent HRP Substrate; Millipore) and scanned using G:BOX Chemi XL 1.4 (SynGene, Frederick, MT, USA) and GeneSys image acquisition software v1.2.8.0 (SynGene). Protein bands from scanned images were quantified through Quantity One v4.6.9 software package (Bio-Rad Laboratories, Inc.). Protein quantification was expressed as the total signal intensity inside the boundary of a band measured in pixel intensity units (density, square millimeter) minus the background signal, considered as 0 (white).

Data were normalized with α-tubulin following the stripping of the same membranes. In brief, membranes were stripped through two incubations at room temperature under agitation for 10 min with a stripping buffer containing 0.02 M glycine, 3% (*w:v*) SDS and 1% (*v:v*) Tween 20 (pH 2.2) in 100 mL Mili-Q water. Thereafter, membranes were washed twice in TBS at room temperature for 10 min and twice with TBST for 5 min. Following this, stripped membranes were incubated with an anti-α-tubulin mouse antibody (anti-α-tubulin antibody clone DM1A, MABT205; Millipore) diluted 1:5000 (*v:v*) in blocking solution at room temperature under agitation for 60 min. After washing with blocking solution twice (5 min each), membranes were incubated with a HRP-conjugated anti-mouse antibody (P0260; Dako Denmark A/S) diluted 1:15,000 (*v:v*) in blocking solution (*v:v*) at room temperature under agitation for 60 min. Reactive bands were visualized and band density calculated as previously described. Relative content of Tyr-P-GSK3α was normalized, dividing the intensity of Tyr-P-GSK3α by that of α-tubulin. Data were corrected to a basal arbitrary value of 100 for the control point, which corresponded to the incubation in the medium containing BSA and 38 mM bicarbonate at 0 min.

2.8. Immunoblotting of DARPP-32 (PKA)

The procedure was similar to that carried out to determine the Tyr-P-GSK3α/α-tubulin ratio. Thus, after the SDS-PAGE/membrane transference procedure conducted as described above, membranes were incubated with an anti-phospho-DARPP-32 (Thr75) primary antibody (Cell Signaling Technology; Leyden; The Netherlands) at 4 °C overnight. The dilution factor was 1:1000 (*v:v*). After three washes, membranes were incubated with a horseradish peroxidase (HRP)-conjugated donkey anti-rabbit secondary antibody (Dako; Glostrup, Denmark) at a dilution of 1:5000 (*v:v*) in blocking solution for 60 min. Membranes were washed six times and then revealed using a chemiluminescent HRP substrate (ImmunoCruz Western Blotting Luminol Reagent; Santa Cruz Biotechnology, Dallas, TX, USA). Protein quantification was expressed as the total signal intensity inside the boundary of a band measured in pixel intensity units (density, square millimeter) minus the background signal, considered as zero (white). Afterwards, the same membranes were reprobed with a specific total PKAα antibody. For this purpose, membranes were stripped through two incubations at room temperature under agitation for 10 min with a stripping buffer containing 0.02 M glycine, 3% (*w:v*) SDS and 1% (*v:v*) Tween 20 (pH 2.2) in 100 mL Mili-Q water. Afterwards, membranes were incubated with the specific anti-PKAα primary antibody (Cell Signaling Technology; final dilution: 1:1000, *v:v*) at 4 °C overnight. After three washes, membranes were incubated with a horseradish peroxidase (HRP)-conjugated donkey anti-rabbit secondary antibody (Dako; Glostrup, Denmark) at a dilution of 1:5000 (*v:v*) in blocking solution for 60 min. Membranes were washed six times and were then revealed using a chemiluminescent HRP substrate (ImmunoCruz Western Blotting Luminol Reagent; Santa Cruz Biotechnology, Dallas, TX, USA). Reactive bands were visualized and their densities were calculated as previously described. At this point, the DARPP-32/PKA ratio was calculated by dividing the intensity of DARPP-32 by that of PKA. Data were corrected to a basal arbitrary value of 100 for the control point, which corresponded to the incubation in the medium containing BSA and 38 mM bicarbonate at 0 min.

2.9. Determination of O_2 Consumption Rate

Determination of O_2 consumption rate was carried out through a SensorDish® Reader (SDR) system (PreSens Gmbh; Regensburg, Germany). Briefly, 1 mL of each sample was incubated in four separate media (No BSA/No Bic, BSA/No Bic, BSA + 38 mM Bic, and No BSA + 38 mM Bic) for 0 min or 240 min. Following this, samples were transferred onto Oxodish® OD24 plates (24 wells/plate) specifically designed for this device. Plates were sealed with Parafilm®, loaded into the SDR system, and incubated at 38.5 °C (controlled atmosphere) for 120 min. During that period, O_2 concentration was recorded in each well at a rate of one reading/min. Results were exported to an Excel file and O_2 consumption rate was normalized against the total number of viable sperm per sample, determined through flow cytometry (SYBR14$^+$/PI$^-$ sperm) using another aliquot. For this experiment, only a single concentration of bicarbonate (38 mM) was used. This was because, from an operative point of view, the analysis of all bicarbonate concentrations at all incubation times was not feasible. Thus, the authors decided to focus on the highest concentration (38 mM) and on the outcomes after 240 min of incubation at 38.5 °C and at a 5% CO_2 atmosphere, as this is the time needed for pig sperm to achieve the capacitated status.

2.10. Statistical Analyses

Data were analyzed with a statistical package (IBM SPSS for Windows, ver. 25.0; Armonk, NY, USA) and are shown as mean ± standard error of the mean (SEM). Data were first checked for normality (Shapiro–Wilk test) and homogeneity of variances (Levene test). Following this, a linear mixed model (repeated measures) was run with the treatment as the fixed-effects inter-subject factor and the incubation time as the intra-subjects factor. Sidak's post-hoc test was run for pair-wise comparisons. The level of significance was set at $p \leq 0.05$.

3. Results

3.1. Effects of Different Concentrations of Bicarbonate and BSA on Plasma Membrane Integrity

Plasma membrane integrity during in vitro capacitation and after the addition of progesterone was evaluated through SYBR14/PI, as shown in Figure 1 (mean ± SEM). A significant ($p < 0.05$) reduction in the percentages of sperm with an intact plasma membrane (SYBR14$^+$/PI$^-$) was observed in treatments containing 38 mM of bicarbonate, with or without BSA, after 120 min of starting the experiment. After 120 min and 240 min of incubation, all media containing either no bicarbonate or bicarbonate at 5 mM presented significantly ($p < 0.05$) higher percentages of sperm with an intact plasma membrane than treatments containing 15 mM or 38 mM bicarbonate. After 300 min of incubation, the medium without BSA and without bicarbonate had a significantly ($p < 0.05$) higher percentage of sperm with an intact plasma membrane than the other treatments.

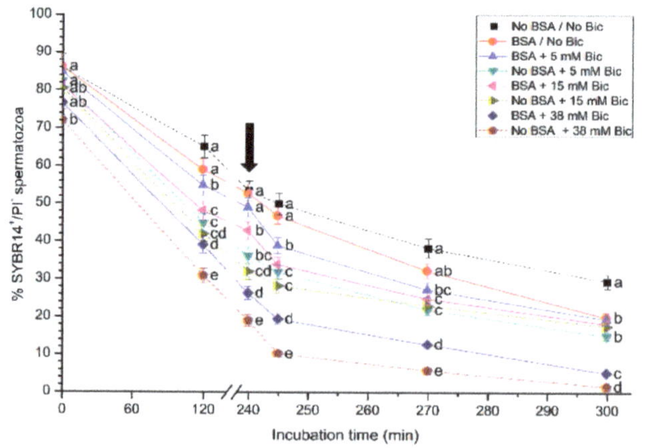

Figure 1. Percentages of sperm with an intact plasma membrane (SYBR14$^+$/PI$^-$) following incubation with different concentrations of bicarbonate (Bic; 0 mM, 5 mM, 15 mM, 38 mM) and BSA (0, 5 mg/mL) for 300 min. The black arrow indicates the time at which 10 µg/mL of progesterone was added (i.e., 240 min) to induce acrosome exocytosis. Different superscripts (a–e) indicate significant differences ($p < 0.05$) between treatments within the same time point. Data are shown as mean ± SEM for 12 independent experiments.

3.2. Effects of Different Concentrations of Bicarbonate and BSA on Membrane Lipid Disorder

The impact of different bicarbonate and BSA concentrations on the lipid disorder of sperm membrane is shown in Figure 2. Percentages of viable sperm with low lipid membrane disorder ($M540^-$/YO-PRO-1^-; Figure 2a) after 120 min of incubation were significantly higher ($p < 0.05$) in the treatments without bicarbonate, either with or without BSA, than in those containing bicarbonate at 5 mM, 15 mM and 38 mM. Additionally, percentages of viable sperm with low membrane lipid disorder in the medium without BSA and with bicarbonate at 38 mM were significantly ($p < 0.05$) lower than in the other treatments throughout the entire period of incubation.

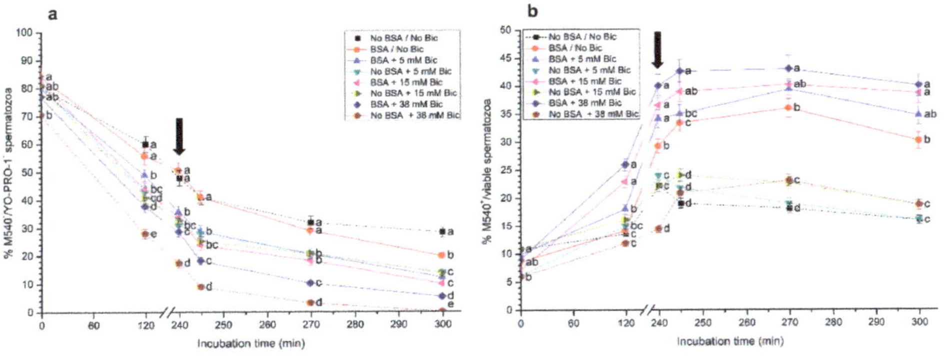

Figure 2. (**a**) Percentages of viable sperm with low membrane lipid disorder ($M540^-$/YO-PRO-1^-), and (**b**) percentages of sperm with high membrane lipid disorder considering the viable sperm population only ($M540^+$/viable sperm) following incubation with different concentrations of bicarbonate (Bic; 0 mM, 5 mM, 15 mM, 38 mM) and BSA (0, 5 mg/mL) for 300 min. The black arrow indicates the time at which 10 µg/mL of progesterone was added (i.e., 240 min) to induce acrosome exocytosis. Different superscripts (a–e) indicate significant differences ($p < 0.05$) between treatments within the same time point. Data are shown as mean ± SEM for 12 independent experiments.

On the other hand, when percentages of sperm with high membrane lipid disorder ($M540^+$) were calculated considering the viable sperm population only (YO-PRO-1^-), the media containing BSA, with or without bicarbonate, showed significantly ($p < 0.05$) higher values of this parameter than those without BSA, from 240 min and until the end of the incubation period (Figure 2b). Remarkably, treatments containing BSA and bicarbonate at 5 mM, 15 mM or 38 mM showed significantly ($p < 0.05$) higher percentages of $M540^+$/viable sperm than those containing BSA and no bicarbonate throughout all the incubation time.

3.3. Effects of Different Concentrations of Bicarbonate and BSA on Acrosome Integrity

As shown in Figure 3a, percentages of viable sperm with an intact acrosome (PNA^+/EthD-1^-) in the medium containing BSA and 38 mM were significantly ($p < 0.05$) lower than in the other treatments throughout the entire incubation period. In contrast, media without bicarbonate, either with or without BSA, showed significantly ($p < 0.05$) higher percentages of viable sperm with an intact acrosome than the other treatments. In addition, it was observed that the higher the concentration of bicarbonate, the lower the percentages of viable sperm with an intact acrosome. With regard to the percentages of viable sperm with an exocytosed acrosome (PNA^-/EthD-1^-), the presence of BSA in the media containing the same concentration of bicarbonate (i.e., 0 mM, 5 mM, 15 mM and 38 mM) led to significantly ($p < 0.05$) higher percentages of this sperm population after 240 min, 245 min, 270 min and 300 min of incubation (Supplementary Table S1).

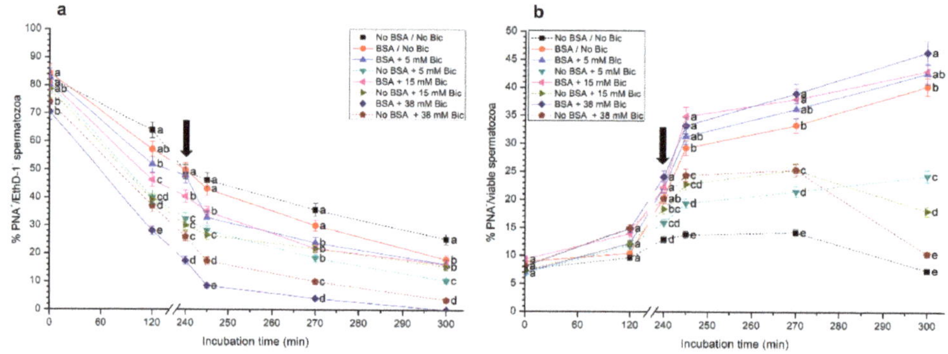

Figure 3. (**a**) Percentages of viable sperm with an intact acrosome ($PNA^+/EthD\text{-}1^-$), and (**b**) percentages of sperm with an exocytosed acrosome considering the viable sperm population only (PNA^-/viable sperm) following incubation with different concentrations of bicarbonate (Bic; 0 mM, 5 mM, 15 mM, 38 mM) and BSA (0, 5 mg/mL) for 300 min. The black arrow indicates the time at which 10 µg/mL of progesterone was added (i.e., 240 min) to induce acrosome exocytosis. Different superscripts (a–e) indicate significant differences ($p<0.05$) between treatments within the same time point. Data are shown as mean ± SEM for 12 independent experiments.

When percentages of sperm with an exocytosed acrosome (PNA^-) were calculated considering the viable sperm population ($EthD\text{-}1^-$), no significant differences between treatments were observed at 0 min and 120 min (Figure 3b). However, treatments containing BSA, with or without bicarbonate, showed significantly ($p < 0.05$) higher percentages of viable sperm with an exocytosed acrosome (PNA^-/viable sperm) than treatments without BSA, right after the addition of progesterone (245 min), and after 270 min and 300 min of incubation. In addition, percentages of viable sperm with an exocytosed acrosome were significantly ($p < 0.05$) higher in the medium with BSA and 38 mM than in that with BSA but without bicarbonate, from 245 min to 300 min of incubation.

3.4. Effects of Different Concentrations of Bicarbonate and BSA on Intracellular Calcium Levels

Percentages of viable sperm with high intracellular calcium levels (%$Fluo3^+/PI^-$ and %$Rhod5^+/YO\text{-}PRO\text{-}1^-$ sperm) are shown in Figure 4. Significantly ($p < 0.05$) higher percentages of $Fluo3^+/PI^-$ sperm were observed in the medium containing BSA and 38 mM bicarbonate from 0 min to 300 min of incubation, with maximum values being observed after 5 min of progesterone addition (i.e., 245 min; Figure 4a). On the other hand, the medium without BSA and with 5 mM bicarbonate showed the lowest percentage of $Fluo3^+/PI^-$ sperm when compared to the other experimental conditions.

In a similar fashion to that observed for Fluo-3/PI staining, percentages of viable sperm with a positive Rhod5 signal ($Rhod5^+/YO\text{-}PRO\text{-}1^-$) progressively increased when sperm were incubated in media containing BSA ($p < 0.05$), especially at 38 mM bicarbonate (Figure 4b). The addition of progesterone induced a rapid increase in these percentages, which reached the highest values 5 min after progesterone addition (i.e., 245 min of incubation). Moreover, the lowest percentages of $Rhod5^+/YO\text{-}PRO\text{-}1^-$ sperm from 0 min to 300 min of incubation were observed in the media without BSA, especially when containing 5 mM, 15 mM and 38 mM bicarbonate.

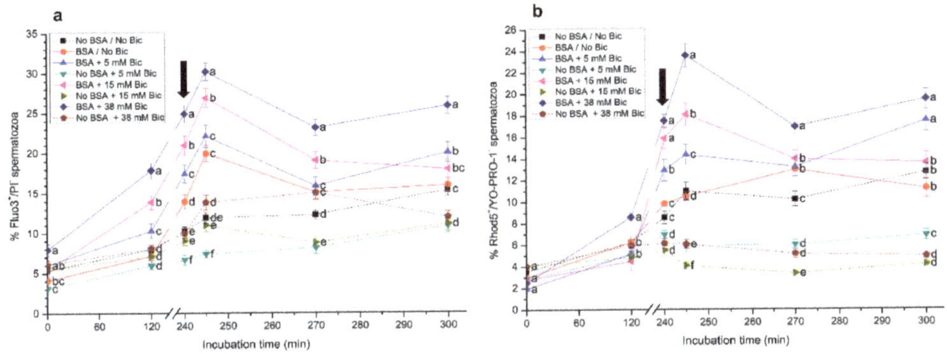

Figure 4. Percentages of viable sperm with high intracellular calcium levels evaluated through Fluo3^{+}/PI^{-} (**a**) and Rhod5^{+}/YO-PRO-1^{-} (**b**) following incubation with different concentrations of bicarbonate (Bic; 0 mM, 5 mM, 15 mM, 38 mM) and BSA (0, 5 mg/mL) for 300 min. The black arrow indicates the time at which 10 µg/mL of progesterone was added (i.e., 240 min) to induce acrosome exocytosis. Different superscripts (a–e) indicate significant differences ($p < 0.05$) between treatments within the same time point. Data are shown as mean ± SEM for 12 independent experiments.

3.5. Effects of Different Concentrations of Bicarbonate and BSA on Mitochondrial Membrane Potential

Incubation of pig spermatozoa with BSA and 38 mM bicarbonate significantly ($p < 0.05$) increased the percentage of spermatozoa with high mitochondrial membrane potential (MMP) from 0 min to 300 min of incubation (Figure 5a). The addition of progesterone after 240 min of incubation induced a rapid increase in this percentage, which was followed by a decrease at 270 min. On the other hand, while incubation of sperm in a medium without BSA and without bicarbonate (No BSA/No Bic) prevented the increase in the percentage of sperm with high MMP from 0 min to 240 min of incubation, the addition of progesterone to this medium after 240 min of incubation also increased that percentage (Figure 5a). Remarkably, percentages of sperm with high MMP did not differ between the medium without BSA/bicarbonate (No BSA/No Bic) and the medium with BSA and 38 mM bicarbonate (BSA + 38 mM Bic) after 270 min of incubation. At 300 min, sperm incubated with BSA and 38 mM bicarbonate showed a significantly ($p < 0.05$) higher percentage of sperm with high MMP than in the other experimental conditions. Furthermore, the addition of progesterone to the medium with BSA and 15 mM bicarbonate after 240 min of incubation also showed an increased percentage of sperm with high MMP. However, the extent of that increase was significantly ($p < 0.05$) lower than that observed in the medium containing BSA and 38 mM bicarbonate. Media without BSA showed, in all cases, significantly ($p < 0.05$) lower percentages of sperm with high MMP than those containing both BSA and bicarbonate (Figure 5a).

Figure 5b shows the FL2:FL1 ratio in the sperm population with high MMP. After 240 min of incubation, these ratios were significantly ($p < 0.05$) higher in the media with BSA than in those without this protein, especially when exogenous bicarbonate was present. Following the addition of progesterone, the treatments containing BSA and bicarbonate at 15 mM (BSA + 15 mM Bic) or 38 mM (BSA + 38 mM Bic) showed significantly ($p < 0.05$) higher FL2:FL1 ratios than the other media. In spite of this, after 270 min of incubation, the media containing BSA either with (38 mM) or without bicarbonate, showed significantly ($p < 0.05$) higher FL2:FL1 ratios than the media without BSA. At 300 min, no significant differences between treatments were observed (Figure 5b).

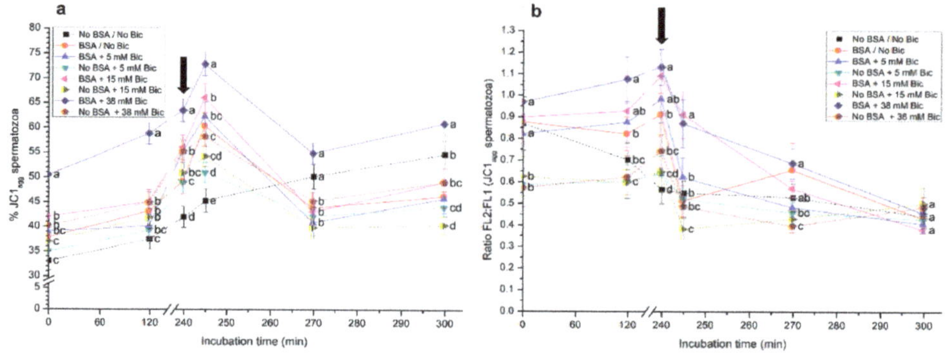

Figure 5. (**a**) Percentages of sperm with high mitochondrial membrane potential (MMP; $JC1_{agg}$), and (**b**) FL2:FL1 (orange:green) ratios in the sperm population with high MMP ($JC1_{agg}$) following incubation with different concentrations of bicarbonate (Bic; 0 mM, 5 mM, 15 mM, 38 mM) and BSA (0, 5 mg/mL) for 300 min. The black arrow indicates the time at which 10 µg/mL of progesterone was added (i.e., 240 min) to induce acrosome exocytosis. Different superscripts (a–d) indicate significant differences ($p < 0.05$) between treatments within the same time point. Data are shown as mean ± SEM for 12 independent experiments.

3.6. Effects of Different Concentrations of Bicarbonate and BSA on Sperm Motility

In all media, total sperm motility decreased throughout incubation time, reaching minimal values at the end of the experiment (Figure 6a). In addition, percentages of total motile spermatozoa incubated in media containing BSA and bicarbonate at either 0 mM or 5 mM were significantly ($p < 0.05$) higher than in the other treatments. In contrast, incubation of spermatozoa in the medium with 38 mM bicarbonate and no BSA (No BSA + 38 mM Bic) showed the lowest percentage of total motile sperm, with almost complete immobilization at the end of the experiment (Figure 6a). Furthermore, sperm agglutination was very high in treatments containing 38 mM, either with or without BSA, from 120 min and 300 min of incubation.

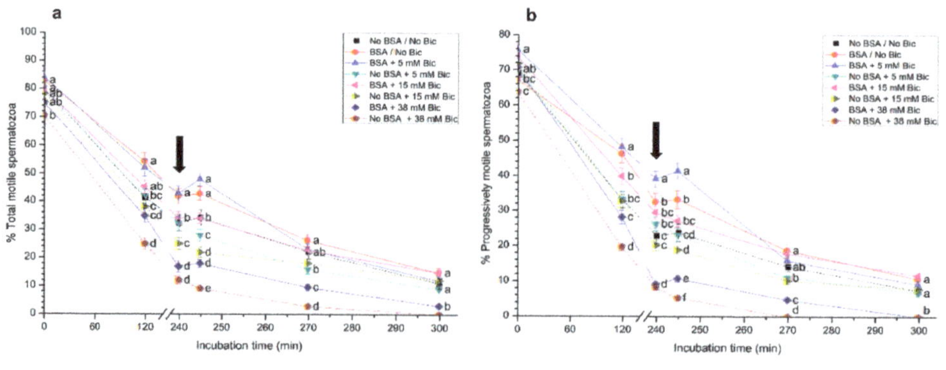

Figure 6. (**a**) Percentages of total and (**b**) progressively motile sperm following incubation with different concentrations of bicarbonate (Bic; 0 mM, 5 mM, 15 mM, 38 mM) and BSA (0, 5 mg/mL) for 300 min. The black arrow indicates the time at which 10 µg/mL of progesterone was added (i.e., 240 min) to induce acrosome exocytosis. Different superscripts (a–f) indicate significant differences ($p < 0.05$) between treatments within the same time point. Data are shown as mean ± SEM for 12 independent experiments.

Similar results were observed for the percentages for progressively motile spermatozoa (Figure 6b). Again, the medium with BSA and 5 mM bicarbonate presented the highest progressive sperm motility, followed by the medium with BSA and without bicarbonate. In contrast, the medium containing 38 mM bicarbonate, either with or without BSA, showed the lowest percentages of progressively motile sperm.

As shown in Table 1, VCL was significantly ($p < 0.05$) higher in treatments containing BSA than in those that did not contain this protein after 240 min, 245 min, 270 min and 300 min of incubation. Although the treatment with BSA and the highest concentration of bicarbonate (i.e., 38 mM) showed the highest VCL after 240 min, 245 min, 270 min and 300 min of incubation, significant ($p < 0.05$) differences were only observed when the medium with BSA and without bicarbonate was compared with that containing BSA and 38 mM bicarbonate. On the other hand, sperm incubated in media with BSA, regardless of whether or not they contained bicarbonate, showed significantly ($p < 0.05$) higher VSL and VAP than those incubated without BSA, from 240 min to 300 min of incubation (Table 1).

Table 1. Curvilinear velocity (VCL), straight-line velocity (VSL) and average path velocity (VAP) of pig sperm following incubation with different concentrations of bicarbonate (Bic; 0 mM, 5 mM, 15 mM, 38 mM) and BSA (0, 5 mg/mL) for 300 min. Progesterone (10 µg/mL) was added at 240 min to induce acrosome exocytosis. Different superscripts (a–d) indicate significant differences ($p < 0.05$) between treatments (rows) within the same time point. Data are shown as mean ± SEM for 12 independent experiments.

Media	0 Min	120 Min	240 Min	245 Min	270 Min	300 Min
			VCL (µm/s)			
No BSA/No Bic	52.5 ± 3.1 ^{a}	49.2 ± 2.8 ^{a,b}	41.8 ± 2.5 ^{a}	36.7 ± 2.1 ^{a}	28.1 ± 1.6 ^{a}	9.2 ± 0.6 ^{a}
BSA/No Bic	63.4 ± 3.5 ^{b}	56.1 ± 2.9 ^{b,c}	54.3 ± 3.1 ^{b}	51.2 ± 3.3 ^{b}	49.5 ± 2.8 ^{b}	47.8 ± 3.0 ^{b}
BSA + 5 mM Bic	74.8 ± 3.7 ^{c}	62.3 ± 3.5 ^{c,d}	56.5 ± 3.1 ^{b}	55.1 ± 3.0 ^{b,c}	52.2 ± 3.2 ^{b,c}	49.7 ± 2.8 ^{b,c}
No BSA + 5 mM Bic	75.2 ± 4.2 ^{c}	44.5 ± 2.8 ^{a}	39.7 ± 2.4 ^{a}	38.1 ± 2.1 ^{a}	35.4 ± 2.3 ^{a}	10.3 ± 0.4 ^{a}
BSA + 15 mM Bic	76.2 ± 4.1 ^{c}	63.7 ± 3.8 ^{c,d}	58.5 ± 3.3 ^{b}	57.8 ± 3.1 ^{b,c}	54.0 ±3.3 ^{b,c}	51.3 ± 2.9 ^{b,c}
No BSA + 15 mM Bic	70.8 ± 4.6 ^{b,c}	52.3 ± 2.9 ^{a,b}	41.2 ± 2.2 ^{a}	39.8 ± 2.3 ^{a}	32.3 ±1.8 ^{a}	8.9 ± 0.4 ^{a}
BSA + 38 mM Bic	75.9 ± 4.3 ^{c}	66.3 ± 4.0 ^{d}	60.5 ± 3.8 ^{b}	62.1 ± 3.5 ^{c}	59.3 ±3.4 ^{c}	55.6 ± 3.8 ^{c}
No BSA + 38 mM Bic	67.4 ± 3.8 ^{b,c}	54.2 ± 3.1 ^{b}	34.8 ± 2.1 ^{a}	32.2 ± 1.9 ^{a}	19.1 ±1.1 ^{d}	3.0 ± 0.2 ^{c}
			VSL (µm/s)			
No BSA/No Bic	25.4 ± 1.6 ^{a}	19.2 ± 1.1 ^{a}	13.4 ± 0.9 ^{a}	10.3 ± 0.6 ^{a}	6.7 ± 0.4 ^{a}	3.8 ± 0.1 ^{a}
BSA/No Bic	33.8 ± 2.0 ^{a}	27.5 ± 1.8 ^{b}	25.6 ± 1.5 ^{b}	25.1 ± 1.5 ^{b}	23.2 ± 1.4 ^{b}	20.9 ± 1.3 ^{b}
BSA + 5 mM Bic	32.5 ± 2.1 ^{a}	29.8 ± 1.9 ^{b}	28.3 ± 1.7 ^{b}	26.4 ± 1.4 ^{b}	22.3 ± 1.3 ^{b}	19.5 ± 1.2 ^{b}
No BSA + 5 mM Bic	30.1 ± 1.8 ^{a}	23.2 ± 1.4 ^{a,b}	18.4 ± 1.2 ^{a}	15.6 ± 0.9 ^{a}	9.6 ± 0.5 ^{a}	3.2 ± 0.2 ^{a}
BSA + 15 mM Bic	31.4 ± 2.1 ^{a}	30.3 ± 1.9 ^{b}	26.2 ± 1.7 ^{b}	25.8 ± 1.8 ^{b}	27.1 ± 1.5 ^{b}	22.9 ± 1.3 ^{b}
No BSA + 15 mM Bic	30.5 ± 2.2 ^{a}	18.3 ± 1.2 ^{a}	17.1 ± 1.0 ^{a}	12.8 ± 0.8 ^{a}	7.2 ± 0.5 ^{a}	1.8 ± 0.1 ^{a}
BSA + 38 mM Bic	33.5 ± 2.1 ^{a}	31.8 ± 2.3 ^{b}	25.6 ± 1.7 ^{b}	23.4 ± 1.4 ^{b}	20.2 ± 1.3 ^{b}	18.3 ± 1.0 ^{b}
No BSA + 38 mM Bic	29.2 ± 2.0 ^{a}	26.7 ± 1.8 ^{a,b}	15.4 ± 1.1 ^{a}	11.3 ± 0.7 ^{a}	9.6 ± 0.6 ^{a}	1.4 ± 0.1 ^{a}
			VAP (µm/s)			
No BSA/No Bic	40.3 ± 2.5 ^{a}	34.8 ± 2.2 ^{a}	26.2 ± 1.8 ^{a}	27.1 ± 1.7 ^{a}	19.2 ± 1.2 ^{a}	5.6 ± 0.4 ^{a}
BSA/No Bic	42.8 ± 2.8 ^{a,b}	41.3 ± 2.7 ^{a,b}	38.9 ± 2.5 ^{b,c}	40.1 ± 2.8 ^{b}	39.5 ± 2.5 ^{b}	37.0 ± 2.6 ^{b}
BSA + 5 mM Bic	46.1 ± 2.6 ^{a,b}	43.5 ± 2.4 ^{b}	40.6 ± 2.2 ^{c}	48.2 ± 2.7 ^{b}	42.5 ± 2.3 ^{b}	43.4 ± 2.4 ^{b}
No BSA + 5 mM Bic	44.1 ± 2.5 ^{a,b}	39.2 ± 2.3 ^{a,b}	30.1 ± 2.1 ^{a}	21.4 ± 1.3 ^{a}	18.5 ± 1.1 ^{a}	7.6 ± 0.5 ^{a}
BSA + 15 mM Bic	46.7 ± 2.8 ^{a,b}	40.1 ± 2.5 ^{a,b}	38.4 ± 2.3 ^{b,c}	43.2 ± 2.6 ^{b}	37.0 ± 2.2 ^{b}	38.1 ± 2.0 ^{b}
No BSA + 15 mM Bic	42.3 ± 2.6 ^{a,b}	38.1 ± 2.4 ^{a,b}	32.5 ± 2.2 ^{a,b}	23.8 ± 1.5 ^{a}	20.7 ± 1.3 ^{a}	5.5 ± 0.3 ^{a}
BSA + 38 mM Bic	49.5 ± 2.8 ^{b}	43.3 ± 2.6 ^{b}	42.8 ± 2.6 ^{c}	48.7 ± 3.0 ^{b}	45.2 ± 2.5 ^{b}	35.8 ± 2.2 ^{b}
No BSA + 38 mM Bic	44.6 ± 2.4 ^{a,b}	40.8 ± 2.3 ^{a,b}	29.1 ± 1.8 ^{a}	22.4 ± 1.4 ^{a}	15.6 ± 1.0 ^{a}	2.2 ± 0.2 ^{a}

Percentages of LIN were significantly ($p < 0.05$) lower in the medium without BSA and without bicarbonate than in those with BSA and/or different concentrations of bicarbonate after 240 min of incubation. At 270 min and 300 min, percentages of LIN were significantly ($p < 0.05$) lower in the treatment containing BSA and 38 mM bicarbonate than in that containing BSA and no bicarbonate (Table 2). With regard to percentages of STR, media that did not contain BSA and 0 mM (No BSA/No Bic) or 15 mM bicarbonate (No BSA + 15 mM Bic) showed significantly ($p < 0.05$) lower values of this parameter than their counterparts containing BSA (BSA/No Bic and BSA + 15 mM Bic) after 120 min of incubation. From 240 min to 300 min of incubation, percentages of STR were significantly ($p < 0.05$) lower in the treatment without BSA and without bicarbonate (No BSA/No Bic) than in that containing BSA and 5 mM bicarbonate (BSA + 5 mM Bic). Finally, percentages of WOB were significantly ($p < 0.05$) higher in the medium without BSA and with 5 mM bicarbonate (No BSA + 5 mM Bic) than in the other treatments after 120 min of incubation. After 245 min and 270 min of incubation, percentages of WOB were significantly ($p < 0.05$) higher in sperm incubated with BSA and 5 mM bicarbonate than in those incubated without BSA and the same concentration of bicarbonate (i.e., BSA + 5 mM Bic vs. No BSA + 5 mM Bic). At 300 min, while treatments containing BSA and bicarbonate at 5 mM or 15 mM showed significantly ($p < 0.05$) higher percentages of WOB than those with no BSA and the same concentrations of bicarbonate (i.e., BSA + 5 mM Bic vs. No BSA + 5 mM Bic; BSA + 15 mM Bic vs. No BSA + 15 mM Bic), those observed in the medium containing BSA and 38 mM bicarbonate were significantly ($p < 0.05$) lower than in those with no BSA and bicarbonate at 38 mM (i.e., BSA + 38 mM Bic vs. No BSA + 38 mM Bic).

Table 2. Percentages of linearity (LIN), straightness (STR) and oscillation (WOB) following incubation of pig sperm with different concentrations of bicarbonate (Bic; 0 mM, 5 mM, 15 mM, 38 mM) and BSA (0, 5 mg/mL) for 300 min. Progesterone (10 µg/mL) was added at 240 min to induce acrosome exocytosis. Different superscripts (a–e) indicate significant differences ($p < 0.05$) between treatments (rows) within the same time point. Data are shown as mean ± SEM for 12 independent experiments.

Media	0 Min	120 Min	240 Min	245 Min	270 Min	300 Min
			LIN (%)			
No BSA/No Bic	48.4 ± 2.9 ^{a,b}	39.0 ± 2.3 ^{a,c}	32.1 ± 1.8 ^{a}	28.1 ± 1.6 ^{a}	23.8 ± 1.4 ^{a}	41.3 ± 2.5 ^{a}
BSA/No Bic	53.3 ± 3.1 ^{a}	49.0 ± 2.8 ^{b}	47.1 ± 2.7 ^{b}	49.0 ± 2.8 ^{b}	46.9 ± 2.7 ^{b}	43.7 ± 2.6 ^{a}
BSA + 5 mM Bic	43.4 ± 2.5 ^{b}	47.8 ± 2.8 ^{a,b}	50.1 ± 2.9 ^{b}	47.9 ± 2.8 ^{b}	42.7 ± 2.4 ^{b,d}	39.2 ± 2.3 ^{a,b}
No BSA + 5 mM Bic	40.0 ± 2.4 ^{b}	52.1 ± 3.0 ^{b}	46.3 ± 2.7 ^{b}	40.9 ± 2.4 ^{b,c}	27.1 ±1.6 ^{a,c}	31.1 ± 1.8 ^{b}
BSA + 15 mM Bic	41.2 ± 2.3 ^{b}	47.6 ± 2.8 ^{a,b}	44.8 ± 2.6 ^{b}	44.6 ± 2.5 ^{b,c}	50.2 ± 2.9 ^{b}	44.6 ± 2.6 ^{a}
No BSA + 15 mM Bic	43.1 ± 2.5 ^{b}	35.0 ± 2.0 ^{c}	41.5 ± 2.4 ^{b}	32.2 ± 1.9 ^{a}	22.3 ± 1.3 ^{a}	20.2 ± 1.2 ^{c}
BSA + 38 mM Bic	44. 1± 2.6 ^{b}	48.0 ± 2.8 ^{a,b}	42.3 ± 2.5 ^{b}	37.7 ± 2.2 ^{a,c}	34.1 ± 2.0 ^{c,d}	32.9 ± 1.9 ^{b}
No BSA + 38 mM Bic	43.3 ± 2.6 ^{b}	49.3 ± 2.9 ^{b}	44.3 ± 2.7 ^{b}	35.1 ± 2.1 ^{a,c}	50.3 ± 3.0 ^{b}	46.7 ± 2.8 ^{a}
			STR (%)			
No BSA/No Bic	63.0 ± 3.5 ^{a}	55.2 ± 3.2 ^{a,b}	51.1 ± 3.0 ^{a}	38.0 ± 2.3 ^{a}	34.9 ± 2.1 ^{a}	67.9 ± 3.6 ^{a}
BSA/No Bic	79.0 ± 4.4 ^{b}	66.6 ± 3.8 ^{c,d,e}	65.8 ± 3.7 ^{b,c}	62.6 ± 3.6 ^{b}	58.7 ± 3.5 ^{b}	56.5 ± 3.2 ^{b,e}
BSA + 5 mM Bic	70.5 ± 4.1 ^{a,b}	68.5 ± 3.9 ^{c,d,e}	69.7 ± 4.0 ^{c}	54.8 ± 3.1 ^{b}	52.5 ± 3.1 ^{b,c}	44.9 ± 2.6 ^{c}
No BSA + 5 mM Bic	68.3 ± 3.8 ^{a}	59.2 ± 3.4 ^{b,e}	61.1 ± 3.5 ^{b,c}	72.9 ± 4.1 ^{c}	51.9 ± 2.9 ^{b,c}	42.1 ± 2.4 ^{c}
BSA + 15 mM Bic	67.2 ± 3.8 ^{a}	75.6 ± 4.2 ^{c}	68.2 ± 3.8 ^{b,c}	59.7 ± 3.4 ^{b}	73.2 ± 4.1 ^{d}	60.1 ± 3.5 ^{a,b,e}
No BSA + 15 mM Bic	72.1 ± 4.2 ^{a,b}	48.0 ± 2.8 ^{a}	52.6 ± 3.1 ^{a}	53.8 ± 3.0 ^{b}	34.8 ± 2.0 ^{a}	32.7 ± 1.9 ^{d}
BSA + 38 mM Bic	67.7 ± 3.9 ^{a}	73.4 ± 4.1 ^{c,d}	59.8 ± 3.4 ^{a,b}	48.0 ± 2.8 ^{a,b}	44.7 ± 2.6 ^{c}	51.1 ± 2.9 ^{b,c}
No BSA + 38 mM Bic	65.5 ± 3.8 ^{a}	65.4 ± 3.3 ^{b,d,e}	52.9 ± 3.0 ^{a}	50.4 ± 2.9 ^{a,b}	61.5 ± 3.5 ^{b}	63.6 ± 3.7 ^{a,e}
			WOB (%)			
No BSA/No Bic	76.8 ± 4.2 ^{a}	70.7 ± 4.1 ^{a,b}	62.7 ± 3.5 ^{a}	73.8 ± 4.3 ^{a}	68.3 ± 3.8 ^{a,b}	60.9 ± 3.4 ^{a}
BSA/No Bic	67.5 ± 3.9 ^{a}	73.6 ± 4.1 ^{a}	71.6 ± 4.2 ^{a,b}	78.3 ± 4.3 ^{a,b}	79.8 ± 4.5 ^{d}	77.4 ± 4.4 ^{b}
BSA + 5 mM Bic	61.6 ± 3.4 ^{b}	69.8 ± 3.9 ^{a,b}	71.9 ± 4.1 ^{a,b}	87.5 ± 4.8 ^{b}	81.4 ± 4.7 ^{d}	87.3 ± 4.8 ^{c}
No BSA + 5 mM Bic	58.6 ± 3.3 ^{b}	88.1 ± 4.8 ^{c}	75.8 ± 4.2 ^{b,c}	56.2 ± 3.1 ^{c}	52.3 ± 2.9 ^{c}	73.8 ± 4.2 ^{b}
BSA + 15 mM Bic	61.3 ± 3.3 ^{b}	63.0 ± 3.6 ^{b}	65.6 ± 3.7 ^{a,b}	74.7 ± 4.2 ^{a}	68.5 ± 3.8 ^{a,b}	74.3 ± 4.1 ^{b}
No BSA + 15 mM Bic	59.7 ± 3.4 ^{b}	72.8 ± 4.1 ^{a,b}	78.9 ± 4.4 ^{c}	59.8 ± 3.4 ^{c}	64.1 ± 3.7 ^{a}	61.8 ± 3.6 ^{a}
BSA + 38 mM Bic	65.2 ± 3.7 ^{a,b}	65.3 ± 3.5 ^{b}	70.7 ± 4.2 ^{a,b}	78.4 ± 4.5 ^{a,b}	76.2 ± 4.3 ^{b,d}	64.4 ± 3.7 ^{a}
No BSA + 38 mM Bic	66.2 ± 3.7 ^{a,b}	75.3 ± 4.2 ^{a}	83.6 ± 4.7 ^{c}	69.6 ± 4.0 ^{a}	81.7 ± 4.6 ^{d}	73.3 ± 4.1 ^{b}

Table 3 shows ALH and BCF of pig sperm incubated with different media. With regard to ALH, although no significant differences between treatments were observed at the beginning of the experiment (0 min), those containing BSA and bicarbonate (5 mM, 15 mM or 38 mM) showed significantly ($p < 0.05$) higher values of this parameter than those without BSA, and the same concentration of bicarbonate (BSA + 5 mM Bic vs. No BSA + 5 mM Bic; BSA + 15 mM Bic vs. No BSA + 15 mM Bic; BSA + 38 mM Bic vs. No BSA + 38 mM Bic) after 120 min of incubation and until the end of the experimental period.

Finally, sperm incubated with BSA, regardless of the concentration of bicarbonate, showed significantly ($p < 0.05$) higher BCF than those incubated without BSA after 245 min, 270 min and 300 min of incubation.

Table 3. Amplitude of lateral head displacement (ALH) and frequency of head displacement (BCF) following incubation of pig sperm with different concentrations of bicarbonate (Bic; 0 mM, 5 mM, 15 mM, 38 mM) and BSA (0, 5 mg/mL) for 300 min. Progesterone (10 µg/mL) was added at 240 min to induce acrosome exocytosis. Different superscripts (a–d) indicate significant differences ($p < 0.05$) between treatments (rows) within the same time point. Data are shown as mean ± SEM for 12 independent experiments.

Media	0 Min	120 Min	240 Min	245 Min	270 Min	300 Min
			ALH (µm)			
No BSA/No Bic	2.5 ± 0.1^{a}	2.4 ± 0.1^{a}	2.6 ± 0.1^{a}	2.1 ± 0.1^{a}	2.2 ± 0.1^{a}	1.9 ± 0.1^{a}
BSA/No Bic	2.7 ± 0.1^{a}	3.2 ± 0.2^{b}	3.7 ± 0.2^{b}	3.9 ± 0.2^{b}	4.1 ± 0.2^{b}	3.8 ± 0.2^{b}
BSA + 5 mM Bic	2.6 ± 0.1^{a}	3.1 ± 0.2^{b}	3.9 ± 0.2^{b}	4.1 ± 0.2^{b}	4.3 ± 0.2^{b}	4.2 ± 0.2^{b}
No BSA + 5 mM Bic	2.3 ± 0.1^{a}	2.2 ± 0.1^{a}	2.4 ± 0.1^{a}	2.3 ± 0.1^{a}	1.9 ± 0.1^{a}	1.7 ± 0.1^{a}
BSA + 15 mM Bic	2.5 ± 0.1^{a}	3.3 ± 0.2^{b}	3.4 ± 0.2^{b}	3.6 ± 0.2^{b}	4.0 ± 0.2^{b}	3.9 ± 0.2^{b}
No BSA + 15 mM Bic	2.2 ± 0.1^{a}	2.4 ± 0.1^{a}	2.6 ± 0.1^{a}	2.5 ± 0.1^{a}	2.3 ± 0.1^{a}	2.0 ± 0.1^{a}
BSA + 38 mM Bic	2.6 ± 0.1^{a}	3.5 ± 0.2^{b}	3.7 ± 0.2^{b}	3.8 ± 0.2^{b}	3.6 ± 0.2^{b}	3.5 ± 0.2^{b}
No BSA + 38 mM Bic	2.1 ± 0.1^{a}	2.4 ± 0.1^{a}	2.5 ± 0.1^{a}	2.3 ± 0.1^{a}	2.1 ± 0.1^{a}	1.8 ± 0.1^{a}
			BCF (Hz)			
No BSA/No Bic	$6.8 \pm 0.4^{a,b}$	5.1 ± 0.3^{a}	4.9 ± 0.3^{a}	5.2 ± 0.2^{a}	4.7 ± 0.2^{a}	3.8 ± 0.2^{a}
BSA/No Bic	6.5 ± 0.3^{a}	6.5 ± 0.3^{b}	7.3 ± 0.4^{b}	7.1 ± 0.4^{b}	6.7 ± 0.3^{b}	6.4 ± 0.3^{b}
BSA + 5 mM Bic	6.1 ± 0.3^{a}	$6.7 \pm 0.3^{b,c}$	7.6 ± 0.4^{b}	7.2 ± 0.4^{b}	6.7 ± 0.3^{b}	6.5 ± 0.3^{b}
No BSA + 5 mM Bic	6.2 ± 0.3^{a}	$5.9 \pm 0.3^{a,b}$	$5.5 \pm 0.3^{a,d}$	5.3 ± 0.3^{a}	5.1 ± 0.3^{a}	4.9 ± 0.2^{c}
BSA + 15 mM Bic	7.2 ± 0.4^{b}	$7.1 \pm 0.4^{c,d}$	6.4 ± 0.3^{c}	6.9 ± 0.3^{b}	6.5 ± 0.3^{b}	6.2 ± 0.3^{b}
No BSA + 15 mM Bic	$6.8 \pm 0.4^{a,b}$	$6.2 \pm 0.3^{b,c}$	$5.8 \pm 0.3^{c,d}$	5.5 ± 0.3^{a}	5.4 ± 0.3^{a}	5.1 ± 0.3^{c}
BSA + 38 mM Bic	7.4 ± 0.4^{b}	7.7 ± 0.4^{d}	7.5 ± 0.4^{b}	7.2 ± 0.4^{b}	6.3 ± 0.3^{b}	6.0 ± 0.3^{b}
No BSA + 38 mM Bic	7.1 ± 0.4^{b}	$6.4 \pm 0.3^{b,c}$	$5.9 \pm 0.3^{c,d}$	5.6 ± 0.3^{a}	5.3 ± 0.3^{a}	5.2 ± 0.3^{c}

3.7. *Effects of Different Concentrations of Bicarbonate and BSA on DARPP-32 Phosphorylation Levels*

Figure 7 shows normalized phosphorylated DARPP-32:PKA ratios after 0 min, 240 min, 245 min and 300 min of incubation in media, with or without BSA, and with or without 38 mM bicarbonate. Only these four treatments and time points are shown, since they were those in which differences were clearer. Sperm incubated in media containing either BSA alone (BSA/No Bic) or BSA and 38 mM bicarbonate (BSA + 38 mM Bic) showed significantly ($p < 0.05$) higher phosphorylated DARPP-32:PKA ratios than the other treatments after 0 min, 240 min and 245 min of incubation (Figure 7a). Remarkably, sperm incubated in the medium containing 38 mM bicarbonate but no BSA (No BSA + 38 mM Bic) showed similar phosphorylated DARPP-32:PKA ratios to those incubated in a medium without BSA and without bicarbonate (No BSA/No Bic) after 0 min and 240 min of incubation. At 300 min, the medium without BSA and with 38 mM bicarbonate (No BSA + 38 mM Bic) showed a significantly ($p < 0.05$) lower phosphorylated DARPP-32:PKA ratio than the other treatments.

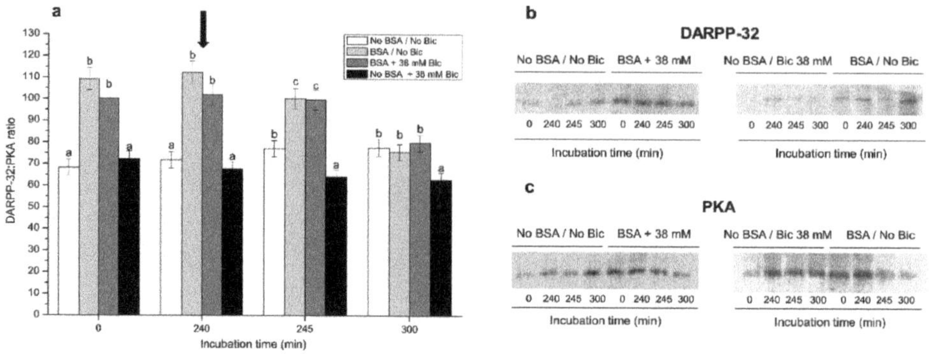

Figure 7. (**a**) Ratios between phosphorylated DARPP-32 and PKA following incubation with different concentrations of bicarbonate (Bic; 0 mM, 38 mM) and BSA (0, 5 mg/mL) for 300 min. The black arrow indicates the time at which 10 µg/mL of progesterone was added (i.e., 240 min) to induce acrosome exocytosis. Different superscripts (a–c) indicate significant differences ($p < 0.05$) between treatments within the same time point. Data were corrected to a basal arbitrary value of 100 for the control point, which corresponded to the incubation in the medium containing BSA and 38 mM bicarbonate at 0 min. Results are shown as mean ± SEM for 12 independent experiments. Representative blots for DARPP-32 (**b**) and PKA (**c**) after 0, 240, 245 and 300 min of incubation with No BSA/No Bic; BSA + 38 mM Bic; No BSA + 38 mM Bic; or BSA/No Bic.

3.8. Effects of Different Concentrations of Bicarbonate and BSA on Tyrosine Phosphorylation Levels of GSK3α

Figure 8 shows phosphorylated GSK3α:α-tubulin ratios after 0 min, 240 min, 245 min and 300 min of incubation in media, with or without BSA, and with or without 38 mM bicarbonate. In a similar fashion to the case of phosphorylated DARPP-32:PKA ratios, only these four treatments and time points are shown, as they were those in which differences were clearer. Incubation of sperm in media containing BSA and/or 38 mM bicarbonate (i.e., BSA/No Bic, BSA + 38 mM, and No BSA + 38 mM) showed significantly ($p < 0.05$) higher phosphorylated GSK3α:α-tubulin ratios after 0 min and 240 min of incubation. At 245 min, sperm incubated in a medium with BSA and 38 mM bicarbonate presented significantly ($p < 0.05$) higher phosphorylated GSK3α:α-tubulin ratios than the other three treatments, and those incubated in a medium with BSA and without bicarbonate exhibited significantly ($p < 0.05$) higher phosphorylated GSK3α:α-tubulin ratios than those incubated in media without BSA (No BSA/No Bic, No BSA + 38 mM Bic). At 300 min, phosphorylated GSK3α:α-tubulin ratios in sperm incubated in media containing BSA, with or without bicarbonate, were significantly ($p < 0.05$) higher than in those incubated in media without BSA.

Finally, sperm incubated with BSA, regardless of the concentration of bicarbonate, showed significantly ($p < 0.05$) higher BCF than those incubated without BSA after 245 min, 270 min and 300 min of incubation.

Table 3. Amplitude of lateral head displacement (ALH) and frequency of head displacement (BCF) following incubation of pig sperm with different concentrations of bicarbonate (Bic; 0 mM, 5 mM, 15 mM, 38 mM) and BSA (0, 5 mg/mL) for 300 min. Progesterone (10 μg/mL) was added at 240 min to induce acrosome exocytosis. Different superscripts (a–d) indicate significant differences ($p < 0.05$) between treatments (rows) within the same time point. Data are shown as mean ± SEM for 12 independent experiments.

Media	0 Min	120 Min	240 Min	245 Min	270 Min	300 Min
			ALH (μm)			
No BSA/No Bic	2.5 ± 0.1 [a]	2.4 ± 0.1 [a]	2.6 ± 0.1 [a]	2.1 ± 0.1 [a]	2.2 ± 0.1 [a]	1.9 ± 0.1 [a]
BSA/No Bic	2.7 ± 0.1 [a]	3.2 ± 0.2 [b]	3.7 ± 0.2 [b]	3.9 ± 0.2 [b]	4.1 ± 0.2 [b]	3.8 ± 0.2 [b]
BSA + 5 mM Bic	2.6 ± 0.1 [a]	3.1 ± 0.2 [b]	3.9 ± 0.2 [b]	4.1 ± 0.2 [b]	4.3 ± 0.2 [b]	4.2 ± 0.2 [b]
No BSA + 5 mM Bic	2.3 ± 0.1 [a]	2.2 ± 0.1 [a]	2.4 ± 0.1 [a]	2.3 ± 0.1 [a]	1.9 ± 0.1 [a]	1.7 ± 0.1 [a]
BSA + 15 mM Bic	2.5 ± 0.1 [a]	3.3 ± 0.2 [b]	3.4 ± 0.2 [b]	3.6 ± 0.2 [b]	4.0 ± 0.2 [b]	3.9 ± 0.2 [b]
No BSA + 15 mM Bic	2.2 ± 0.1 [a]	2.4 ± 0.1 [a]	2.6 ± 0.1 [a]	2.5 ± 0.1 [a]	2.3 ± 0.1 [a]	2.0 ± 0.1 [a]
BSA + 38 mM Bic	2.6 ± 0.1 [a]	3.5 ± 0.2 [b]	3.7 ± 0.2 [b]	3.8 ± 0.2 [b]	3.6 ± 0.2 [b]	3.5 ± 0.2 [b]
No BSA + 38 mM Bic	2.1 ± 0.1 [a]	2.4 ± 0.1 [a]	2.5 ± 0.1 [a]	2.3 ± 0.1 [a]	2.1 ± 0.1 [a]	1.8 ± 0.1 [a]
			BCF (Hz)			
No BSA/No Bic	6.8 ± 0.4 [a,b]	5.1 ± 0.3 [a]	4.9 ± 0.3 [a]	5.2 ± 0.2 [a]	4.7 ± 0.2 [a]	3.8 ± 0.2 [a]
BSA/No Bic	6.5 ± 0.3 [a]	6.5 ± 0.3 [b]	7.3 ± 0.4 [b]	7.1 ± 0.4 [b]	6.7 ± 0.3 [b]	6.4 ± 0.3 [b]
BSA + 5 mM Bic	6.1 ± 0.3 [a]	6.7 ± 0.3 [b,c]	7.6 ± 0.4 [b]	7.2 ± 0.4 [b]	6.7 ± 0.3 [b]	6.5 ± 0.3 [b]
No BSA + 5 mM Bic	6.2 ± 0.3 [a]	5.9 ± 0.3 [a,b]	5.5 ± 0.3 [a,d]	5.3 ± 0.3 [a]	5.1 ± 0.3 [a]	4.9 ± 0.2 [c]
BSA + 15 mM Bic	7.2 ± 0.4 [b]	7.1 ± 0.4 [c,d]	6.4 ± 0.3 [c]	6.9 ± 0.3 [b]	6.5 ± 0.3 [b]	6.2 ± 0.3 [b]
No BSA + 15 mM Bic	6.8 ± 0.4 [a,b]	6.2 ± 0.3 [b,c]	5.8 ± 0.3 [c,d]	5.5 ± 0.3 [a]	5.4 ± 0.3 [a]	5.1 ± 0.3 [c]
BSA + 38 mM Bic	7.4 ± 0.4 [b]	7.7 ± 0.4 [d]	7.5 ± 0.4 [b]	7.2 ± 0.4 [b]	6.3 ± 0.3 [b]	6.0 ± 0.3 [b]
No BSA + 38 mM Bic	7.1 ± 0.4 [b]	6.4 ± 0.3 [b,c]	5.9 ± 0.3 [c,d]	5.6 ± 0.3 [a]	5.3 ± 0.3 [a]	5.2 ± 0.3 [c]

3.7. *Effects of Different Concentrations of Bicarbonate and BSA on DARPP-32 Phosphorylation Levels*

Figure 7 shows normalized phosphorylated DARPP-32:PKA ratios after 0 min, 240 min, 245 min and 300 min of incubation in media, with or without BSA, and with or without 38 mM bicarbonate. Only these four treatments and time points are shown, since they were those in which differences were clearer. Sperm incubated in media containing either BSA alone (BSA/No Bic) or BSA and 38 mM bicarbonate (BSA + 38 mM Bic) showed significantly ($p < 0.05$) higher phosphorylated DARPP-32:PKA ratios than the other treatments after 0 min, 240 min and 245 min of incubation (Figure 7a). Remarkably, sperm incubated in the medium containing 38 mM bicarbonate but no BSA (No BSA + 38 mM Bic) showed similar phosphorylated DARPP-32:PKA ratios to those incubated in a medium without BSA and without bicarbonate (No BSA/No Bic) after 0 min and 240 min of incubation. At 300 min, the medium without BSA and with 38 mM bicarbonate (No BSA + 38 mM Bic) showed a significantly ($p < 0.05$) lower phosphorylated DARPP-32:PKA ratio than the other treatments.

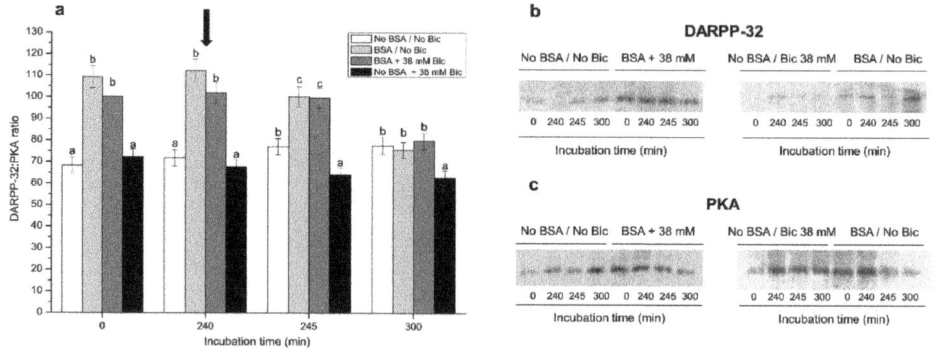

Figure 7. (**a**) Ratios between phosphorylated DARPP-32 and PKA following incubation with different concentrations of bicarbonate (Bic; 0 mM, 38 mM) and BSA (0, 5 mg/mL) for 300 min. The black arrow indicates the time at which 10 μg/mL of progesterone was added (i.e., 240 min) to induce acrosome exocytosis. Different superscripts (a–c) indicate significant differences ($p < 0.05$) between treatments within the same time point. Data were corrected to a basal arbitrary value of 100 for the control point, which corresponded to the incubation in the medium containing BSA and 38 mM bicarbonate at 0 min. Results are shown as mean ± SEM for 12 independent experiments. Representative blots for DARPP-32 (**b**) and PKA (**c**) after 0, 240, 245 and 300 min of incubation with No BSA/No Bic; BSA + 38 mM Bic; No BSA + 38 mM Bic; or BSA/No Bic.

3.8. Effects of Different Concentrations of Bicarbonate and BSA on Tyrosine Phosphorylation Levels of GSK3α

Figure 8 shows phosphorylated GSK3α:α-tubulin ratios after 0 min, 240 min, 245 min and 300 min of incubation in media, with or without BSA, and with or without 38 mM bicarbonate. In a similar fashion to the case of phosphorylated DARPP-32:PKA ratios, only these four treatments and time points are shown, as they were those in which differences were clearer. Incubation of sperm in media containing BSA and/or 38 mM bicarbonate (i.e., BSA/No Bic, BSA + 38 mM, and No BSA + 38 mM) showed significantly ($p < 0.05$) higher phosphorylated GSK3α:α-tubulin ratios after 0 min and 240 min of incubation. At 245 min, sperm incubated in a medium with BSA and 38 mM bicarbonate presented significantly ($p < 0.05$) higher phosphorylated GSK3α:α-tubulin ratios than the other three treatments, and those incubated in a medium with BSA and without bicarbonate exhibited significantly ($p < 0.05$) higher phosphorylated GSK3α:α-tubulin ratios than those incubated in media without BSA (No BSA/No Bic, No BSA + 38 mM Bic). At 300 min, phosphorylated GSK3α:α-tubulin ratios in sperm incubated in media containing BSA, with or without bicarbonate, were significantly ($p < 0.05$) higher than in those incubated in media without BSA.

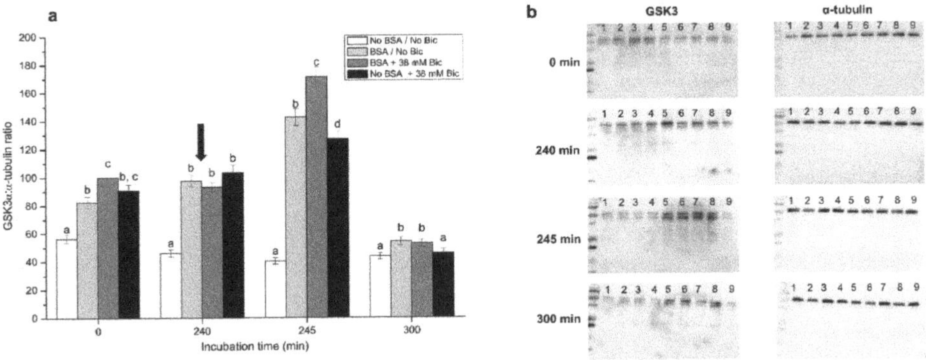

Figure 8. (**a**) Ratios between tyrosine-phosphorylated GSK3α and α-tubulin following incubation with different concentrations of bicarbonate (Bic; 0 mM, 38 mM) and BSA (0, 5 mg/mL) for 300 min. The black arrow indicates the time at which 10 μg/mL of progesterone was added (i.e., 240 min) to induce acrosome exocytosis. Different superscripts (a–d) indicate significant differences ($p < 0.05$) between treatments within the same time point. Data were corrected to a basal arbitrary value of 100 for the control point, which corresponded to the incubation in the medium containing BSA and 38 mM bicarbonate at 0 min. Results are shown as mean ± SEM for 12 independent experiments. (**b**) Representative blots for tyrosine-phosphorylated GSK3 and α-tubulin. Lane 1: No BSA/No Bic; Lane 2: BSA + 38 mM Bic; Lane 3: No BSA + 38 mM Bic; Lane 4: BSA/No Bic; Lane 5: No BSA + 5 mM Bic; Lane 6: BSA + 5 mM Bic; Lane 7: No BSA + 15 mM Bic; Lane 8: BSA + 15 mM Bic; Lane 9: No BSA/No Bic.

3.9. Effects of Bicarbonate and BSA on O_2 Consumption Rate

Figure 9 shows the O_2 consumption rate in pig sperm incubated in media containing BSA and/or 38 mM bicarbonate (i.e., No BSA/No Bic, BSA/No Bic, BSA + 38 mM Bic, and No BSA + 38 mM Bic).

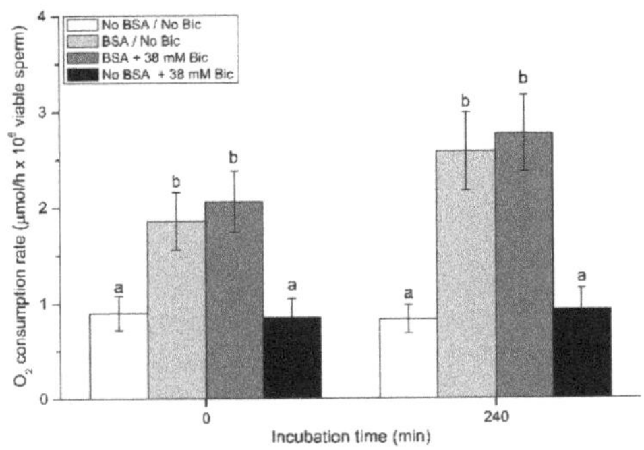

Figure 9. O_2 consumption rate of pig sperm in the presence of different concentrations of bicarbonate (Bic; 0 mM, 38 mM) and BSA (0, 5 mg/mL) at 0 min and 240 min. Different superscripts (a,b) indicate significant differences ($p < 0.05$) between treatments within the same time point. Results are shown as mean ± SEM for 12 independent experiments.

At 0 min, the O_2 consumption rate in sperm diluted in the medium without BSA and bicarbonate (No BSA/No Bic) was low (0.90 ± 0.18 μmoL/h × 10^6 viable sperm). These ratios did not significantly increase after 240 min of incubation. Similar results were observed in the treatment containing 38 mM bicarbonate and no BSA (No BSA + 38 mM Bic). In contrast, sperm incubated in the presence of BSA, either with or without bicarbonate, showed significantly ($p < 0.05$) higher O_2 consumption rates than those incubated without BSA, at both 0 min (2.06 ± 0.32 μmoL/h × 10^6 viable sperm) and 240 min (2.77 ± 0.40 μmoL/h × 10^6 viable sperm).

4. Discussion

Our findings clearly indicate that the presence of BSA is essential for pig sperm to elicit in vitro capacitation. Moreover, bicarbonate is not essential to launch the in vitro capacitation process, but it modulates its efficiency. Therefore, the current work supports that in vitro capacitation of pig sperm can be achieved in a medium without bicarbonate, but not without BSA, although the efficiency of the process is not optimal in the absence of the former. This conclusion is not only reached from the changes in sperm function parameters related to capacitation, such as membrane lipid disorder, acrosome exocytosis and intracellular calcium levels, but also from the increase observed in phosphorylation levels of Tyr-GSK3α and DARPP-32. In fact, previous works already demonstrated that in vitro capacitation in pig sperm can be efficiently achieved in a medium with BSA but without bicarbonate [7,29,30,32,40]. In this context, it is worth noting that these observations are somewhat different from those observed in other mammalian species, in which bicarbonate is considered to be an essential component to elicit sperm capacitation [41]. These discrepancies do highlight the existence of species-specific mechanisms that are regulating the achievement of sperm capacitation in vitro. In relation, while different studies were conducted to induce in vitro capacitation in pig sperm, no consensus was reached on what the best medium is. For instance, high concentrations of bicarbonate (i.e., 37.6 mM) in in vitro capacitation media induce strong agglutination in pig sperm [9], thereby greatly impairing their recovery. While high concentrations of bicarbonate also induce sperm agglutination in sheep [42] and horses [43], their impact is not so apparent in hamsters [41] and cattle [44]. In addition, experimental conditions also differ between studies evaluating the response of pig sperm to in vitro capacitation. In effect, whereas some studies, including the present one, incubated sperm cells for 240 min prior to adding progesterone [45–47], other authors used shorter incubation periods (60 min [22], 120 min [28], and 180 min [48,49]). The fact that the optimal concentration of bicarbonate in the in vitro capacitation medium of mammalian sperm differs between species should not be considered as exceptional, since, for example, other components such as heparin are required for bovine, but not for porcine sperm [44]. Therefore, although the general mechanism of mammalian sperm capacitation is known [5], it seems that species differ in particular events triggered by definite components.

Focusing on the addition of bicarbonate to capacitation medium, the results of our work have elucidated some interesting aspects. With regard to membrane integrity and lipid disorder, capacitating media containing high concentrations of bicarbonate (15 mM and 38 mM) led to the lowest percentages of membrane-intact sperm and the highest percentages of viable sperm with high membrane lipid disorder. This was expected, since bicarbonate is known to be an important regulator of phospholipid scrambling, thereby affecting membrane fluidity [50]. In effect, high intracellular levels of bicarbonate induce plasma membrane asymmetry, which activates scramblase enzymes that translocate membrane phospholipids, such as phosphatidylserine and phosphatidylethanolamine, thereby reducing membrane stability and making cholesterol available to external receptors [50,51]. Cholesterol can then be removed by BSA, which leads to the rapid collapse of membrane asymmetry [5,12,52]. Based on our data, the addition of bicarbonate to capacitation medium does accelerate the increase in the percentage of sperm with high membrane lipid disorder, which takes more time when sperm are incubated without bicarbonate in an atmosphere of 5% CO_2. The fact that pig sperm do not require exogenous bicarbonate to elicit in vitro capacitation can be related to their specific membrane structure. It is well known that the proportion of unsaturated:saturated fatty acids in pig sperm membrane is higher, and that of cholesterol:phospholipid lower, than in other species, such as the human, horse and cattle [53–56]. In this way, it is likely that the specific composition of pig sperm plasma membrane makes these cells more prone to increase their membrane lipid disorder in the presence of BSA under a 5% CO_2 atmosphere, which would be sufficient to induce these capacitation-related changes. A concentration of bicarbonate as low as 2 mM has also been reported by other authors to induce in vitro capacitation in pig sperm [45].

The aforementioned changes in sperm plasma membrane appeared to be linked with the results observed in sperm motility. Sperm incubated in in vitro capacitation media containing BSA and either 5 mM or 15 mM bicarbonate showed higher total and progressive motility, and VCL, VSL and VAP. In addition, our sperm motility data agreed with the significant changes observed in the sperm population with high MMP, matching with previous works reporting a significant correlation between MMP and sperm motility [57–59]. It is well known that mitochondrial activity increases during in vitro capacitation [60]; this activity further raises after the induction of acrosome exocytosis, which is likely to be related to the increase in the rate of mitochondrial oxidative phosphorylation [7]. In addition, we observed a mitochondria-stored calcium peak, indicated by Fluo3-staining, immediately after the induction of acrosome exocytosis with progesterone. In this context, it is worth bearing in mind the close relationship between sperm calcium metabolism and mitochondrial function. Calcium is a key regulator for sperm hyperactivation in the mouse [61], which is an essential step in the achievement of sperm capacitation. This motile pattern is characterized by an asymmetrical and wide amplitude flagellar movement [62]. In our conditions, the highest concentration of bicarbonate led to the lowest values of both total and progressive motility, despite the existence of an acrosome exocytosis-related calcium peak. These results, combined with the observed changes in the percentages of both plasma membrane and acrosome integrity, suggest that high concentrations of bicarbonate cause damage to the sperm cell. Although these detrimental effects seem to be related to capacitation-linked changes, our results suggest that adding bicarbonate in the range of 30–40 mM should be avoided if pig sperm are capacitated under a 5% CO_2 atmosphere for 240 min.

In this study, we used two different fluorochromes (Fluo3 and Rhod5) that stain calcium stores of both sperm head and mid-piece [7]. It is well understood that in vitro capacitation and the subsequent progesterone-induced acrosome exocytosis increase calcium influx through CatSper channels, which is required for sperm to acquire the ability to fertilize the oocyte [63,64]. In pigs, previous research showed that calcium plays a vital role during sperm capacitation and acrosome reaction [7,65]. Herein, we observed that the presence of BSA in the capacitation medium is essential for the calcium peak to occur after the induction of acrosome exocytosis with progesterone. These data are in agreement with Espinosa et al. [66], who found that BSA in the medium is responsible for calcium to enter mouse spermatogenic cells. In addition, the calcium peak was proportional to bicarbonate concentration, with the highest peaks being observed at 38 mM, followed by 15 mM and 5 mM bicarbonate. These results indicate that the entry of calcium into the sperm is a result of the joint action of BSA and bicarbonate. As stated previously, removal of cholesterol from sperm plasma membrane during capacitation is induced by BSA, which leads to the displacement and rearrangement of phospholipids and increases membrane fluidity [67,68]. Cholesterol efflux triggers a cascade of events, such as the increase of calcium levels via CatSper channels [69] or the activation, through bicarbonate, of a soluble, non-membrane bound adenylyl cyclase (sAC) [5,70]. This latter point is important, since activation of sAC leads to an increase in intracellular cAMP levels, which in turn activate protein kinase A, a crucial protein for the signaling pathway of mammalian sperm capacitation [17,71]. The fact that low concentrations (5 mM) or even the complete absence of bicarbonate in in vitro capacitation media also allows pig sperm to increase their intracellular calcium levels suggests that, under a 5% CO_2 atmosphere, the amount of bicarbonate required to activate sAC does not need to be high.

On the other hand, we also analyzed the activity of PKA through tyrosine phosphorylation levels of DARPP-32. DARPP-32 is involved in the modulation of the activity of several capacitation-involved protein kinases and phosphatases [72]. The activity of DARPP-32 is regulated by a complex system of intracellular signaling pathways involving PKA and calcium [73,74]. Remarkably, our results showed that the highest phosphorylation levels of DARPP-32, which indicate its maximal activity, are reached in the media containing BSA, regardless of the presence of bicarbonate. Conversely, the lowest phosphorylation levels of DARPP-32 occurred in the medium without BSA and with 38 mM bicarbonate. These findings match with the results obtained from sperm function parameters, including membrane lipid disorder and acrosome exocytosis, and support that PKA activation, which is essential for sperm capacitation, can be achieved without adding bicarbonate to the capacitation medium. In contrast, BSA must be present, since the lack of that protein in the medium does not allow DARPP-32 to be phosphorylated.

The results obtained from the analysis of both MMP and O_2 consumption rates suggest that mitochondria are activated at the start of the incubation in capacitation media, but only if BSA is present. This immediate effect on mitochondria in the presence of BSA takes place together with the activation of PKA and GSK3α. In this context, it is worth highlighting that phosphorylation of specific sperm proteins is crucial to reach the capacitated status [14]. In pigs, glycogen synthase kinase-3 (GSK3) is present in sperm cells and is phosphorylated at Ser9 and Ser21 in response to the activation of PKA, which completely activates GSK3 [75,76]. It has been reported that GSK3 activity is involved in the modulation of sperm motility, acrosome exocytosis and, ultimately, fertilization [75,77,78]. In this way, the results obtained from both PKA and GSK3α activities highlight how relevant BSA is for pig sperm to elicit in vitro capacitation and progesterone-induced acrosome exocytosis. In this respect, the mechanism through which BSA modulates the activities of PKA and GSK3α has been posited to be related to the activation of a CatSper-modulated extracellular Ca^{2+}-influx [69,79]. Our results are in agreement with this mechanism, thus highlighting the relevance of BSA as an effector of capacitation in pig sperm. Furthermore, our data also suggest that serine as well astyrosine-phosphorylation regulate kinase activity of GSK3α and GSK3β in mammalian sperm [78,80–83]. Related to the specific role of serine residues, Rival et al. recently demonstrated that phosphatidylserine in mouse sperm plasmalemma is progressively exposed on the head region of viable spermatozoa during epididymal maturation [84]. Phosphatidylserine is recognized by specific oocyte membrane receptors, such as BAI1, CD36, TIM4, and MERTK, and when these phospholipids are masked, fertilization is inhibited.

Finally, it is worth mentioning that while this work aimed to address the effects of different concentrations of bicarbonate on the ability of pig sperm to elicit in vitro capacitation, a constant concentration of BSA was used. That concentration was chosen based on previous works, in which 5 mg/mL BSA was found to elicit in vitro capacitation of pig sperm [7,9,29,30]. Therefore, while this work specifically focused on the effects of bicarbonate, further studies should aim to question how different BSA concentrations modulate the response of pig sperm to the induction of in vitro capacitation and acrosome reaction. In addition, further research involving other indicators related to the underlying molecular mechanism, such as the phosphorylation of other sperm proteins and changes in intracellular cAMP levels, is likely to help gain new insights into how BSA and bicarbonate modulate PKA activation at different incubation times. On the other hand, although we did not find much variation between the pH of the different media throughout incubation (data not shown), we did not evaluate the potential impact that the different media composition could have on their osmolality. Hence, further studies investigating the effects of different BSA concentrations on how pig sperm elicit in vitro capacitation should also take this matter into consideration.

5. Conclusions

In conclusion, our results indicate that BSA is a crucial factor for pig sperm to elicit in vitro capacitation. In contrast, while the addition of bicarbonate to the medium is not a limiting factor to induce capacitation, the presence of high levels of bicarbonate reduces the time required for sperm to reach that status. In addition, based on all sperm parameters evaluated and on the phosphorylation levels of DARPP-32 and GSK3α, media containing

BSA and either no or low levels of bicarbonate are the most suitable for inducing pig sperm capacitation. In effect, despite some capacitation events taking longer, sperm cells do not die as fast as the medium containing BSA and 38 mM bicarbonate when evaluated after 300 min of incubation.

Supplementary Materials: The following are available online at https://www.mdpi.com/article/10.3390/biology10111105/s1, Supplementary Table S1 and Supplementary File 1 (File S1).

Author Contributions: Conceptualization, S.B., J.E.R.-G., M.G.Z. and M.Y.; methodology, B.R.C., A.P.P.P. and O.B.-P.; investigation, B.R.C., A.P.P.P., O.B.-P., E.P., J.E.R.-G. and M.Y.; data curation, B.R.C., O.B.-P. and J.E.R.-G.; writing—original draft preparation, B.R.C.; review and editing, E.P., S.B., M.G.Z., J.E.R.-G. and M.Y.; funding acquisition, S.B., M.G.Z., J.E.R.-G. and M.Y. All authors have read and agreed to the published version of the manuscript.

Funding: This research was supported by the Coordenação de Aperfeiçoamento de Pessoal de Nível Superior (CAPES), Brazil (Grants: 88881.135379/2016-01), the Ministry of Science, Innovation and Universities, Spain (Grants: RYC-2014-15581 and AGL2017- 88329-R), and the Regional Government of Catalonia (2017-SGR-1229).

Institutional Review Board Statement: Not applicable. Semen samples were purchased from a local farm, which operates under standard commercial conditions. Therefore, authors did not manipulate any animal and specific approval from an Ethics Commission was not required.

Informed Consent Statement: Not applicable.

Data Availability Statement: The data presented in this study are available in the article and Supplementary File S1.

Acknowledgments: The authors would like to thank the support from TechnoSperm staff (University of Girona) for their technical support.

Conflicts of Interest: The authors declare no conflict of interest.

References

1. Gervasi, M.G.; Visconti, P.E. Chang's meaning of capacitation: A molecular perspective. *Mol. Reprod. Dev.* **2016**, *83*, 860–874. [CrossRef]
2. Austin, C. Observations on the penetration of the sperm into the mammalian egg. *Aust. J. Biol. Sci.* **1951**, *4*, 581–596. [CrossRef]
3. Chang, M.C. Fertilizing capacity of spermatozoa deposited into the fallopian tubes. *Nature* **1951**, *168*, 697–698. [CrossRef]
4. Buffone, M.G.; Wertheimer, E.V.; Visconti, P.E.; Krapf, D. Central role of soluble adenylyl cyclase and cAMP in sperm physiology. *Biochim. Biophys. Acta* **2014**, *1842*, 2610–2620. [CrossRef]
5. Visconti, P.E. Understanding the molecular basis of sperm capacitation through kinase design. *Proc. Natl. Acad. Sci. USA* **2009**, *106*, 667–668. [CrossRef] [PubMed]
6. Ferramosca, A.; Zara, V. Bioenergetics of mammalian sperm capacitation. *Biomed Res. Int.* **2014**, *2014*, 902953. [CrossRef]
7. Yeste, M.; Fernández-Novell, J.M.; Ramió-Lluch, L.; Estrada, E.; Rocha, L.G.; Cebrián-Pérez, J.A.; Muiño-Blanco, T.; Concha, I.I.; Ramírez, A.; Rodríguez-Gil, J.E. Intracellular calcium movements of boar spermatozoa during 'in vitro' capacitation and subsequent acrosome exocytosis follow a multiple-storage place, extracellular calcium-dependent model. *Andrology* **2015**, *3*, 729–747. [CrossRef]
8. Betarelli, R.P.; Rocco, M.; Yeste, M.; Fernández-Novell, J.M.; Placci, A.; Azevedo-Pereira, B.; Castillo-Martín, M.; Estrada, E.; Peña, A.; Zangeronimo, M.G.; et al. The achievement of boar sperm in vitro capacitation is related to an increase of disrupted disulphide bonds and intracellular reactive oxygen species levels. *Andrology* **2018**, *6*, 781–797. [CrossRef]
9. Rocco, M.; Betarelli, R.; Placci, A.; Fernández-Novell, J.M.; Spinaci, M.; Casao, A.; Muiño-Blanco, T.; Cebrián-Pérez, J.A.; Peña, A.; Rigau, T.; et al. Melatonin affects the motility and adhesiveness of in vitro capacitated boar spermatozoa via a mechanism that does not depend on intracellular ROS levels. *Andrology* **2018**, *6*, 720–736. [CrossRef]
10. Yeste, M.; Llavanera, M.; Scornik, F.; Pérez, G.; Puig-Parri, J.; Brugada, R.; Bonet, S.; Pinart, E. Elucidating the role of K+ channels during in vitro capacitation of boar spermatozoa: Do SLO1 channels play a crucial role? *Int. J. Mol. Sci.* **2019**, *20*, 6330. [CrossRef]
11. Yeste, M. Boar spermatozoa within the oviductal environment (II): Sperm capacitation. In *Boar Reproduction*; Bonet, S., Casas, I., Holt, W.V., Yeste, M., Eds.; Springer: Heidelberg, Germany, 2013; pp. 347–405.
12. Salicioni, A.M.; Platt, M.D.; Wertheimer, E.V.; Arcelay, E.; Allaire, A.; Sosnik, J.; Visconti, P.E. Signalling pathways involved in sperm capacitation. *Soc. Reprod. Fertil. Suppl.* **2007**, *65*, 245–259.
13. Stival, C.; Molina, L.D.C.P.; Paudel, B.; Buffone, M.G.; Visconti, P.E.; Krapf, D. Sperm capacitation and acrosome reaction in mammalian sperm. *Adv. Anat. Embryol. Cell. Biol.* **2016**, *220*, 93–106.
14. Visconti, P.E.; Kopf, G.S. Regulation of protein phosphorylation during sperm capacitation. *Biol. Reprod.* **1998**, *59*, 1–6. [CrossRef]

15. Flesch, F.M.; Wijnand, E.; van de Lest, C.H.; Colenbrander, B.; van Golde, L.M.; Gadella, B.M. Capacitation dependent activation of tyrosine phosphorylation generates two sperm head plasma membrane proteins with high primary binding affinity for the zona pellucida. *Mol. Reprod. Dev.* **2001**, *60*, 107–115. [CrossRef]
16. Gadella, B.M.; Van Gestel, R.A. Bicarbonate and its role in mammalian sperm function. *Anim. Reprod. Sci.* **2004**, *82–83*, 307–319. [CrossRef]
17. Gangwar, D.K.; Atreja, S.K. Signalling events and associated pathways related to the mammalian sperm capacitation. *Reprod. Domest. Anim.* **2015**, *50*, 705–711. [CrossRef]
18. Naz, R.K.; Rajesh, P.B. Role of tyrosine phosphorylation in sperm capacitation/acrosome reaction. *Reprod. Biol. Endocrinol.* **2004**, *2*, 75. [CrossRef]
19. Rodriguez-Martinez, H. Role of the oviduct in sperm capacitation. *Theriogenology* **2007**, *68* (Suppl. 1), S138–S146. [CrossRef]
20. Tsai, P.S.; De Vries, K.J.; De Boer-Brouwer, M.; Garcia-Gil, N.; Van Gestel, R.A.; Colenbrander, B.; Gadella, B.M.; Van Haeften, T. Syntaxin and VAMP association with lipid rafts depends on cholesterol depletion in capacitating sperm cells. *Mol. Membr. Biol.* **2007**, *24*, 313–324. [CrossRef]
21. Rodriguez-Martinez, H.; Ekstedt, E.; Einarsson, S. Acidification of epididymal fluid in the boar. *Int. J. Androl.* **1990**, *13*, 238–243. [CrossRef]
22. Soriano-Úbeda, C.; Romero-Aguirregomezcorta, J.; Matás, C.; Visconti, P.E.; García-Vázquez, F.A. Manipulation of bicarbonate concentration in sperm capacitation media improvesin vitro fertilisation output in porcine species. *J. Anim. Sci. Biotechnol.* **2019**, *10*, 19. [CrossRef]
23. Tardif, S.; Dubé, C.; Chevalier, S.; Bailey, J.L. Capacitation is associated with tyrosine phosphorylation and tyrosine kinase-like activity of pig sperm proteins. *Biol. Reprod.* **2001**, *65*, 784–792. [CrossRef]
24. Holt, W.V.; Harrison, R.A. Bicarbonate stimulation of boar sperm motility via a protein kinase A-dependent pathway: Between-cell and between-ejaculate differences are not due to deficiencies in protein kinase A activation. *J. Androl.* **2002**, *23*, 557–565.
25. Fàbrega, A.; Puigmulé, M.; Yeste, M.; Casas, I.; Bonet, S.; Pinart, E. Impact of epididymal maturation, ejaculation and in vitro capacitation on tyrosine phosphorylation patterns exhibited of boar (*Sus domesticus*) spermatozoa. *Theriogenology* **2001**, *76*, 1356–1366. [CrossRef] [PubMed]
26. Puigmulé, M.; Fàbrega, A.; Yeste, M.; Bonet, S.; Pinart, E. Study of the proacrosin–acrosin system in epididymal, ejaculated and in vitro capacitated boar spermatozoa. *Reprod. Fertil. Dev.* **2011**, *23*, 837–845. [CrossRef] [PubMed]
27. Yeste, M.; Codony, F.; Estrada, E.; Lleonart, M.; Balasch, S.; Peña, A.; Bonet, S.; Rodríguez-Gil, J.E. Specific LED-based red light photo-stimulation procedures improve overall sperm function and reproductive performance of boar ejaculates. *Sci. Rep.* **2016**, *6*, 22569. [CrossRef]
28. Breininger, E.; Dubois, D.; Pereyra, V.E.; Rodriguez, P.C.; Satorre, M.M.; Cetica, P.D. Participation of phosphofructokinase, malate dehydrogenase and isocitrate dehydrogenase in capacitation and acrosome reaction of boar spermatozoa. *Reprod. Domest. Anim.* **2017**, *52*, 731–740. [CrossRef]
29. Ramió-Lluch, L.; Fernández-Novell, J.M.; Peña, A.; Colás, C.; Cebrián-Pérez, J.A.; Muiño-Blanco, T.; Ramírez, A.; Concha, I.I.; Rigau, T.; Rodríguez-Gil, J.E. 'In vitro' capacitation and acrosome reaction are concomitant with specific changes in mitochondrial activity in boar sperm: Evidence for a nucleated mitochondrial activation and for the existence of a capacitation-sensitive subpopulational structure. *Reprod. Domest. Anim.* **2011**, *46*, 664–673. [CrossRef] [PubMed]
30. Ramió-Lluch, L.; Yeste, M.; Fernández-Novell, J.M.; Estrada, E.; Rocha, L.; Cebrián-Pérez, J.A.; Muiño-Blanco, T.; Concha, I.I.; Ramírez, A.; Rodríguez-Gil, J.E. Oligomycin A-induced inhibition of mitochondrial ATP-synthase activity suppresses boar sperm motility and in vitro capacitation achievement without modifying overall sperm energy levels. *Reprod. Fertil. Dev.* **2014**, *26*, 883–897. [CrossRef] [PubMed]
31. Jiménez, I.; González-Márquez, H.; Ortiz, R.; Herrera, J.A.; García, A.; Betancourt, M.; Fierro, R. Changes in the distribution of lectin receptors during capacitation and acrosome reaction in boar spermatozoa. *Theriogenology* **2003**, *59*, 1171–1180. [CrossRef]
32. Ramió, L.; Rivera, M.M.; Ramírez, A.; Concha, I.I.; Peña, A.; Rigau, T.; Rodríguez-Gil, J.E. Dynamics of motile-sperm subpopulation structure in boar ejaculates subjected to "in vitro" capacitation and further "in vitro" acrosome reaction. *Theriogenology* **2008**, *69*, 501–512. [CrossRef]
33. Yeste, M.; Lloyd, R.E.; Badia, E.; Briz, M.; Bonet, S.; Holt, W.V. Direct contact between boar spermatozoa and porcine oviductal epithelial cell (OEC) cultures is needed for optimal sperm survival in vitro. *Anim. Reprod. Sci.* **2009**, *113*, 263–278. [CrossRef]
34. Lee, J.A.; Spidlen, J.; Boyce, K.; Cai, J.; Crosbie, N.; Dalhin, M.; Furlong, J.; Gasparetto, M.; Goldberg, M.; Goralczyk, E.M.; et al. MIFlowCyt: The minimum information about a flow cytometry experiment. *Cytometry A* **2008**, *73*, 926–930. [CrossRef] [PubMed]
35. Yeste, M.; Flores, E.; Estrada, E.; Bonet, S.; Rigau, T.; Rodríguez-Gil, J.E. Reduced glutathione and procaine hydrochloride protect the nucleoprotein structure of boar spermatozoa during freeze thawing by stabilising disulfide bonds. *Reprod. Fertil. Dev.* **2013**, *25*, 1036–1050. [CrossRef] [PubMed]
36. Petrunkina, A.M.; Waberski, D.; Bollwein, H.; Sieme, H. Identifying non-sperm particles during flow cytometric physiological assessment: A simple approach. *Theriogenology* **2010**, *73*, 995–1000. [CrossRef] [PubMed]
37. Garner, D.L.; Johnsons, L.A. Viability assessment of mammalian sperm using SYBR-14 and propidium iodide. *Biol. Reprod.* **1995**, *53*, 276–284. [CrossRef]
38. Harrison, R.A.P.; Ashworth, P.J.C.; Miller, N.G.A. Bicarbonate/CO_2, an effector of capacitation, induces a rapid and reversible change in the lipid architecture of boar sperm plasma membranes. *Mol. Reprod. Dev.* **1996**, *45*, 378–391. [CrossRef]

39. Guthrie, H.D.; Welch, G.R. Determination of intracellular reactive oxygen species and high mitochondrial membrane potential in Percoll treated viable boar sperm using fluorescence-activated flow cytometry. *J. Anim. Sci.* **2006**, *84*, 2089–2100. [CrossRef] [PubMed]
40. Ramió-Lluch, L.; Blanco Prieto, O.; Ramírez, A.; Fernández-Novell, J.M.; Peña, A.; Rodríguez-Gil, J.E. "In vitro" capacitation and further progesterone-induced acrosome exocytosis are linked to specific changes in the expression and location of threonine phosphorylation of boar spermatozoa. *Reprod. Domest. Anim.* **2019**, *54*, 1085–1094. [CrossRef]
41. Boatman, D.E.; Robbins, R.S. Bicarbonate: Carbon-dioxide regulation of sperm capacitation, hyperactivated motility, and acrosome reactions. *Biol. Reprod.* **1991**, *44*, 806–813. [CrossRef]
42. Leahy, T.; Rickard, J.P.; Aitken, R.J.; de Graaf, S.P. Penicillamine prevents ram sperm agglutination in media that support capacitation. *Reproduction* **2016**, *151*, 167–177. [CrossRef]
43. Leemans, B.; Gadella, B.M.; Stout, T.A.; Sostaric, E.; De Schauwer, C.; Nelis, H.; Hoogewijs, M.; Van Soom, A. Combined albumin and bicarbonate induces head-to-head sperm agglutination which physically prevents equine sperm-oviduct binding. *Reproduction* **2016**, *151*, 313–330. [CrossRef]
44. Alonso, C.A.I.; Osycka-Salut, C.E.; Castellano, L.; Cesari, A.; Di Siervi, N.; Mutto, A.; Johannisson, A.; Morrell, J.M.; Davio, C.; Perez-Martinez, S. Extracellular cAMP activates molecular signalling pathways associated with sperm capacitation in bovines. *Mol. Hum. Reprod.* **2017**, *23*, 521–534. [CrossRef] [PubMed]
45. Kerns, K.; Zigo, M.; Drobnis, E.Z.; Sutovsky, M.; Sutovsky, P. Zinc ion flux during mammalian sperm capacitation. *Nat. Commun.* **2018**, *9*, 2061. [CrossRef]
46. Ded, L.; Dostalova, P.; Zatecka, E.; Dorosh, A.; Komrskova, K.; Peknicova, J. Fluorescent analysis of boar sperm capacitation process in vitro. *Reprod. Biol. Endocrinol.* **2019**, *17*, 109. [CrossRef] [PubMed]
47. Hamilton, L.E.; Zigo, M.; Mao, J.; Xu, W.; Sutovsky, P.; O'Flaherty, C.; Oko, R. GSTO2 Isoforms Participate in the Oxidative Regulation of the Plasmalemma in Eutherian Spermatozoa during Capacitation. *Antioxidants* **2019**, *8*, 601. [CrossRef] [PubMed]
48. Waberski, D.; Luther, A.M.; Grünther, B.; Jäkel, H.; Henning, H.; Vogel, C.; Peralta, W.; Weitze, K.F. Sperm function in vitro and fertility after antibiotic-free, hypothermic storage of liquid preserved boar semen. *Sci. Rep.* **2019**, *9*, 14748. [CrossRef] [PubMed]
49. Zapata-Carmona, H.; Soriano-Úbeda, C.; París-Oller, E.; Matás, C. Periovulatory oviductal fluid decreases sperm PKA activity, tyrosine phosphorylation and in vitro fertilization in pig. *Andrology* **2019**, *8*, 756–768. [CrossRef] [PubMed]
50. Harrison, R.A.; Gadella, B.M. Bicarbonate-induced membrane processing in sperm capacitation. *Theriogenology* **2005**, *63*, 342–351. [CrossRef]
51. Gadella, B.M.; Harrison, R.A. The capacitating agent bicarbonate induces protein kinase A-dependent changes in phospholipid transbilayer behavior in the sperm plasma membrane. *Development* **2000**, *127*, 2407–2420. [CrossRef]
52. Lee, S.H.; Park, C.K. Antioxidative effects of magnetized extender containing bovine serum albumin on sperm oxidative stress during long-term liquid preservation of boar semen. *Biochem. Biophys. Res. Commun.* **2015**, *464*, 467–472. [CrossRef]
53. Parks, J.E.; Lynch, D.V. Lipid composition and thermotropic phase behavior of boar, bull, stallion, and rooster sperm membranes. *Cryobiology* **1992**, *29*, 255–266. [CrossRef]
54. Waterhouse, K.E.; Hofmo, P.O.; Tverdal, A.; Miller, R.R., Jr. Within and between breed differences in freezing tolerance and plasma membrane fatty acid composition of boar sperm. *Reproduction* **2006**, *131*, 887–894. [CrossRef]
55. Martínez-Soto, J.C.; Landeras, J.; Gadea, J. Spermatozoa and seminal plasma fatty acids as predictors of cryopreservation success. *Andrology* **2013**, *1*, 365–375. [CrossRef]
56. Mandal, R.; Badyakar, D.; Chakrabarty, J. Role of membrane lipid fatty acids in sperm cryopreservation. *Adv. Androl.* **2014**, *2014*, 190542. [CrossRef]
57. Gallon, F.; Marchetti, C.; Jouy, N.; Marchetti, P. The functionality of mitochondria differentiates human spermatozoa with high and low fertilizing capability. *Fertil. Steril.* **2006**, *86*, 1526–1530. [CrossRef] [PubMed]
58. Paoli, D.; Gallo, M.; Rizzo, F.; Baldi, E.; Francavilla, S.; Lenzi, A.; Lombardo, F.M.D.; Gandini, L. Mitochondrial membrane potential profile and its correlation with increasing sperm motility. *Fertil. Steril.* **2011**, *95*, 2315–2319. [CrossRef] [PubMed]
59. Wang, M.J.; Ou, J.X.; Chen, G.W.; Wu, J.P.; Shi, H.J.; Wai-Sum, O.; Martin-DeLeon, P.A.; Chen, H. Does prohibitin expression regulate sperm mitochondrial membrane potential, sperm motility, and male fertility? *Antioxid. Redox Signal.* **2012**, *17*, 513–519. [CrossRef]
60. Hurtado de Llera, A.; Martin-Hidalgo, D.; Rodriguez-Gil, J.E.; Gil, M.C.; Garcia-Marin, L.J.; Bragado, M.J. AMP-activated kinase, AMPK, is involved in the maintenance of plasma membrane organization in boar spermatozoa. *Biochim. Biophys. Acta* **2013**, *1828*, 2143–2151. [CrossRef]
61. Sánchez-Cárdenas, C.; Montoya, F.; Navarrete, F.A.; Hernández-Cruz, A.; Corkidi, G.; Visconti, P.E.; Darszon, A. Intracellular Ca2+ threshold reversibly switches flagellar beat off and on. *Biol. Reprod.* **2018**, *99*, 1010–1021. [CrossRef]
62. Singh, A.P.; Rajender, S. CatSper channel, sperm function and male fertility. *Reprod. Biomed. Online* **2015**, *30*, 28–38. [CrossRef]
63. Lishko, P.V.; Kirichok, Y.; Ren, D.; Navarro, B.; Chung, J.J.; Clapham, D.E. The control of male fertility by spermatozoan ion channels. *Annu. Rev. Physiol.* **2012**, *74*, 453–475. [CrossRef] [PubMed]
64. Brown, S.G.; Costello, S.; Kelly, M.C.; Ramalingam, M.; Drew, E.; Publicover, S.J.; Barratt, C.L.R.; Da Silva, S.M. Complex CatSper-dependent and independent [Ca2+]i signalling in human spermatozoa induced by follicular fluid. *Hum. Reprod.* **2017**, *32*, 1995–2006. [CrossRef]

65. Gadella, B.M.; Boerke, A. An update on post-ejaculatory remodeling of the sperm surface before mammalian fertilization. *Theriogenology* **2016**, *85*, 113–124. [CrossRef] [PubMed]
66. Espinosa, F.; López-González, I.; Muñoz-Garay, C.; Felix, R.; De la Vega-Beltrán, J.L.; Kopf, G.S.; Visconti, P.E.; Darszon, A. Dual regulation of the T-type Ca2+ current by serum albumin and β-estradiol in mammalian spermatogenic cells. *FEBS Lett.* **2000**, *475*, 251–256. [CrossRef]
67. Flesch, F.M.; Gadella, B.M. Dynamics of the mammalian sperm plasma membrane in the process of fertilization. *Biochim. Biophys. Acta* **2000**, *1469*, 197–235. [CrossRef]
68. Boerke, A.; van der Lit, J.; Lolicato, F.; Stout, T.A.; Helms, J.B.; Gadella, B.M. Removal of GPI-anchored membrane proteins causes clustering of lipid microdomains in the apical head area of porcine sperm. *Theriogenology* **2014**, *81*, 613–624. [CrossRef]
69. Xia, J.; Ren, D. The BSA-induced Ca (2+) influx during sperm capacitation is CATSPER channel-dependent. *Reprod. Biol. Endocrinol.* **2009**, *7*, 119. [CrossRef]
70. Tardif, S.; Lefièvre, L.; Gagnon, C.; Bailey, J.L. Implication of cAMP during porcine sperm capacitation and protein tyrosine phosphorylation. *Mol. Reprod. Dev.* **2004**, *69*, 428–435. [CrossRef]
71. Puga Molina, L.C.; Pinto, N.A.; Torres Rodríguez, P.; Romarowski, A.; Vicens Sanchez, A.; Visconti, P.E.; Darszon, A.; Treviño, C.L.; Buffone, M.G. Essential Role of CFTR in PKA-Dependent Phosphorylation, Alkalinization, and Hyperpolarization During Human Sperm Capacitation. *J. Cell. Physiol.* **2017**, *232*, 1404–1414. [CrossRef]
72. Córdova, A.; Strobel, P.; Ulloa, O.; Urra, J.; Contreras, C.; Valenzuela, P.; Carrillo, A.; Burgos, R.; Menarim, B.; Ratto, M.; et al. Development of stallion sperm cryopreservation extenders by analysis of semen quality and new cryocapacitation markers. *Cryobiology* **2012**, *65*, 352. [CrossRef]
73. Nishi, A.; Bibb, J.A.; Matsuyama, S.; Hamada, M.; Higashi, H.; Nairn, A.C.; Greengard, P. Regulation of DARPP-32 dephosphorylation at PKA- and Cdk5-sites by NMDA and AMPA receptors: Distinct roles of calcineurin and protein phosphatase-2A. *J. Neurochem.* **2002**, *81*, 832–841. [CrossRef]
74. Lindskog, M.; Kim, M.; Wikström, M.A.; Blackwell, K.T.; Kotaleski, J.H. Transient calcium and dopamine increase PKA activity and DARPP-32 phosphorylation. *PLoS Comput. Biol.* **2006**, *2*, e119. [CrossRef] [PubMed]
75. Aparicio, I.M.; Bragado, M.J.; Gil, M.C.; Garcia-Herreros, M.; Gonzalez-Fernandez, L.; Tapia, J.A.; Garcia-Marin, L.J. Phosphatidylinositol 3-kinase pathway regulates sperm viability but not capacitation on boar spermatozoa. *Mol. Reprod. Dev.* **2007**, *74*, 1035–1042. [CrossRef]
76. Hurtado de Llera, A.; Martin-Hidalgo, D.; Gil, M.C.; Garcia-Marin, L.J.; Bragado, M.J. New insights into transduction pathways that regulate boar sperm function. *Theriogenology* **2016**, *85*, 12–20. [CrossRef]
77. Reid, A.T.; Anderson, A.L.; Roman, S.D.; McLaughlin, E.A.; McCluskey, A.; Robinson, P.J.; Aitken, R.J.; Nixon, B. Glycogen synthase kinase 3 regulates acrosomal exocytosis in mouse spermatozoa via dynamin phosphorylation. *FASEB J.* **2015**, *29*, 2872–2882. [CrossRef]
78. Belenky, M.; Breitbart, H. Role and regulation of Glycogen synthase kinase-3 beta in bovine spermatozoa. *Mol. Reprod. Dev.* **2017**, *84*, 8–18. [CrossRef] [PubMed]
79. Jin, S.K.; Yang, W.X. Factors and pathways involved in capacitation: How are they regulated? *Oncotarget* **2017**, *8*, 3600–3627. [CrossRef]
80. Stambolic, V.; Woodgett, J.R. Mitogen inactivation of glycogen synthase kinase-3 beta in intact cells via serine 9 phosphorylation. *Biochem. J.* **1994**, *303*, 701–704. [CrossRef] [PubMed]
81. Frame, S.; Cohen, P.; Biondi, R.M. A common phosphate binding site explains the unique substrate specificity of GSK3 and its inactivation by phosphorylation. *Mol. Cell* **2001**, *7*, 1321–1327. [CrossRef]
82. Somanath, P.R.; Jack, S.L.; Vijayaraghavan, S. Changes in sperm glycogen synthase kinase-3 serine phosphorylation and activity accompany motility initiation and stimulation. *J. Androl.* **2004**, *25*, 605–617. [CrossRef]
83. Aparicio, I.M.; Bragado, M.J.; Gil, M.C.; Garcia-Herreros, M.; Gonzalez-Fernandez, L.; Tapia, J.A.; Garcia-Marin, L.J. Porcine sperm motility is regulated by serine phosphorylation of the glycogen synthase kinase-3α. *Reproduction* **2007**, *134*, 435–444. [CrossRef] [PubMed]
84. Rival, C.M.; Xu, W.; Shankman, L.S.; Morioka, S.; Arandjelovic, S.; Lee, C.S.; Wheeler, K.M.; Smith, R.P.; Haney, L.B.; Isakson, B.E.; et al. Phosphatidylserine on viable sperm and phagocytic machinery in oocytes regulate mammalian fertilization. *Nat. Commun.* **2019**, *10*, 4456. [CrossRef] [PubMed]

biology

Review

Sperm Phosphoproteome: Unraveling Male Infertility

Rebeca Serrano, Luis J. Garcia-Marin and Maria J. Bragado *

Grupo de Investigación en Señalización Intracelular y Tecnología de la Reproducción (SINTREP), Instituto de Investigación INBIO G+C, Universidad de Extremadura, 10003 Caceres, Spain; rebecasp@unex.es (R.S.); ljgarcia@unex.es (L.J.G.-M.)
* Correspondence: jbragado@unex.es

Simple Summary: Approximately 24% of men referred to assisted reproductive technology (ART) present with idiopathic male infertility. The current standard analysis of human semen does not allow for an accurate diagnosis of this infertility with unknown etiology. Spermatozoa cellular development and maturation, as well the acquisition of suitable motility and capacitation, are tightly coordinated by sperm protein phosphorylation, among other protein post-translational modifications. Extraordinary advances have been achieved in the field of spermatozoa using proteomics methodology in combination with bioinformatics. The aim of this work is to review, using a proteomic and phosphoproteomic analysis, the updated knowledge about proteins and phosphoproteins of spermatozoa that regulate cell processes necessary to achieve a proper fertilization. The phosphorylation of sperm proteins involved in spermatogenesis, in sperm capacitation, and in the maintenance of correct sperm motility, and consequently in sperm quality, is focused on in this review. Further investigations of protein spermatozoa in larger populations combined with other multi-omics technologies would offer a precise perspective of male fertility and would be especially relevant for those cases involving repeated failures in ART linked to idiopathic infertility.

Abstract: Infertility affects approximately 15% of couples worldwide of childbearing age, and in many cases the etiology of male infertility is unknown. The current standard evaluation of semen is insufficient to establish an accurate diagnosis. Proteomics techniques, such as phosphoproteomics, applied in this field are a powerful tool to understand the mechanisms that regulate sperm functions such as motility, which is essential for successful fertilization. Among the post-translational modifications of sperm proteins, this review summarizes, from a proteomic perspective, the updated knowledge of protein phosphorylation, in human spermatozoa, as a relevant molecular mechanism involved in the regulation of sperm physiology. Specifically, the role of sperm protein phosphorylation in motility and, consequently, in sperm quality is highlighted. Additionally, through the analysis of published comparative phosphoproteomic studies, some candidate human sperm phosphoproteins associated with low sperm motility are proposed. Despite the remarkable advances in phosphoproteomics technologies, the relatively low number of studies performed in human spermatozoa suggests that phosphoproteomics has not been applied to its full potential in studying male infertility yet. Therefore, further studies will improve the application of this procedure and overcome the limitations, increasing the understanding of regulatory mechanisms underlying protein phosphorylation in sperm motility and, consequently, in male fertility.

Keywords: human spermatozoa; sperm proteins; PTM; phosphorylation; phosphoproteomics; sperm motility; male infertility

Citation: Serrano, R.; Garcia-Marin, L.J.; Bragado, M.J. Sperm Phosphoproteome: Unraveling Male Infertility. *Biology* **2022**, *11*, 659. https://doi.org/10.3390/biology11050659

Academic Editors: Pilar Santolaria Blasco, Rosaura Pérez-Pé and Jessica Rickard

Received: 7 April 2022
Accepted: 22 April 2022
Published: 25 April 2022

Publisher's Note: MDPI stays neutral with regard to jurisdictional claims in published maps and institutional affiliations.

1. Introduction

Our understanding of sperm physiology remains relatively superficial despite the large number of manuscripts focused on spermatozoa [1]. Indeed, a global estimation calculates that 1 in 15 men of reproductive age are infertile, and the diagnosis of idiopathic male infertility, for which the cause is unknown, is a reality for 24% of men referred to assisted reproductive technology (ART) [2]. To date, basic semen analysis, or seminogram, is the best predictive test used routinely in laboratories for the assessment of male partner fertility, where it is analyzed if the semen samples meet the macroscopic (volume, pH, color, and viscosity) and microscopic characteristics (sperm concentration, total motility, progressive motility, and sperm morphology) established by the World Health Organization (WHO), which published a set of guidelines for evaluating semen quality 40 years ago. The last update, in 2010, includes the current reference parameters to make the first prognosis of male infertility [3].

Nevertheless, the seminogram is not the most suitable analysis to obtain an accurate diagnosis because semen parameters within the reference interval do not guarantee fertility, nor do values outside those limits necessarily imply male infertility or pathology [4]. Several studies have shown that men with sperm parameters (sperm number, morphology, and motility) below the thresholds outlined by the WHO can be fertile [5–8]. Additionally, in the same way, there are cases of men with normal sperm parameters that are infertile [9]. Therefore, more in-depth analysis and understanding of spermatozoa physiology at the molecular level are necessary to improve the current evaluation of male fertility by the routine semen analysis.

Fertilization might be considered the endpoint of sperm function. To get it successfully, spermatozoon, a highly specialized haploid cell that contains exceptionally condensed chromatin and will deliver the paternal DNA to the oocyte, must be completely functional. For that, spermatozoa undergo a series of physiological and biochemical changes from their developmental stages and during their transit through the male first and female reproductive tract later, which occur apparently in the complete absence of simultaneous gene transcription and protein translation. Although several coding and non-coding RNAs exist in human spermatozoa [10], which may play a role in gene silencing or heterochromatinization, their transcriptional and translational activities are nearly silent. Thus, sperm proteins of mature spermatozoa might undergo different post-translational modifications (PTM), such as phosphorylation or acetylation among, and become very important molecular mechanisms by which spermatozoa acquire functionality [11]. For this reason, the alteration of sperm status (for example, by errors in spermatogenesis or maturation) may be accompanied by a distinctive pattern of PTMs, characteristic of spermatozoa, in particular with low quality or motility and, therefore, low male reproductive prognostic.

In this regard, it is necessary to focus our attention on sperm motile capacity, and the abnormal content or presence of PTMs, such as the reduced abundance of lysine glutarylation in several proteins located in the tail of human spermatozoa, just as a diminished quantity of S-sulfhydrated H3 and H3.3 histones positively correlate with sperm progressive motility [12,13]; conversely the level of SUMO1-positive spermatozoa and the quantity of lysine 2-hydroxyisobutyrylation in sperm proteins trend towards higher grades in asthenozoospermic men compared with normozoospermic ones, indicating a negative association with the motility of human spermatozoa in this case [14,15]. Additionally, lysine acetylation seems to be essential for human sperm motility and fertilization [16].

The most extensively studied PTM in human spermatozoa is phosphorylation [17]. The phosphorylation of specific sperm proteins plays an important role in regulating sperm processes essential for fertilization, such as sperm motility, capacitation, or acrosome reactions [18–20]. Extraordinary advances have been achieved in the field of male infertility in recent decades, especially with the use of proteomics techniques and the bioinformatic analysis of human sperm proteomic data. However, there are many well-recognized causes of male infertility in humans whose molecular basis is only just beginning to be understood. The study of the global protein phosphorylation landscape of spermatozoa in different species proposes wide phosphoregulation in other processes such as sperm formation [21,22], maturation [23], capacitation [24–28], and motility [18,29–32]. Nevertheless, it remains to be fully explained which are ultimately the molecular mechanisms responsible for spermatozoa motility and, therefore, for sperm quality. Advances in global and quantitative methods to elucidate dynamic phosphorylation events in spermatozoa will be essential for a systematic understanding of their functional behavior. They will allow for a more comprehensive analysis of the biochemical basis of defective semen quality and identify possible biomarkers for different pathologies and conditions related to infertility. A few studies applying quantitative mass spectrometry (MS)-based proteomics have proposed some molecular mechanisms through which protein phosphorylation might affect sperm motility in humans [18,29–31].

From a proteomic perspective, this review summarizes the current knowledge of protein phosphorylation in human spermatozoa as a molecular mechanism responsible for the regulation of spermatozoa motility, and subsequently of sperm quality.

2. Proteomics and Sperm Physiology

According to the recent data, the proteomic approach is a powerful tool to identify human sperm proteins as biomarkers of fertility [33]. For example, a recent seminal plasma proteomic-based study proposes the HSPA2 protein, a molecular chaperone mediating protein folding, as a possible biomarker of spermatogenesis status. Azoospermic men (who have a complete absence of spermatozoa in their ejaculate) lack HSPA2, which is present as additional protein isoforms in cryptozoospermia (<0.1 million spermatozoa mL^{-1}) [34]. In addition, a study comparing high- and low-quality sperm nuclear extracts by proteomic analysis recently showed that the presence of Topoisomerase 2A in the human spermatozoa head is highly correlated to poor head morphology. So, Topoisomerase 2A, a protein normally involved in the alteration of DNA topology, may also be considered a potential biomarker to confirm male infertility in clinical practice [35]. The first study that examined the potential variability of the proteome in different semen samples and proposed proteomics as a useful tool for studying defects in sperm function was published almost 2 decades ago by Pixton and collaborators [36]. Since then, many proteomics studies have performed a comparative proteomics analysis between sperm cells from infertile patients and healthy donors [37–46].

On the other hand, sperm motility is essential for successful fertilization, so low sperm motility is highly associated with male infertility [47,48]. In fact, this defect has been the subject of research for years because it is frequently observed in andrology laboratories. In this sense, a retrospective study based on a large population reveals that about 82% of infertile men had impaired sperm motility [48]. Asthenozoospermia (AS), characterized by normal concentrations of spermatozoa (>15 million spermatozoa mL^{-1}) and sperm progressive motility <32% [3], is one of the major causes of male infertility, which approximately accounts for 20% of infertility among men. The etiology of AS is varied and can be seen as a unique condition in isolated disease, associated with other sperm anomalies or as part of a syndromic association. In some cases, routine clinical examinations do not find clear causes, leading to so-called idiopathic asthenozoospermia [41]. Although the lower expression of several proteins might cause spermatozoa with poor motility, the molecular basis of AS is difficult to establish. Nonetheless, proteomic studies on asthenozoospermic individuals have increased in recent years, promoting the idea that the number of identified proteins related to sperm motility is rising [43].

For instance, in four different proteome analyses comparing sperm samples from asthenozoospermic vs. normozoospermic men, the altered expression of the HSPA2 protein was found. Interestingly, increased expression was observed in two studies [38,42], whereas HSPA2 expression decreased in the other two works [39,43]. Other chaperones HSPs (HSPA5, HSPA9, and HSPA1L) were also found downregulated in asthenozoospermic men [41,43]. Additionally, in a recent study comparing proteomes of high or low-motility human spermatozoa, HSPA1L and HSPA9 were also significantly decreased in low-motility spermatozoa [49]. Conversely, in other comparative sperm proteomics studies, the expression level of chaperones did not indicate significant differences [37,44–46]. It has to be mentioned that the lack of agreement or the opposite expression differences between the studies published may be due to several factors such as different quantitative technologies, different sample sources, low sample size, or even ethnic differences, among others. HSPs have a potential relationship with sperm quality, and they are important in normal sperm physiology [50], spermatogenesis, and sperm maturation [51–53], although the association between their altered expression and impaired motility is not yet fully understood. It is established that the role of the HSPs is to ensure the correct folding of proteins, their re-folding of misfolded protein, and the orientation control of tagged proteins for subsequent degradation [54]. Thus, reduced HSPs expression might be associated with a decrease in sperm motility due to the accumulation of misfolded protein [55].

In addition, a higher level of triosephosphate isomerase (TPI), an extremely efficient metabolic enzyme in glycolysis and gluconeogenesis, is also associated with a reduced sperm motility phenotype [37,39,42]. Other proteins responsible for energy metabolism that could play an important role in spermatozoa motility maintenance are COX proteins (COX5B, COX6B, COX20, and COX41), which are involved in the oxidative phosphorylation (OXPHOS) pathway and showed lower levels in AS [38,42,43,45,46]. In conclusion, these are some examples that highlight the importance of glycolysis and OXPHOS as major metabolic pathways that provide energy to support human sperm motility.

Moreover, a reduction in sperm motility may be affected by other proteins, such as SEMG1 and SEMG2, which work as seminal plasma motility inhibitor proteins and are found up-regulated in asthenozoospermic men [37,38,46]. Another group of proteins with altered expression in spermatozoa with impaired motility involves different subunits of the proteasome such as PSMA3, PSMB3, PSMB4, PSMB5, PSMB6, PSMC2, PSMC6, and PSMD11 [37–39,42,45,49]. The proteasome plays a key role in the formation of condensed spermatozoa because it mediates the protein turnover of ubiquitinated proteins during spermatogenesis, when many proteins and organelles are degraded [55]. So, defects in the proteasome system might lead to the accumulation of ubiquitinated molecules and be related to sperm motility [39]. It is worth noting that SEMG1 and PSMB5 are also downregulated in a proteomic study that compares the proteomic profiles of human sperm samples that had or had not achieved a previous pregnancy via ART [56]. In addition, low levels of the major cytoskeleton components in spermatozoa flagella such as tektins (TEK1, TEK4, and TEK5) [39,41,42], outer dense fibers (ODF2) [41,43,49], or tubulin proteins (TUBB2C, TUBB2B, and TUBA3C) [39,43,49] are also associated with reduced sperm motility in comparative proteomics studies. On the other hand, altered levels of another protein that plays a role in the movement and structural organization of cells, such as CLU, have been found. CLU expression is decreased in some analyses [40,49], whereas it is increased in others [38,43,46].

Proteomics is also a current methodology used to study variations in sperm proteins that are altered in some disorders. For example, the SPEF2 protein is widely expressed in cilia-related organs such as the lung, spleen, trachea, brain, and testis [57], and its encoding gene is involved in a genetically heterogeneous disorder such as the so-called multiple morphological abnormalities of the sperm flagella (MMAF) [58]. Moreover, spermiogenesis failure by a deficiency in SPEF2 causes severe asthenoteratozoospermia, characterized by reduced sperm motility and abnormal sperm morphology [58]. Recently, an MS-proteomic analysis of human spermatozoa from three individuals with SPEF2 mutations compared with normal controls showed that this protein regulates the expression of various proteins involved in the flagellar assembly with which it interacts [59]. This methodology allows one to understand the protein networks from the whole sperm proteome, being especially useful in the study of sperm tail development since sperm flagellum is composed of more than 1000 proteins in the case of humans [60].

Human sperm cryopreservation plays an important role in assisted reproductive technology for male fertility preservation and the treatment of infertile couples. In this regard, proteomics approaches have also been useful to study the pathogenesis of sperm cryo-damage during the process of cryopreservation, comparing the proteomic differences between fresh and cryopreserved human sperm [61,62]. Fu's lab, using MS and a novel proteomics technology named data-independent acquisition (DIA), identified 174 proteins significantly deregulated, including four enzymes involved in glycolysis (GPI, LDHB, ADH5, and PGAM1) and other proteins related to propanoate, glyoxylate, pyruvate, and dicarboxylate metabolism and gluconeogenesis [62]. Five years before, another proteomic analysis had found that 37% of the proteins involved in the metabolism are differentially expressed between freeze-thawed and fresh sperm samples [61]. So, both studies, although using different proteomic strategies, conclude that metabolic pathways play an important role during sperm cryo-preservation. Interestingly, phosphoglycerate mutase proteins, PGAM1 and PGAM2 [46] and the glucose 6-phosphate isomerase (GPI), evaluated by proteomic approaches, are also found to be significantly decreased in AS [42,46]. Furthermore, the supplement with the product of GPI, fructose-6-phosphate, significantly promotes human spermatozoa motility in vitro [46]. So, it can be postulated that during spermatozoa cryopreservation, when a marked reduction in sperm motility occurs, the supplement with fructose-6-phosphate could also help to recuperate the rates of spermatozoa motility.

Altogether, there are a set of proteins related to sperm quality in the literature. However, the internal relationship and the mechanisms underlying abnormal protein expressions and defective sperm function are not clear yet [49].

3. Phosphoproteomics Technique in Male Fertility

3.1. Phosphorylation as Post-Translational Modification of Sperm Proteins

Defects in PTMs have been linked to numerous human diseases and disorders, so the importance of PTMs in maintaining normal cellular states is essential [63]. Hence, previous and emerging data indicate that some male reproduction diseases, including the failure of sperm motility, arise through the deregulation of PTMs in spermatozoa. Despite that, more than 431 reversible and irreversible PTM mechanisms exist in the cell [64], and we know that protein phosphorylation affects an estimated one-third of all cellular proteins [65], with most proteins phosphorylated at one or more sites in a mammalian cell [66]. However, we know only a small subset of the in vivo phosphorylation sites described. Most studies have focused on Ser, Thr, and Tyr phosphorylation (canonical phosphorylation), but there are other amino acid residues that are less common, including His, Lys, Arg, Asp, Glu, and Cys, that can also be phosphorylated (noncanonical phosphorylation) [67]. This variability of possibilities further complicates the effort in studying protein phosphorylation, whose consequences may affect its activation status [68].

Nonetheless, protein phosphorylation is not permanent due to the activity of phosphatases [69]. Consequently, the deregulation of kinases and phosphatases pathways is linked to many diseases, including infertility. So, deciphering the molecular elements that determine the biochemical balance of phosphorylation/dephosphorylation is essential for correct reproductive function.

3.2. Phosphoproteomics Technique

Phosphoproteomics is a large-scale analysis that identifies and quantifies the phosphorylated proteins in addition to the mapping of the phosphorylation sites in a complex biological sample using MS [70]. Briefly, an MS-based phosphoproteomics study on the role of in vivo phosphorylation in sperm physiology starts with isolating sperm cells from the seminal plasma and other cells coexisting in semen by swim-up procedure, density centrifugation, or different techniques. The purity of the sperm preparation and the removal of interfering compounds are critical steps in the process because any minor contamination could result in a false-positive identification [71]. Sperm proteins are then extracted and protein mixtures are digested with a specific protease, typically trypsin.

Once proteins are extracted, carrying out a phosphopeptides enrichment procedure before experimental analysis is necessary, given that, for example, almost 30% of all human proteins may be phosphorylated and that each phosphoprotein may exist as multiple phospho-isoforms with different relative abundances and stoichiometries [70]. This procedure allows/has its purpose the characterization from low femtomole level phosphorylated proteins and the improvement of selectivity by reducing the unspecific binding of non-phosphorylated peptides [72].

Among the wide selection of methodologies developed for phosphopeptides enrichment, the most extensively used in the study of sperm cells is immobilized metal ion affinity chromatography (IMAC) [17], which is based upon the affinity that phosphate exhibits towards immobilized metal ions and forms relatively stable complexes with these. So, the nature of the chromatographic stationary phase is of extreme importance [73]. Accordingly, titanium dioxide (TiO2) resin has been one of the most widespread methods for phosphopeptides enrichment from complex biological samples because it has a very high affinity for phosphopeptides, is extremely tolerant towards most buffers used in biological experiments, and is optimal for large-scale phosphoproteomics studies [74].

Later, the sperm phosphopeptides are detected using both conventional and advanced proteomic techniques. Two-dimensional (2D) gel electrophoresis separates sperm proteins based on peptides' isoelectric focusing properties and molecular weight. A modified version named difference gel electrophoresis (DIGE) identifies differentially expressed proteins (DEPs) [75]. The analysis of advanced high-throughput techniques such as MALDI-TOF (matrix-assisted laser desorption/ionization time-of-flight) and LC-MS/MS (liquid chromatography-tandem mass spectrometry) detect low abundance peptides present in a sample with low protein concentration. Therefore, they overcome the limitations of conventional proteomics techniques. In addition, advances in chromatography techniques such as nano HPLC (high-performance liquid chromatography) or UPLC (ultra-performance liquid chromatography) methods enable the decrease of the internal diameter of the LC column to analyze low amounts of a sample with none or very low dilution and with increased sensitivity, which allows for higher sample throughput [73]. Those aspects are fundamental in phosphoproteomics studies.

In addition, MS offers numerous advantages for studying protein phosphorylation, enabling its quantitative, sensitive, and site-specific measure [76]. MS-based quantification strategies rely on light/heavy peptide intensities and can be divided into label-based and label-free approaches. Label-based quantitation methods utilize stable isotope labels by chemical, metabolic, or proteolytic labeling strategies. These are incorporated within the peptides, introducing an expectable mass difference within the two or more experimental conditions. The quantitation is based on comparing the peak intensity ratio of the labeled peptide pairs [77]. In contrast, label-free quantitation compares both relative and absolute protein quantity by utilizing signal intensity and spectral counting of the same peptide [77]. It is also one of the methods of choice in human sperm phosphoproteome studies [27,31] and has gained more acceptance because it shows the highest proteome coverage and is cost-efficient without adding additional steps to labeling samples with alternative differential mass tags [78].

Finally, mass spectral data interpretation is carried out using the different platforms, databases, and software programs available, which allow for the identification and quantification of the assignments of peptides and proteins that make up the sperm phosphoproteome. Bioinformatics methods are indispensable for proteomics-based studies and are helping scientists to interpret the integration of large datasets from proteomics studies [79]. The specific workflow involving the processing of semen samples for sperm phosphoproteomics analysis is shown in Figure 1.

Furthermore, phosphoproteomics, in combination with these high-throughput techniques, is one of the most potent techniques nowadays for the global analysis of signaling networks in defined biological systems [80], such as human spermatozoa [27,31].

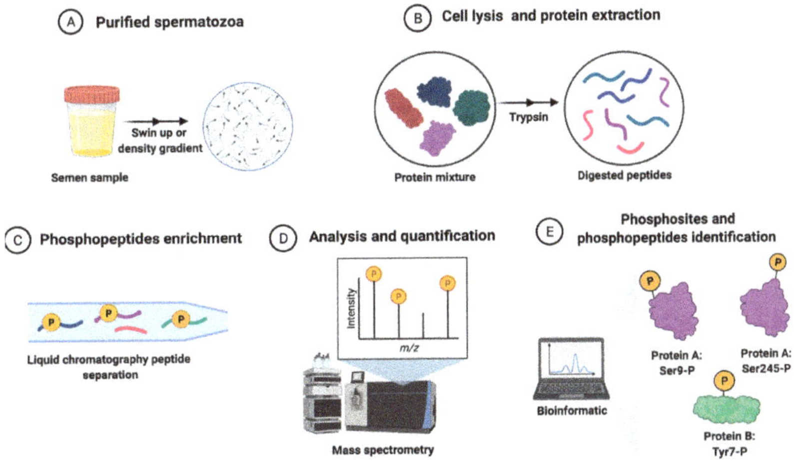

Figure 1. The general workflow of the quantitative phosphoproteomics strategy for human spermatozoa samples analysis from sperm donors. (**A**) Purified sperm cells from semen samples (spermatozoa isolated from other cells and the seminal plasma). (**B**) Sperm protein extraction and digestion after cellular lysis. (**C**) Phosphopeptides enrichment and peptide separation. (**D**) The analysis and quantification of peptides. (**E**) The collection and analysis of sperm phosphoproteome data by bioinformatics tools.

4. Phosphoproteomics and Spermatogenesis

As mentioned before, phosphorylation of sperm proteins is linked with male fertility because this PTM is extremely important in all stages of sperm cell development, being essential for sperm differentiation, maturation, and function [17]. To date, it is the most extensively studied PTM in mammalian spermatozoa, as previously mentioned. Besides, spermatozoa are an excellent cell model for proteomic analysis because they are purified in large numbers and are reliably and robustly driven into different functional states using various incubation media or validated pharmacological manipulations [81]. In fact, in the study of male reproduction, there is research identifying phosphoproteins and their phosphorylated sites through differential phosphoproteomics analysis of sperm cells from infertile men or under experimental or physiological conditions [18,24,29,31,32]. To include a summary of the use of phosphoproteomics studies carried out in human spermatozoa to study sperm biological processes, Table 1 is incorporated.

Table 1. A summary of phosphoproteomics studies in human spermatozoa biological processes.

Biological Process	Study	Type of Samples	Sperm Preparation	Phosphoproteomic Method
Spermatogenesis	Castillo et al. 2019	Testicular tissue		LC-MS/MS
Sperm motility	Chan et al. 2009	Normozoospermic vs. asthenozoospermic spermatozoa	Percoll fractionation	2DE-MALDI-TOF MS
	Parte et al. 2012	Normozoospermic vs. asthenozoospermic spermatozoa	Washing	Nano UPLC-MS
	Martin-Hidalgo et al. 2020	High-mobility vs. low-mobility sperm subpopulations	PureSperm fractionation	Nano HPLC-MS/MS Triple TOF
Sperm capacitation	Ficarro et al. 2003	Capacitated vs. non-capacitated spermatozoa	Percoll fractionation	2DE-anti-phosphotyrosine Immunoblots MS/MS
	Wang et al. 2015	Capacitated vs. non-capacitated spermatozoa	Percoll fractionation	LC-MS/MS

Defects in spermatogenesis are the most common factors for male infertility [4]. In this sense, phosphoproteomics tools are helpful to understand the origin of different causes of infertility because phosphoregulation is highly active during sperm differentiation [21,82,83]. Thanks to the technology for analyzing kinase-substrate relations (KSRs) in coordination with the exploration of the phosphoproteome, a pattern of consistently high activity for many kinases has been elucidated during spermatogenesis in mice, including the MAPKs, CDKs, and especially the POLO-like kinases (PLKs) [83]. In the same way, to decipher the most relevant signaling pathways during the development of male gametes in the human testes, Castillo and coworkers performed global phosphoproteomics on human testicular tissue with full spermatogenesis using a TiO_2 method coupled to MS [21]. They identified 2661 proteins, and 174 of them were different phosphorylated kinases covering 32% of the human kinome, including MAPK1, MAPK3, CDK12, CDK13, and PAK4. Curiously, unlike in mice, PLKs are not among the most active kinases regulating human spermatogenesis [21]. In the same year, using proteomics and phosphoproteomics analyses of mouse testes, Wei's lab suggested that WIP1 phosphatase is involved in maintaining the integrity of the blood–testicular barrier [84]. Therefore, as this organ is essential for spermatogenesis to progress correctly, its alteration results in male subfertility or infertility [84]. Besides, this phosphatase WIP1 seems to be an important regulator of global heterochromatin silencing and is critical in maintaining genome integrity [85], a basic regulatory mechanism for spermatozoa function. Although there are no studies that associate the role of WIP1 with male fertility in humans, this link has been observed previously in the mouse testis phosphoproteome [83].

During mammalian spermiogenesis, the last phase of the spermatogenesis, haploid round spermatids are differentiated into spermatozoa undergoing remarkable morphological changes, chromatin condensation, the biogenesis of the acrosome, the migration of mitochondria to the intermediate piece, and flagellum formation. A large-scale phosphoproteome analysis performed from purified mouse spermatids undergoing spermiogenesis described 735 testis-specific proteins phosphorylated and expressed at high levels. These phosphoproteins are implied in histone modifications and chromosome and cilium organization [22]. Nonetheless, defects at any level of the spermatogenesis, where protein phosphorylation is essential, would be inevitably associated with fertility impairments (such as azoospermia, teratozoospermia, oligozoospermia, or asthenozoospermia).

Evidence indicates that correct epididymal sperm maturation also requires protein phosphorylation and dephosphorylation events in spermatozoa. This epididymal process takes approximately two weeks in humans [86] and is necessary for generating fertile spermatozoa. When the spermatozoa leave the testes, they are immotile, and it is during their transit and storage through the epididymis when they acquire progressive motility and functional capacity for their interaction with the oocyte. During epididymal sperm maturation, besides phosphorylation, ubiquitination, a frequent PTM in regulating many sperm biological processes, is important as it eliminates defective spermatozoa (mostly with defects in morphology) by phagocytosis [87]. Interestingly, more and more scientific evidence shows that many proteins have different types of PTM simultaneously, which all together help to regulate protein stability and activity. The imbalance of theses PTMs' crosstalk may be highly associated with male infertility; however, there are few studies about multiple PTMs co-occurring in sperm proteins (except for histones and protamines) [88]. Zhang and collaborators recently combined the phosphoproteome with the ubiquitylome to study the physiological mechanisms underlying sperm maturation in epididymal spermatozoa of buffalo [23]. Since a few years ago, sperm ubiquitination has been a marker of defective spermatozoon in humans [89]. For these reasons, mistakes in the post-testicular maturation context will affect male fertility. Based on these strategies, new biomarkers of sperm quality or semen abnormalities will help establish a precise landscape of PTMs that features high-quality spermatozoa.

5. Phosphoproteomics and Sperm Motility

In male reproduction, the use of phosphoproteomics is mainly focused on unraveling the molecular mechanisms underlying the regulation of sperm motility [71]. The first evidence about the control of sperm motility by processes that include the regulation of protein phosphorylation was described in dog spermatozoa in 1982 [90]. Later, two protein kinases, PI3K and AKAP3, were demonstrated to be involved in the phosphoregulation of human sperm motility [91]. On the other hand, as mentioned before, the phosphorylation status of a protein depends on the opposing activities of protein kinases and phosphatases. Accordingly, immotile mammalian spermatozoa contain higher activity levels of serine/threonine phosphatase 1 isoform gamma 2 (PP1γ2) compared with motile ones [92]. Moreover, the inhibition of PP1γ2 causes motility initiation in immature spermatozoa, whereas it leads to motility stimulation and changes in flagellar beat parameters in mature spermatozoa [93], indicating that protein phosphatases also regulate flagellar motility. Nowadays, high-throughput techniques are used to precisely decipher the phosphoproteome with functional importance for sperm motility, and the phosphoproteomic profiles of spermatozoa in different functional states (uncapacitated vs. capacitated; normal vs. defective, high vs. low-mobility).

Focusing on phosphorylation at tyrosine residues, phosphotyrosine-containing proteins are present in the spermatozoa of different species. For decades, it has been known that an increase in the protein tyrosine phosphorylation of spermatozoa plays a critical role in regulating sperm motility [94], especially in those processes related to hyperactivated motility. Moreover, deficiencies in the tyrosine-phosphorylated proteins of the sperm tail are associated with AS in human [95]. Later studies have focused on the differential analysis of the phosphorylation status of human sperm proteins by large-scale phosphoproteomics techniques in spermatozoa from healthy and asthenozoospermic donors [18,29] and from two sperm populations with different sperm motility degrees isolated from normozoospermic healthy donors [31]. Despite the different strategies of phosphoproteomics used, changes in sperm motility patterns correlate with the differential phosphorylation of proteins. Using 2D-gel electrophoresis MALDI-TOF/MS, the first study identified 12 proteins exhibiting differential phosphorylation. There was a relatively lower phosphorylation level in asthenozoospermic spermatozoa for 10 proteins, while in 2 of them, the level was higher [18]. In the same way, comparing healthy and asthenozoospermic semen, Parte and collaborators, using IMAC nano UPLC/MS, detected 66 sperm phosphoproteins with altered expression (39 were up and 27 were hypophosphorylated) in asthenozoospermic donors [29].

Moreover, a recent study of human spermatozoa using nano HPLC-MS/MS triple TOF confirms that the sperm proteins phosphorylation level is involved in sperm motility regulation. In fact, human sperm subpopulations with low and high motility statistically differ in up to 119 sperm phosphoproteins [31]. The constructed networks by the STRING database of protein–protein interactions of human spermatozoa from the phosphoproteins identified by the studies of Chan et al. (2009), Parte et al. (2012), and Martin-Hidalgo et al. (2020) (Table 1) and mentioned in the following paragraphs are shown in Figure 2. Combining overlapping data in these studies focused on human sperm motility, the most differentially expressed phosphoproteins were mainly involved in sperm metabolism. For example, GSTM3, a protein important in glutathione metabolism and cellular detoxification, was hypophosphorylated in asthenozoospermic donors [18,29]. Moreover, Ras-related proteins such as RAB2A, RAB2B, or RAB4A, which are involved in vesicle trafficking and ribosomal proteins such as RPLP2, RPL15, or DAP3, involved in the metabolism of proteins, also showed differential protein phosphorylation [18,29,31]. In addition, those proteins, related to the degradation processes in the protein ubiquitination pathway, such as the ubiquitin-protein ligases, UBR4 and HUWE1; polyubiquitin-C; and the ribonucleoprotein ubiquitin-40S ribosomal proteins S27a and RPS27A, were found to be hyperphosphorylated in low-mobility spermatozoa [29,31]. Other phosphoproteins involved in carbohydrate metabolism and energy production, such as pro-glucagon, fructose-bisphosphate aldolase A (ALDOA), glyceraldehyde-3-phosphate dehydrogenase-S (GAPDHS), mannose-6-phosphate isomerase (MPI), or the subunit alpha of the pyruvate dehydrogenase E1 component (PDHA1), also showed altered phosphorylation levels between two sperm subpopulations with very different motility degrees [18,29,31]. Specifically, MPI and PDHA1 were hypophosphorylated in spermatozoa with poor motility patterns [31], while the phosphorylation level of pro-glucagon, ALDOA, and GAPDHS was higher [18,29]. Interestingly, mice sperm lacking GAPDHS show deficient phosphorylation levels of sperm protein phosphatase 1 (PP1) accompanied by defects in sperm motility and male fertility [96]. However, changes in sperm motility do not correlate with differential PP1 phosphorylation in these phosphoproteomic studies. Regarding lipids metabolism, phosphoproteins such as ATP-citrate synthase (ACLY), dihydroxyacetone phosphate acyltransferase (GNPAT), ethanolamine-phosphate cytidylyltransferase (PCYT2), choline-phosphate cytidylyltransferase B (PCYT1B), or phospholipid hydroperoxide glutathione peroxidase (GPX4) were found to be altered according to the motility status of each spermatozoon [29,31]. While the phosphorylation levels of ACLY, GNPAT, PCYT2, and PCYT1B were increased, GPX4 was hypophosphorylated in low-mobility spermatozoa. These results highlight the importance of metabolism in different motility patterns in human spermatozoa. In fact, up to 40% of human proteins that are hyperphosphorylated in low-mobility spermatozoa are involved in metabolism [31]. This interesting finding is consistent with the differences observed by the proteomic approach in the expression levels of sperm proteins between healthy and asthenozoospermic individuals [45,46].

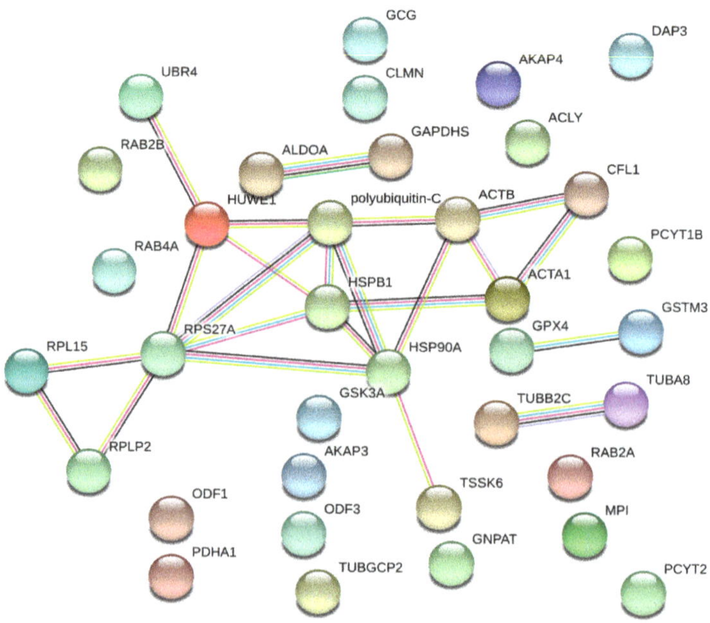

Figure 2. The protein–protein interactions (PPI) network of some phosphoproteins identified by comparative phosphoproteomics studies between human spermatozoa with different levels of motility. PPI are generated by the STRING database (http://string-db.org/, accessed on 14 April 2022). Bubbles show phosphoproteins involved in sperm motility, and lines represent both functional and physical protein associations. The colors of the lines indicate the type of interaction evidenced from different sources: databases (blue), protein homology (grey), high-throughput experiments (pink), co-expression experiments (black), and prior knowledge from research publications (yellow). The absence of a line indicates that no interaction has been detected. Only those interactions with a high confidence interaction score (score $\geq$0.7 according to STRING indications) are shown.

Other matching human phosphoproteins that are also found at altered levels in spermatozoa with impaired motility include those associated with flagellum assembly and motility [18,29,31]. The flagellum, a fundamental structure for spermatozoa motility, requires a specific organization of microtubules. The protein components of the tubulin superfamily (TUBB2C, TUBA8, or TUBGCP2, among others), together with actin and acting-binding proteins such as ACTA1, ACTB, CFL1, or CLMN, play an important role in the assembly of microtubules and consequently in sperm motility. These cytoskeletal proteins showed altered phosphorylation levels in AS vs. normal donors and high vs. low-mobility human spermatozoa [18,29,31]. In such a way, four of them were hyperphosphorylated (TUBB2C, ACTB, CFL1, and CLMN) in spermatozoa with poor motility [29,31]. Furthermore, lower phosphorylation levels are important to regulate flagellum functions in proteins such as components of the fibrous sheath such as AKAP3, AKAP4, or FSIP2, and elements of the outer dense fibers, such as ODF1 and ODF3, located in the midpiece and principal piece of the tail spermatozoa. All of them are related to sperm motility showing hypophosphorylated levels in AS and low-mobility human spermatozoa [29,31]. These data agree with the significantly increased phosphorylation of AKAP3 and AKAP4 observed during hyperactivated motility (movement with high amplitude and asymmetric thrashing of the sperm tail) associated with capacitated human spermatozoa [24,27]. These two PKA-anchoring proteins, which mediate the PKA activity by localizing this kinase to specific cellular structures and organelles, are postulated to coordinate sperm capacitation events, including motility hyperactivation [97].

On the other hand, the proteins of the heat shock family HSP—HSPB1 and HSP90A—were hyperphosphorylated in low-mobility human spermatozoa [29,31]. These HSP mediate protein folding and signal transduction and prevent protein aggregation predominantly; their role in human sperm function and male fertility is not clear yet, although a potential correlation with sperm quality has been described [98].

Other phosphoproteomics techniques to study protein changes associated with sperm motility were applied during sperm cryopreservation, when a marked reduction in sperm motility after freezing and thawing occurs. Decreased motility parameters are the most significant phenotype of cryodamage. Indeed, the cutbacks of human sperm motility were between 25% and 75% of the total in an evaluation of freeze-thawed sperm samples relative to that of fresh sperm samples from the same normozoospermic donors [99]. The level of tyrosine phosphorylated proteins in the freeze-thawed group [61], as well as the number of several kinases required for sperm production and function, such as testis-specific serine/threonine-protein kinase 6 (TSSK6) [100], was higher than those in the fresh group. Recently, several phosphoproteomics studies have been performed to investigate the molecular differences between fresh and cryopreserved (post-thaw) human spermatozoa. Wang's lab constructed a quantitative phosphoproteome to investigate the expression change of phosphorylated sites during sperm cryopreservation [32], and they identified glycogen synthase kinase 3A, GSK3A, as a key kinase that may play an important role in regulating human sperm motility. The A isoform of the serine threonine kinase GSK3 contains at least five phosphorylation sites, including phosphorylation at Ser21, which inhibits its kinase activity and negatively correlates with human sperm motility [31,101]. The regulatory role of GSK3A is also required for proper motility in other mammalian spermatozoa [102–105]. Therefore, low GSK3 activity (more phosphorylated at Ser21), together with PP1γ2 activity, might be a prerequisite for the optimum function of mammalian spermatozoa [93].

STRING analysis revealed 23 high-confidence PPI collected from direct (physical) and indirect (functional) associations between 18 of the 36 phosphoproteins involved in sperm motility that were mentioned in previous paragraphs. So, 50% of the phosphoproteins identified by comparative phosphoproteomics studies between human spermatozoa with different motility levels are associated with one another (Figure 2). Moreover, focusing on protein functions, up to 56% are involved in sperm metabolism. These studies provide important information about proteins or their molecular mechanisms associated with male infertility, related to low-mobility, overall. However, they did not provide any evidence on the spermatozoa phosphoproteomics based on the type of infertility. A recent study combining phosphoproteomics results with functional analysis in human spermatozoa to analyze the role of G-protein-coupled receptors (GPCRs) in spermatozoa physiology identified phosphorylation changes in sperm-specific proteins downstream of the kappa-opioid receptor, which modulates human sperm motility [30].

Despite the revelations of these studies, molecular characteristics associated with the ability to fertilize the oocyte are still poorly understood in spermatozoa. Up to now, few candidate phosphoproteins have been associated with AS or low sperm motility. Still, there are many uncharacterized phosphoproteins and undescribed phosphorylation residues in spermatozoa. Therefore, functional studies in the future should elucidate their importance in sperm physiology.

6. Phosphoproteomics and Sperm Capacitation

The other two crucial stages in male fertility are sperm capacitation and acrosomal reaction, with protein phosphorylation also being crucially involved. Sperm capacitation to acquire the ability to fertilize occurs during sperm transit through the female reproductive tract for a specific time [106,107]. This process can be achieved in vitro, and its physical manifestation is hyperactivated motility, a powerful and asymmetric sperm tail movement [108]. After ejaculation, mammalian spermatozoa move actively, but they must undergo capacitation to fertilize the ovum. This process requires a sequence of physiological and biochemical changes, including increases in tyrosine phosphorylation, mostly in tail proteins. The first study to identify proteins phosphorylated during human sperm capacitation by a proteomic technique used two-dimensional polyacrylamide gel electrophoresis (PAGE), anti-phosphotyrosine antibody labeling, and MS/MS [24]. In that study, during human capacitation, the tyrosine phosphorylation sites of AKAP3 and AKAP4 proteins were detected, which are the major structural component of sperm fibrous sheath [24]. Later, another quantitative phosphoproteomics analysis comparing uncapacitated vs. capacitated human spermatozoa supported these results about AKAPs using IMAC-TiO2 (LC)-MS/MS [27]. These mentioned functional assays to study signaling pathways during human sperm capacitation, where phosphoproteomics analysis was combined with a prediction of cellular kinase-substrate relationships, indicated that the insulin growth factor 1 receptor, IGF1R, is an enriched tyrosine phosphorylation kinase. This receptor interacts during the capacitation process with the up-regulated phosphorylation sites for AKAP3 or AKAP4 proteins. As a result, the phosphorylation levels of AKAP3 and AKAP4, which include six and two phosphotyrosine sites, respectively, are significantly increased during human capacitation [24,27]. This tyrosine phosphorylation pathway mediated by IGF1R is crucial for human sperm capacitation and hyperactivated motility [27].

7. Outlook

Definitely, phosphorylation, among the PTMs, plays a vital role in regulating sperm processes that are essential for fertilization. Therefore, all these reviewed data together are a strong indicator of the contributions of sperm phosphoproteins in the most important male reproductive functions. So, the end goal of using proteomics technologies in male reproduction is to generate relevant data to increase the knowledge and the discovery of noninvasive predictive biomarkers of the prognosis and diagnoses of male infertility.

The precise cause(s) of male infertility remains elusive, although the correlation between immotile or poorly motile spermatozoa and infertility is scientifically strong. PTMs by phosphorylation allow eukaryotic cells and spermatozoa to dynamically regulate their intracellular signal integration and physiological states. Up to now, there have been some candidate human sperm phosphoproteins associated with AS or low sperm motility, but many uncharacterized sperm phosphoproteins and undescribed phosphorylation sites still exist. Nonetheless, providing clarification through functional studies would provide unprecedented insights into the regulatory tasks of phosphorylation and the molecular networks that govern spermatozoa function. The application of global phosphoproteomics profiling technology in evaluating sperm quality associated with male infertility would allow for significant advances to be made in identifying male infertility biomarkers. Therefore, more studies will improve the application of this procedure and overcome the limitations, providing data that will contribute to new knowledge and abundant resources for the screening of these molecular biomarkers in correlation with sperm quality. In addition, phosphoproteomics has found differences in spermatozoa from a distinct breed of pigs presenting different sperm qualities [109]. This research, which elucidates the mechanisms of regulation of male reproduction in other mammals, might be extrapolated to humans. In line with this, future studies of phosphoproteomics in a larger population could help explain whether the different results observed in some proteomics studies can be attributed to biological variation in various ethnic groups, as Siva et al. (2010) has speculated. Hence, further investigations combined with other single-omics fields (such as genomics, epigenomics, transcriptomics, and metabolomics) or with multi-omics contributions (proteogenomics, proteotranscriptomics, or reproductomics) are necessary. Findings from new studies would broaden our understanding of the mechanisms underlying the role of protein phosphorylation in sperm motility and male fertility. They would offer a unique perspective for future research into male fertility, especially for repeated failures in ART linked to unknown infertility causes.

Author Contributions: R.S. contributed to the writing and original draft preparation; R.S., L.J.G.-M. and M.J.B. contributed to the review of the manuscript, editing, visualization, and supervision; L.J.G.-M. and M.J.B. contributed to project administration and to funding acquisition. All authors have read and agreed to the published version of the manuscript.

Funding: This work was supported by Junta de Extremadura and European Social Funds (Grants: IB16184, IB20154 and GR18094). Rebeca Serrano is recipient of a research contract granted by Consejería de Economía, Ciencia y Agenda Digital, Junta de Extremadura, and by the European Social Funds (IB20154).

Institutional Review Board Statement: Not applicable.

Informed Consent Statement: Not applicable.

Data Availability Statement: Not applicable.

Conflicts of Interest: The authors declare no conflict of interest.

Patents: This research/review did not generate patents.

References

1. Varner, D.D. Odyssey of the spermatozoon. *Asian J. Androl.* **2015**, *17*, 522–528. [CrossRef] [PubMed]
2. Hetherington, L.; Schneider, E.K.; Scott, C.; DeKretser, D.; Muller, C.H.; Hondermarck, H.; Velkov, T.; Baker, M.A. Deficiency in Outer Dense Fiber 1 Is a Marker and Potential Driver of Idiopathic Male Infertility. *Mol. Cell. Proteom.* **2016**, *15*, 3685–3693. [CrossRef]
3. World Human Organization. *WHO Laboratory Manual for the Examination and Processing of Human Semen*, 5th ed.; WHO Press: Geneva, Switzerland, 2010; p. 287.
4. Esteves, S.C.; Miyaoka, R.; Agarwal, A. An update on the clinical assessment of the infertile male. *Clinics* **2011**, *66*, 691–700. [CrossRef] [PubMed]
5. Eliasson, R. Semen analysis with regard to sperm number, sperm morphology and functional aspects. *Asian J. Androl.* **2010**, *12*, 26–32. [CrossRef] [PubMed]
6. Sánchez, V.; Wistuba, J.; Mallidis, C. Semen analysis: Update on clinical value, current needs and future perspectives. *Reproduction* **2013**, *146*, R249–R258. [CrossRef]
7. Wang, C.; Swerdloff, R.S. Limitations of semen analysis as a test of male fertility and anticipated needs from newer tests. *Fertil. Steril.* **2014**, *102*, 1502–1507. [CrossRef]
8. Papillon-Smith, J.; Baker, S.E.; Agbo, C.; Dahan, M.H. Pregnancy rates with intrauterine insemination: Comparing 1999 and 2010 World Health Organization semen analysis norms. *Reprod. Biomed. Online* **2015**, *30*, 392–400. [CrossRef]
9. Stern, J.E.; Luke, B.; Hornstein, M.D.; Cabral, H.; Gopal, D.; Diop, H.; Kotelchuck, M. The effect of father's age in fertile, subfertile, and assisted reproductive technology pregnancies: A population based cohort study. *J. Assist. Reprod. Genet.* **2014**, *31*, 1437–1444. [CrossRef]

10. Santiago, J.; Silva, J.V.; Howl, J.; Santos, M.A.S.; Fardilha, M. All you need to know about sperm RNAs. *Hum. Reprod. Update* **2021**, *28*, 67–91. [CrossRef]
11. Samanta, L.; Swain, N.; Ayaz, A.; Venugopal, V.; Agarwal, A. Post-Translational Modifications in sperm Proteome: The Chemistry of Proteome diversifications in the Pathophysiology of male factor infertility. *Biochim. Biophys. Acta* **2016**, *1860*, 1450–1465. [CrossRef]
12. Cheng, Y.M.; Hu, X.N.; Peng, Z.; Pan, T.T.; Wang, F.; Chen, H.Y.; Chen, W.Q.; Zhang, Y.; Zeng, X.H.; Luo, T. Lysine glutarylation in human sperm is associated with progressive motility. *Hum. Reprod.* **2019**, *34*, 1186–1194. [CrossRef]
13. Qi, Q.; Pan, H.; Jiang, N.; Zhang, M.; Sun, S.; Wan, X.; Zhang, F.; Zhang, L.; Diao, H.; Wang, J.; et al. A novel posttranslational modification of histone, H3 S-sulfhydration, is down-regulated in asthenozoospermic sperm. *J. Assist. Reprod. Genet.* **2021**, *38*, 3175–3193. [CrossRef]
14. Marchiani, S.; Tamburrino, L.; Giuliano, L.; Nosi, D.; Sarli, V.; Gandini, L.; Piomboni, P.; Belmonte, G.; Forti, G.; Baldi, E.; et al. Sumo1-ylation of human spermatozoa and its relationship with semen quality. *Int. J. Androl.* **2011**, *34*, 581–593. [CrossRef] [PubMed]
15. Cheng, Y.M.; Peng, Z.; Chen, H.Y.; Pan, T.T.; Hu, X.N.; Wang, F.; Luo, T. Posttranslational lysine 2-hydroxyisobutyrylation of human sperm tail proteins affects motility. *Hum. Reprod.* **2020**, *35*, 494–503. [CrossRef]
16. Sun, G.; Jiang, M.; Zhou, T.; Guo, Y.; Cui, Y.; Guo, X.; Sha, J. Insights into the lysine acetylproteome of human sperm. *J. Proteom.* **2014**, *109*, 199–211. [CrossRef] [PubMed]
17. Porambo, J.R.; Salicioni, A.M.; Visconti, P.E.; Platt, M.D. Sperm phosphoproteomics: Historical perspectives and current methodologies. *Expert Rev. Proteom.* **2012**, *9*, 533–548. [CrossRef]
18. Chan, C.C.; Shui, H.A.; Wu, C.H.; Wang, C.Y.; Sun, G.H.; Chen, H.M.; Wu, G.J. Motility and protein phosphorylation in healthy and asthenozoospermic sperm. *J. Proteome Res.* **2009**, *8*, 5382–5386. [CrossRef] [PubMed]
19. Barbonetti, A.; Vassallo, M.R.; Fortunato, D.; Francavilla, S.; Maccarrone, M.; Francavilla, F. Energetic metabolism and human sperm motility: Impact of CB_1 receptor activation. *Endocrinology* **2010**, *151*, 5882–5892. [CrossRef]
20. Visconti, P.E.; Krapf, D.; de la Vega-Beltrán, J.L.; Acevedo, J.J.; Darszon, A. Ion channels, phosphorylation and mammalian sperm capacitation. *Asian J. Androl.* **2011**, *13*, 395405. [CrossRef]
21. Castillo, J.; Knol, J.C.; Korver, C.M.; Piersma, S.R.; Pham, T.V.; de Goeij-de Haas, R.R.; van Pelt, A.M.M.; Jimenez, C.R.; Jansen, B.J.H. Human Testis Phosphoproteome Reveals Kinases as Potential Targets in Spermatogenesis and Testicular Cancer. *Mol. Cell. Proteom.* **2019**, *18* (Suppl. S1), S132–S144. [CrossRef]
22. Li, Y.; Cheng, Y.; Zhu, T.; Zhang, H.; Li, W.; Guo, Y.; Qi, Y.; Chen, X.; Zhang, J.; Sha, J.; et al. The Protein Phosphorylation Landscape of Mouse Spermatids during Spermiogenesis. *Proteomics* **2019**, *19*, e1900055. [CrossRef] [PubMed]
23. Zhang, P.F.; Huang, Y.L.; Fu, Q.; He, W.T.; Xiao, K.; Zhang, M. Integrated analysis of phosphoproteome and ubiquitylome in epididymal sperm of buffalo (Bubalus bubalis). *Mol. Reprod. Dev.* **2021**, *88*, 15–33. [CrossRef] [PubMed]
24. Ficarro, S.; Chertihin, O.; Westbrook, V.A.; White, F.; Jayes, F.; Kalab, P.; Marto, J.A.; Shabanowitz, J.; Herr, J.C.; Hunt, D.F.; et al. Phosphoproteome analysis of capacitated human sperm. Evidence of tyrosine phosphorylation of a kinase-anchoring protein 3 and valosin-containing protein/p97 during capacitation. *J. Biol. Chem.* **2003**, *278*, 11579–11589. [CrossRef]
25. Bailey, J.L.; Tardif, S.; Dubé, C.; Beaulieu, M.; Reyes-Moreno, C.; Lefièvre, L.; Leclerc, P. Use of phosphoproteomics to study tyrosine kinase activity in capacitating boar sperm. Kinase activity and capacitation. *Theriogenology* **2005**, *63*, 599–614. [CrossRef] [PubMed]
26. (Kota, V.; Dhople, V.M.; Shivaji, S. Tyrosine phosphoproteome of hamster spermatozoa: Role of glycerol-3-phosphate dehydrogenase 2 in sperm capacitation. *Proteomics* **2009**, *9*, 1809–1826. [CrossRef]
27. Wang, J.; Qi, L.; Huang, S.; Zhou, T.; Guo, Y.; Wang, G.; Guo, X.; Zhou, Z.; Sha, J. Quantitative phosphoproteomics analysis reveals a key role of insulin growth factor 1 receptor (IGF1R) tyrosine kinase in human sperm capacitation. *Mol. Cell. Proteom.* **2015**, *14*, 1104–1112. [CrossRef]
28. Syifa, N.; Yang, J.T.; Wu, C.S.; Lin, M.H.; Wu, W.L.; Lai, C.W.; Ku, S.H.; Liang, S.Y.; Hung, Y.C.; Chou, C.T.; et al. Phosphoproteomics and Bioinformatics Analyses Reveal Key Roles of GSK-3 and AKAP4 in Mouse Sperm Capacitation. *Int. J. Mol. Sci.* **2020**, *21*, 7283. [CrossRef]
29. Parte, P.P.; Rao, P.; Redij, S.; Lobo, V.; D'Souza, S.J.; Gajbhiye, R.; Kulkarni, V. Sperm phosphoproteome profiling by ultra performance liquid chromatography followed by data independent analysis (LC-MS(E)) reveals altered proteomic signatures in asthenozoospermia. *J. Proteom.* **2012**, *75*, 5861–5871. [CrossRef]
30. Urizar-Arenaza, I.; Osinalde, N.; Akimov, V.; Puglia, M.; Candenas, L.; Pinto, F.M.; Muñoa-Hoyos, I.; Gianzo, M.; Matorras, R.; Irazusta, J.; et al. Phosphoproteomic and Functional Analyses Reveal Sperm-specific Protein Changes Downstream of Kappa Opioid Receptor in Human Spermatozoa. *Mol. Cell Proteom.* **2019**, *18* (Suppl. S1), S118–S131. [CrossRef]
31. Martin-Hidalgo, D.; Serrano, R.; Zaragoza, C.; Garcia-Marin, L.J.; Bragado, M.J. Human sperm phosphoproteome reveals differential phosphoprotein signatures that regulate human sperm motility. *J. Proteom.* **2020**, *215*, 103654. [CrossRef]
32. Wang, J.; Wang, M.; Hong, R.; Tang, S.; Xu, Y.; Zhao, X.; Zhou, T.; Wang, Z.; Huang, S. Quantitative phosphoproteomics reveals GSK3A substrate network is involved in the cryodamage of sperm motility. *Biosci. Rep.* **2021**, *41*, BSR20211326. [CrossRef] [PubMed]
33. Intasqui, P.; Agarwal, A.; Sharma, R.; Samanta, L.; Bertolla, R.P. Towards the identification of reliable sperm biomarkers for male infertility: A sperm proteomic approach. *Andrologia* **2018**, *50*, e12919. [CrossRef] [PubMed]

34. Nowicka-Bauer, K.; Malcher, A.; Włoczkowska, O.; Kamieniczna, M.; Olszewska, M.; Kurpisz, M.K. Evaluation of seminal plasma HSPA2 protein as a biomarker of human spermatogenesis status. *Reprod. Biol.* **2022**, *22*, 100597. [CrossRef] [PubMed]
35. Netherton, J.; Ogle, R.A.; Hetherington, L.; Silva Balbin Villaverde, A.I.; Hondermarck, H.; Baker, M.A. Proteomic Analysis Reveals that Topoisomerase 2A is Associated with Defective Sperm Head Morphology. *Mol. Cell. Proteom.* **2020**, *19*, 444–455. [CrossRef]
36. Pixton, K.L.; Deeks, E.D.; Flesch, F.M.; Moseley, F.L.; Björndahl, L.; Ashton, P.R.; Barratt, C.L.; Brewis, I.A. Sperm proteome mapping of a patient who experienced failed fertilization at IVF reveals altered expression of at least 20 proteins compared with fertile donors: Case report. *Hum. Reprod.* **2004**, *19*, 1438–1447. [CrossRef]
37. Zhao, C.; Huo, R.; Wang, F.Q.; Lin, M.; Zhou, Z.M.; Sha, J.H. Identification of several proteins involved in regulation of sperm motility by proteomic analysis. *Fertil. Steril.* **2007**, *87*, 436–438. [CrossRef]
38. Martínez-Heredia, J.; de Mateo, S.; Vidal-Taboada, J.M.; Ballescà, J.L.; Oliva, R. Identification of proteomic differences in asthenozoospermic sperm samples. *Hum. Reprod.* **2008**, *23*, 783–791. [CrossRef]
39. Siva, A.B.; Kameshwari, D.B.; Singh, V.; Pavani, K.; Sundaram, C.S.; Rangaraj, N.; Deenadayal, M.; Shivaji, S. Proteomics-based study on asthenozoospermia: Differential expression of proteasome alpha complex. *Mol. Hum. Reprod.* **2010**, *16*, 452–462. [CrossRef]
40. Thacker, S.; Yadav, S.P.; Sharma, R.K.; Kashou, A.; Willard, B.; Zhang, D.; Agarwal, A. Evaluation of sperm proteins in infertile men: A proteomic approach. *Fertil. Steril.* **2011**, *95*, 2745–2748. [CrossRef]
41. Shen, S.; Wang, J.; Liang, J.; He, D. Comparative proteomic study between human normal motility sperm and idiopathic asthenozoospermia. *World J. Urol.* **2013**, *31*, 1395–1401. [CrossRef]
42. Amaral, A.; Paiva, C.; Attardo Parrinello, C.; Estanyol, J.M.; Ballescà, J.L.; Ramalho- Santos, J.; Oliva, R. Identification of proteins involved in human sperm motility using highthroughput differential proteomics. *J. Proteome Res.* **2014**, *13*, 5670–5684. [CrossRef] [PubMed]
43. Hashemitabar, M.; Sabbagh, S.; Orazizadeh, M.; Ghadiri, A.; Bahmanzadeh, M. A proteomic analysis on human sperm tail: Comparison between normozoospermia and asthenozoospermia. *J. Assist. Reprod. Genet.* **2015**, *32*, 853–863. [CrossRef] [PubMed]
44. Saraswat, M.; Joenväärä, S.; Jain, T.; Tomar, A.K.; Sinha, A.; Singh, S.; Yadav, S.; Renkonen, R. Human Spermatozoa Quantitative Proteomic Signature Classifies Normo- and Asthenozoospermia. *Mol. Cell. Proteom.* **2017**, *16*, 57–72. [CrossRef] [PubMed]
45. Moscatelli, N.; Lunetti, P.; Braccia, C.; Armirotti, A.; Pisanello, F.; De Vittorio, M.; Zara, V.; Ferramosca, A. Comparative Proteomic Analysis of Proteins Involved in Bioenergetics Pathways Associated with Human Sperm Motility. *Int. J. Mol. Sci.* **2019**, *20*, 3000. [CrossRef]
46. Guo, Y.; Jiang, W.; Yu, W.; Niu, X.; Liu, F.; Zhou, T.; Zhang, H.; Li, Y.; Zhu, H.; Zhou, Z.; et al. Proteomics analysis of asthenozoospermia and identification of glucose-6-phosphate isomerase as an important enzyme for sperm motility. *J. Proteom.* **2019**, *208*, 103478. [CrossRef]
47. Aitken, R.J.; Sutton, M.; Warner, P.; Richardson, D.W. Relationship between the movement characteristics of human spermatozoa and their ability to penetrate cervical mucus and zona-free hamster oocytes. *J. Reprod. Fertil.* **1985**, *73*, 441–449. [CrossRef]
48. Curi, S.M.; Ariagno, J.I.; Chenlo, P.H.; Mendeluk, G.R.; Pugliese, M.N.; Sardi Segovia, L.M.; Repetto, H.E.; Blanco, A.M. Asthenozoospermia: Analysis of a large population. *Arch. Androl.* **2003**, *49*, 343–349. [CrossRef]
49. Liu, X.; Teng, Z.; Wang, Z.; Zhu, P.; Song, Z.; Liu, F. Expressions of HSPA1L and HSPA9 are associated with poor sperm quality of low-motility spermatozoa in fertile men. *Andrologia* **2022**, *54*, e14321. [CrossRef]
50. Meccariello, R.; Chianese, R.; Ciaramella, V.; Fasano, S.; Pierantoni, R. Molecular chaperones, cochaperones, and ubiquitination/deubiquitination system: Involvement in the production of high quality spermatozoa. *Biomed. Res. Int.* **2014**, *2014*, 561426. [CrossRef]
51. Dun, M.D.; Aitken, R.J.; Nixon, B. The role of molecular chaperones in spermatogenesis and the post-testicular maturation of mammalian spermatozoa. *Hum. Reprod. Update* **2012**, *18*, 420–435. [CrossRef]
52. Liu, X.X.; Shen, X.F.; Liu, F.J. Screening targeted testis-specific genes for molecular assessment of aberrant sperm quality. *Mol. Med. Rep.* **2016**, *14*, 1594–1600. [CrossRef] [PubMed]
53. Nixon, B.; Bromfield, E.G.; Cui, J.; De Iuliis, G.N. Heat Shock Protein A2 (HSPA2): Regulatory Roles in Germ Cell Development and Sperm Function. *Adv. Anat. Embryol. Cell Biol.* **2017**, *222*, 67–93. [CrossRef] [PubMed]
54. Radons, J. The human HSP70 family of chaperones: Where do we stand? *Cell Stress Chaperones* **2016**, *21*, 379–404. [CrossRef]
55. Bracke, A.; Peeters, K.; Punjabi, U.; Hoogewijs, D.; Dewilde, S. A search for molecular mechanisms underlying male idiopathic infertility. *Reprod. Biomed. Online* **2018**, *36*, 327–339. [CrossRef]
56. Légaré, C.; Droit, A.; Fournier, F.; Bourassa, S.; Force, A.; Cloutier, F.; Tremblay, R.; Sullivan, R. Investigation of male infertility using quantitative comparative proteomics. *J. Proteome Res.* **2014**, *13*, 5403–5414. [CrossRef]
57. Ostrowski, L.E.; Andrews, K.; Potdar, P.; Matsuura, H.; Jetten, A.; Nettesheim, P. Cloning and characterization of KPL2, a novel gene induced during ciliogenesis of tracheal epithelial cells. *Am. J. Respir. Cell Mol. Biol.* **1999**, *20*, 675–683. [CrossRef] [PubMed]
58. Tu, C.; Nie, H.; Meng, L.; Wang, W.; Li, H.; Yuan, S.; Cheng, D.; He, W.; Liu, G.; Du, J.; et al. Novel mutations in SPEF2 causing different defects between flagella and cilia bridge: The phenotypic link between MMAF and PCD. *Hum. Genet.* **2020**, *139*, 257–271. [CrossRef]
59. Li, D.Y.; Yang, X.X.; Tu, C.F.; Wang, W.L.; Meng, L.L.; Lu, G.X.; Tan, Y.Q.; Zhang, Q.J.; Du, J. Sperm flagellar 2 (SPEF2) is essential for sperm flagellar assembly in humans. *Asian J. Androl.* **2021**, *23*, 1–8. [CrossRef]

60. Lehti, M.S.; Sironen, A. Formation and function of sperm tail structures in association with sperm motility defects. *Biol. Reprod.* **2017**, *97*, 522–536. [CrossRef]
61. Wang, S.; Wang, W.; Xu, Y.; Tang, M.; Fang, J.; Sun, H.; Sun, Y.; Gu, M.; Liu, Z.; Zhang, Z.; et al. Proteomic characteristics of human sperm cryopreservation. *Proteomics* **2014**, *14*, 298–310. [CrossRef]
62. Fu, L.; An, Q.; Zhang, K.; Liu, Y.; Tong, Y.; Xu, J.; Zhou, F.; Wang, X.; Guo, Y.; Lu, W.; et al. Quantitative proteomic characterization of human sperm cryopreservation: Using data independent acquisition mass spectrometry. *BMC Urol.* **2019**, *19*, 133. [CrossRef] [PubMed]
63. Wang, Y.C.; Peterson, S.E.; Loring, J.F. Protein post-translational modifications and regulation of pluripotency in human stem cells. *Cell Res.* **2014**, *24*, 143–160. [CrossRef] [PubMed]
64. Khoury, G.A.; Baliban, R.C.; Floudas, C.A. Proteome-wide post-translational modification statistics: Frequency analysis and curation of the swiss-prot database. *Sci. Rep.* **2011**, *1*, 90. [CrossRef] [PubMed]
65. Cohen, P. The role of protein phosphorylation in human health and disease. The Sir Hans Krebs Medal Lecture. *Eur. J. Biochem.* **2001**, *268*, 5001–5010. [CrossRef]
66. Olsen, J.V.; Blagoev, B.; Gnad, F.; Macek, B.; Kumar, C.; Mortensen, P.; Mann, M. Global, in vivo, and site-specific phosphorylation dynamics in signaling networks. *Cell* **2006**, *127*, 635–648. [CrossRef]
67. Adam, K.; Hunter, T. Histidine kinases and the missing phosphoproteome from prokaryotes to eukaryotes. *Lab. Investig.* **2018**, *98*, 233–247. [CrossRef]
68. Beltrao, P.; Bork, P.; Krogan, N.J.; van Noort, V. Evolution and functional cross-talk of protein post-translational modifications. *Mol. Syst. Biol.* **2013**, *9*, 714. [CrossRef]
69. Brohi, R.D.; Huo, L.J. Posttranslational Modifications in Spermatozoa and Effects on Male Fertility and Sperm Viability. *OMICS* **2017**, *21*, 245–256. [CrossRef]
70. Low, T.Y.; Mohtar, M.A.; Lee, P.Y.; Omar, N.; Zhou, H.; Ye, M. Widening the Bottleneck of phosphoproteomics: Evolving strategies for Phosphopeptide enrichment. *Mass Spectrom. Rev.* **2021**, *40*, 309–333. [CrossRef]
71. Amaral, A.; Castillo, J.; Ramalho-Santos, J.; Oliva, R. The combined human sperm proteome: Cellular pathways and implications for basic and clinical science. *Hum. Reprod. Update* **2014**, *20*, 40–62. [CrossRef]
72. Larsen, M.R.; Thingholm, T.E.; Jensen, O.N.; Roepstorff, P.; Jørgensen, T.J. Highly selective enrichment of phosphorylated peptides from peptide mixtures using titanium dioxide microcolumns. *Mol. Cell. Proteom.* **2005**, *4*, 873–886. [CrossRef]
73. Mitulovic, G.; Mechtler, K. HPLC techniques for proteomics analysis—A short overview of latest developments. *Brief. Funct. Genom. Proteomic* **2006**, *5*, 249–260. [CrossRef]
74. Thingholm, T.E.; Larsen, M.R. The Use of Titanium Dioxide for Selective Enrichment of Phosphorylated Peptides. *Methods Mol. Biol.* **2016**, *1355*, 135–146. [CrossRef]
75. Agarwal, A.; Panner Selvam, M.K.; Baskaran, S. Proteomic Analyses of Human Sperm Cells: Understanding the Role of Proteins and Molecular Pathways Affecting Male Reproductive Health. *Int. J. Mol. Sci.* **2020**, *21*, 1621. [CrossRef] [PubMed]
76. Needham, E.J.; Parker, B.L.; Burykin, T.; James, D.E.; Humphrey, S.J. Illuminating the dark phosphoproteome. *Sci. Signal.* **2019**, *12*, 565. [CrossRef] [PubMed]
77. Anand, S.; Samuel, M.; Ang, C.S.; Keerthikumar, S.; Mathivanan, S. Label-Based and Label-Free Strategies for Protein Quantitation. *Methods Mol. Biol.* **2017**, *1549*, 31–43. [CrossRef]
78. Megger, D.A.; Pott, L.L.; Ahrens, M.; Padden, J.; Bracht, T.; Kuhlmann, K.; Eisenacher, M.; Meyer, H.E.; Sitek, B. Comparison of label-free and label-based strategies for proteome analysis of hepatoma cell lines. *Biochim. Biophys. Acta* **2014**, *1844*, 967–976. [CrossRef] [PubMed]
79. Agarwal, A.; Samanta, L.; Bertolla, R.P.; Durairajanayagam, D.; Intasqui, P. *Proteomics in Human Reproduction: Biomarkers for Millennials*; Springer: Berlin/Heidelberg, Germany, 2016.
80. Macek, B.; Mann, M.; Olsen, J.V. Global and site-specific quantitative phosphoproteomics: Principles and applications. *Annu Rev. Pharmacol. Toxicol.* **2009**, *49*, 199–221. [CrossRef]
81. Aitken, R.J.; Baker, M.A. The role of proteomics in understanding sperm cell biology. *Int. J. Androl.* **2008**, *31*, 295–302. [CrossRef]
82. Jordan, P.W.; Karppinen, J.; Handel, M.A. Polo-like kinase is required for synaptonemal complex disassembly and phosphorylation in mouse spermatocytes. *J. Cell Sci.* **2012**, *125*, 5061–5072. [CrossRef]
83. Qi, L.; Liu, Z.; Wang, J.; Cui, Y.; Guo, Y.; Zhou, T.; Zhou, Z.; Guo, X.; Xue, Y.; Sha, J. Systematic analysis of the phosphoproteome and kinase-substrate networks in the mouse testis. *Mol. Cell. Proteom.* **2014**, *13*, 3626–3638. [CrossRef] [PubMed]
84. Wei, Y.; Gao, Q.; Niu, P.; Xu, K.; Qiu, Y.; Hu, Y.; Liu, S.; Zhang, X.; Yu, M.; Liu, Z.; et al. Integrative Proteomic and Phosphoproteomic Profiling of Testis from Wip1 Phosphatase- Knockout Mice: Insights into Mechanisms of Reduced Fertility. *Mol. Cell. Proteom.* **2019**, *18*, 216–230. [CrossRef] [PubMed]
85. Filipponi, D.; Muller, J.; Emelyanov, A.; Bulavin, D.V. Wip1 controls global heterochromatin silencing via ATM/BRCA1-dependent DNA methylation. *Cancer Cell* **2013**, *24*, 528–541. [CrossRef] [PubMed]
86. Heller, C.G.; Clermont, Y. Spermatogenesis in man: An estimate of its duration. *Science* **1963**, *140*, 184–186. [CrossRef] [PubMed]
87. Sutovsky, P.; Moreno, R.; Ramalho-Santos, J.; Dominko, T.; Thompson, W.E.; Schatten, G. A putative, ubiquitin-dependent mechanism for the recognition and elimination of defective spermatozoa in the mammalian epididymis. *J. Cell Sci.* **2001**, *114*, 1665–1675. [CrossRef]

88. Wang, Y.; Wan, J.; Ling, X.; Liu, M.; Zhou, T. The human sperm proteome 2.0: An integrated resource for studying sperm functions at the level of posttranslational modification. *Proteomics* **2016**, *16*, 2597–2601. [CrossRef]
89. Eskandari-Shahraki, M.; Tavalaee, M.; Deemeh, M.R.; Jelodar, G.A.; Nasr-Esfahani, M.H. Proper ubiquitination effect on the fertilisation outcome post-ICSI. *Andrologia* **2013**, *45*, 204–210. [CrossRef]
90. Tash, J.S.; Means, A.R. Regulation of protein phosphorylation and motility of sperm by cyclic adenosine monophosphate and calcium. *Biol. Reprod.* **1982**, *26*, 745–763. [CrossRef]
91. Luconi, M.; Baldi, E. How do sperm swim? Molecular mechanisms underlying sperm motility. *Cell Mol. Biol.* **2003**, *49*, 357–369.
92. Huang, Z.; Khatra, B.; Bollen, M.; Carr, D.W.; Vijayaraghavan, S. Sperm PP1gamma2 is regulated by a homologue of the yeast protein phosphatase binding protein sds22. *Biol. Reprod.* **2002**, *67*, 1936–1942. [CrossRef]
93. Chakrabarti, R.; Cheng, L.; Puri, P.; Soler, D.; Vijayaraghavan, S. Protein phosphatase PP1 gamma 2 in sperm morphogenesis and epididymal initiation of sperm motility. *Asian J. Androl.* **2007**, *9*, 445–452. [CrossRef] [PubMed]
94. Visconti, P.E.; Bailey, J.L.; Moore, G.D.; Pan, D.; Olds-Clarke, P.; Kopf, G.S. Capacitation of mouse spermatozoa. I. Correlation between the capacitation state and protein tyrosine phosphorylation. *Development* **1995**, *121*, 1129–1137. [CrossRef] [PubMed]
95. Yunes, R.; Doncel, G.F.; Acosta, A.A. Incidence of sperm-tail tyrosine phosphorylation and hyperactivated motility in normozoospermic and asthenozoospermic human sperm samples. *Biocell* **2003**, *27*, 29–36. [CrossRef] [PubMed]
96. Huang, Z.; Danshina, P.V.; Mohr, K.; Qu, W.; Goodson, S.G.; O'Connell, T.M.; O'Brien, D.A. Sperm function, protein phosphorylation, and metabolism differ in mice lacking successive sperm-specific glycolytic enzymes. *Biol. Reprod.* **2017**, *97*, 586–597. [CrossRef]
97. Morgan, D.J.; Weisenhaus, M.; Shum, S.; Su, T.; Zheng, R.; Zhang, C.; Shokat, K.M.; Hille, B.; Babcock, D.F.; McKnight, G.S. Tissue-specific PKA inhibition using a chemical genetic approach and its application to studies on sperm capacitation. *Proc. Natl. Acad. Sci. USA* **2008**, *105*, 20740–20745. [CrossRef]
98. Liu, X.; Wang, X.; Liu, F. Decreased expression of heat shock protein A4L in spermatozoa is positively related to poor human sperm quality. *Mol. Reprod. Dev.* **2019**, *86*, 379–386. [CrossRef]
99. Nijs, M.; Ombelet, W. Cryopreservation of human sperm. *Hum. Fertil.* **2001**, *4*, 158–163. [CrossRef]
100. Bogle, O.A.; Kumar, K.; Attardo-Parrinello, C.; Lewis, S.E.; Estanyol, J.M.; Ballescà, J.L.; Oliva, R. Identification of protein changes in human spermatozoa throughout the cryopreservation process. *Andrology* **2017**, *5*, 10–22. [CrossRef]
101. Freitas, M.J.; Silva, J.V.; Brothag, C.; Regadas-Correia, B.; Fardilha, M.; Vijayaraghavan, S. Isoform-specific GSK3A activity is negatively correlated with human sperm motility. *Mol. Hum. Reprod.* **2019**, *25*, 171–183. [CrossRef]
102. Vijayaraghavan, S.; Mohan, J.; Gray, H.; Khatra, B.; Carr, D.W. A role for phosphorylation of glycogen synthase kinase-3alpha in bovine sperm motility regulation. *Biol Reprod.* **2000**, *62*, 1647–1654. [CrossRef]
103. Aparicio, I.M.; Bragado, M.J.; Gil, M.C.; Garcia-Herreros, M.; Gonzalez-Fernandez, L.; Tapia, J.A.; Garcia-Marin, L.J. Porcine sperm motility is regulated by serine phosphorylation of the glycogen synthase kinase-3alpha. *Reproduction* **2007**, *134*, 435–444. [CrossRef] [PubMed]
104. Bhattacharjee, R.; Goswami, S.; Dudiki, T.; Popkie, A.P.; Phiel, C.J.; Kline, D.; Vijayaraghavan, S. Targeted disruption of glycogen synthase kinase 3A (GSK3A) in mice affects sperm motility resulting in male infertility. *Biol. Reprod.* **2015**, *92*, 65. [CrossRef]
105. Zhu, Z.; Li, R.; Wang, L.; Zheng, Y.; Hoque, S.A.M.; Lv, Y.; Zeng, W. Glycogen Synthase Kinase-3 Regulates Sperm Motility and Acrosome Reaction via Affecting Energy Metabolism in Goats. *Front. Physiol.* **2019**, *10*, 968. [CrossRef] [PubMed]
106. Chang, M.C. Fertilizing capacity of spermatozoa deposited into the fallopian tubes. *Nature* **1951**, *168*, 697–698. [CrossRef]
107. Austin, C.R. The capacitation of the mammalian sperm. *Nature* **1952**, *170*, 326. [CrossRef] [PubMed]
108. Yanagimachi, R. Fertility of mammalian spermatozoa: Its development and relativity. *Zygote* **1994**, *2*, 371–372. [CrossRef] [PubMed]
109. Xu, Y.; Han, Q.; Ma, C.; Wang, Y.; Zhang, P.; Li, C.; Cheng, X.; Xu, H. Comparative Proteomics and Phosphoproteomics Analysis Reveal the Possible Breed Difference in Yorkshire and Duroc Boar Spermatozoa. *Front. Cell Dev. Biol.* **2021**, *9*, 652809. [CrossRef]

biology

MDPI

Article

Sperm Selection by Magnetic-Activated Cell Sorting before Microinjection of Autologous Oocytes Increases Cumulative Live Birth Rates with Limited Clinical Impact: A Retrospective Study in Unselected Males

María Gil Juliá [1,*], Irene Hervás [1], Ana Navarro-Gómez Lechón [1], Fernando Quintana [2], David Amorós [3], Alberto Pacheco [4], Cristina González-Ravina [5], Rocío Rivera-Egea [6] and Nicolás Garrido [1]

1 Andrology and Male Infertility Research Group, IVI Foundation—IIS La Fe Health Research Institute, Av. Fernando Abril Martorell, 106. Tower A, 1st Floor, 46026 Valencia, Spain; irene.hervas@ivirma.com (I.H.); ana.navarro@ivirma.com (A.N.-G.L.); nicolas.garrido@ivirma.com (N.G.)
2 IVIRMA Bilbao, Landabarri Bidea, 1–3, 2nd Floor, 48940 Leioa Bizkaia, Spain; fernando.quintana@ivirma.com
3 IVIRMA Barcelona, Ronda del General Mitre, 14, 08017 Barcelona, Spain; david.amoros@ivirma.com
4 IVIRMA Madrid, Av. del Talgo 68–70, 28023 Madrid, Spain; alberto.pacheco@ivirma.com
5 IVIRMA Sevilla, Av. de la República Argentina, 58, 41011 Sevilla, Spain; cristina.gonzalez@ivirma.com
6 IVIRMA Valencia, Pl. de la Policía Local, 3, 46015 Valencia, Spain; rocio.rivera@ivirma.com
* Correspondence: maria.gil@ivirma.com

Citation: Gil Juliá, M.; Hervás, I.; Navarro-Gómez Lechón, A.; Quintana, F.; Amorós, D.; Pacheco, A.; González-Ravina, C.; Rivera-Egea, R.; Garrido, N. Sperm Selection by Magnetic-Activated Cell Sorting before Microinjection of Autologous Oocytes Increases Cumulative Live Birth Rates with Limited Clinical Impact: A Retrospective Study in Unselected Males. *Biology* **2021**, *10*, 430. https://doi.org/10.3390/biology10050430

Academic Editors: Pilar Santolaria Blasco, Rosaura Pérez-Pé and Tamara Leahy

Received: 15 April 2021
Accepted: 11 May 2021
Published: 12 May 2021

Simple Summary: Many couples attending infertility clinics still need to repeat treatments and undergo several failed attempts before achieving a healthy newborn, which leaves room for improvement in the techniques we currently use in the clinic. Among the different procedures susceptible to improvement, the selection of the most adequate sperm to be injected inside the egg is crucial to the cycle's success. Magnetic-activated cell sorting (MACS) is a technique that removes physiologically abnormal sperm that have started a programmed cell death (apoptotic) process from a semen sample. However, it is not recommended to all patients because there is no agreement between the published literature on whether it improves reproductive outcomes. This study used data from all intracytoplasmic sperm injection cycles performed using the patient's own oocytes in our clinics from January 2008 to February 2020. Our findings support that MACS should not be recommended to all infertile couples, since there was no significant difference in results compared to treatments in which MACS was not used. This study provides clinicians and patients with more accurate information on how MACS will impact their chances of pregnancy, and it will lead to studies focused on specific populations to which the technique can be particularly helpful.

Abstract: The application of MACS non-apoptotic sperm selection in infertility clinics is controversial since the published literature does not agree on its effect on reproductive outcomes. Therefore, it is not part of the routine clinical practice. Classical measures of reproductive success (pregnancy or live birth rates per ovarian stimulation) introduce a bias in the evaluation of a technique's effect, since only the best embryo is transferred. This retrospective, multicenter, observational study evaluated the impact of MACS on reproductive outcomes, measuring results in classical parameters and cumulative live birth rates (CLBR). Data from ICSI cycles using autologous oocyte in Spanish IVIRMA fertility clinics from January 2008 to February 2020 were divided into two groups according to their semen processing: standard practice (reference: 46,807 patients) versus an added MACS sperm selection (1779 patients). Only when measured as CLBR per embryo transferred and per MII oocyte used was the difference between groups statistically significant. There were no significant differences between MACS and reference groups on pregnancy and live birth rates. In conclusion, results suggest that non-apoptotic sperm selection by MACS on unselected males prior to ICSI with autologous oocytes has limited clinical impact, showing a subtle increase in CLBR per embryo transferred.

Keywords: MACS; sperm; sperm selection; ICSI; cumulative live birth rate; pregnancy rate

1. Introduction

Although the scientific literature provides embryologists and andrologists with morphological criteria to select the *a priori* most appropriate spermatozoon in the lab [1], this evaluation overlooks the unique molecular and genetic aptitude of each cell. Choosing an inadequate fertilizing sperm can lead to fertilization failure, incorrect embryo development, failed implantation, or miscarriage. Hence, sperm selection is crucial to ensure that the oocyte is correctly fertilized by the most competent [2]. Among the physiological properties involved in sperm function, apoptosis has been proposed as one of the more detrimental to spermatozoa's fertilization potential. An increased presence of apoptotic markers activated caspase-3, externalized phosphatidylserine, or fragmented spermatic DNA has been linked to abnormal sperm motility or morphology [3–8], a decrease in fertilization rate and optimal quality embryos in couples with normozoospermic men [9], a decrease in fertilization potential, and a reduced ability to trigger acrosome reaction [10]. Therefore, the separation of sperm with initiated apoptosis is interesting to ensure the selection of the most physiologically competent sperm.

Magnetic-activated cell sorting (MACS) is a non-destructive cell separation technique that allows for the retention of apoptotic sperm cells expressing phosphatidylserine in their external membrane inside the column [5,9,11]. The eluted sample is enriched with non-apoptotic sperm, ready to be used in assisted reproduction technologies (ART) [12]. Despite not being performed routinely in the clinic, it is suggested to patients with high spermatic DNA fragmentation index, more than two unexplained intracytoplasmic sperm injection (ICSI) failures, and, in certain cases, more than two miscarriages with an unknown female cause. MACS combined with density gradient centrifugation (DGC) has been associated with a higher recovery of sperm with progressive motility (68%) when compared to neat ejaculate (39%), as well as lower DNA fragmentation index (4% MACS-DGC versus 24% in the reference) [3], and with improving the percentage of sperm with normal morphology [13]. In some studies, sperm selection via MACS showed a reduction of spermatic DNA fragmentation (fDNA) when compared to the neat ejaculate from asthenoteratozoospermic, teratozoospermic [14], and normozoospermic men [15]. However, one of these studies reported that the reduction of fDNA was not complete and not significant in all patients, since it was only substantial when samples had an initial fragmentation index $\geq$30% (7.1% after MACS versus 41.4% in the ejaculate) [16]. Another study reported no significant improvement in sperm morphology, motility, fDNA, or markers of fertilization capacity Izumo-1 and PLC-ζ comparing MACS combined with swim-up or DGC capacitation against controls [17]. Besides the effect it may have on enhancing sperm parameters, there is a significant lack of consensus on the extent to which sperm selection by MACS improves outcomes of standard ART cycles, as highlighted by recent meta-analyses [18,19].

Accordingly, this study aimed to retrospectively evaluate the effect of MACS sperm processing prior to ICSI in a large sample size to clarify the controversy surrounding its use. Reproductive success was measured by cumulative live birth rates (CLBR) per embryo transfer (ET), per embryo replaced, and per metaphase II (MII) oocyte until a live birth was achieved. By this approach, every embryo was considered a unique opportunity for pregnancy, providing a more realistic view of the impact of the intervention, eliminating the biases associated with measuring success in parameters that only contemplate the contribution of the best embryo in the cohort [20–22].

2. Materials and Methods

2.1. Study Design

This was a retrospective multicentric observational cohort study. Data were included from ICSI cycles using patients' autologous semen samples and oocytes, performed at 15 Spanish IVIRMA clinics from January 2008 to February 2020, using semen samples from unselected males who underwent standard semen preparation (reference group) or an added sperm selection via MACS (study group).

2.2. IVF Procedures

Patient semen samples were collected, prepared, and examined as previously reported [23,24]. After this, capacitation via swim up [25] or density gradient centrifugation [26] was performed. Samples in the reference group were then used for ICSI, according to routine clinical practice. Samples in the MACS group were added with annexin-V-coated microbeads, incubated for their binding to apoptotic sperm with externalized phosphatidylserine, and processed through the column [25,26].

Ovarian stimulation and endometrial preparation were carried out as previously described [27]. Once ovarian follicles gained ≥17 mm in diameter, triggering was performed using recombinant human chorionic gonadotropin (hCG) or a single dose of GnRH agonist. Oocytes were retrieved 36 h after triggering and were then denudated [28]. After four hours, ICSI was performed and the resulting embryos were cultured [25], scored [29], and transferred. Pre-implantation genetic testing for aneuploidies (PGT-A) was performed on some of the embryos according to standard procedure [30]. Due to the vast time this study encompassed, ETs were performed either on day 2–3 of development or on day 5–6 at the blastocyst stage.

2.3. Database

An Excel database was created, containing information on patient and cycle characteristics as well as their outcomes. Prior to statistical analysis, an exploratory analysis was performed to identify outliers and discrepancies between the database and the exported data from the informatic platform used in the clinics. Data from 48,586 patients, 62,070 cycles, 389,212 embryos, and 500,260 oocytes were included.

2.4. Outcome Measures

The primary outcomes in this study were CLBR per ET, per embryo replaced (referring to the total number of embryos transferred to the same patient in consecutive cycles, not in the same transfer procedure), and per MII oocyte used in consecutive cycles until abandoning treatment or achieving a live birth. As commented in the introduction, this measurement was considered a more realistic approach to evaluate the effectiveness of a treatment or technique on the reproductive outcome. Reproductive success was also measured according to more classical outcomes: the biochemical pregnancy rate per ET, understood as the measure for beta hCG in blood serum higher than 10 IU/L at 14–16 days after ICSI, as well as the clinical pregnancy rate, the detection of a positive beta-hCG test result at 21–23 days after microinjection a week after a positive result in the biochemical pregnancy test, or confirmation via ultrasound of development of the fetal pole and heartbeat in weeks 6.5 to 7 of pregnancy. The ongoing pregnancy rate, the confirmation of the positive result of the clinical pregnancy test via ultrasound at week 12 of development, was also calculated per ET. The live birth rate (LBR) was calculated per transfer and per started controlled ovarian stimulation cycle, and the clinical miscarriage rate was expressed per transfer.

2.5. Statistical Analysis

All statistical analyses was performed using R version 4.0.0.

2.5.1. Descriptive Analysis

A descriptive analysis was performed to study the behavior and distribution of variables referring to the patients' and cycles' characteristics and to evaluate the quality of the data and detect possible anomalies within them. For quantitative variables, the usual summary statistics where calculated, as well as 95% confidence intervals (95% CI) for the mean. Categorical variables were expressed as proportions. Means for the quantitative descriptive variables for both groups were compared by using paired *t*-tests to identify possible differences between the reference and the study groups, due to the retrospective nature of the study. For categorical variables, odds ratios (OR) were obtained and expressed with their 95% CI. The Chi-squared test was used to compare proportions. A p-value of <0.05 was considered statistically significant.

2.5.2. Univariate Analysis

For the outcome rates per transfer and per cycle, Fisher's exact test was used to compare both groups. For cumulative rates per transfer, per embryo replaced, and per MII oocyte consumed, Kaplan–Meier curves were plotted and compared via the Mantel–Cox test. A p-value of <0.05 was considered statistically significant.

2.5.3. Multivariate Adjusted Analysis

A mixed-effects logistic regression model was developed to evaluate the association of variables of clinical impact in the main outcome, the ET resulting in a live birth or not. To correct the coefficient estimates of the fixed effects in the model, the patient identification number and the clinic in which the transfer was performed were chosen as random effects. The logistic model for live birth rate per cycle was adjusted for variables that were statistically significantly different between non-MACS and MACS groups, such as the age and BMI of the female patient, the presence or absence of male factor infertility determined by the semen samples' conformance to the WHO 2010 guidelines for normality, and the transfer of the embryo at the blastocyst stage (over day 5 of embryo development), as well as variables considered of clinical relevance to the outcome based on previous experience of the group, such as the age of the male patient, last recorded endometrial lining, and whether or not the embryos of that cycle had been analyzed by PGT-A. Moreover, two separate models were created, dividing cycles into two populations: those who had the embryos analyzed by PGT-A and those who did not. To confirm the results, and control for potential confounders in the computation of CLBR, a Cox regression was performed considering the female patient's age and BMI.

3. Results

3.1. Descriptive Variables

Summary statistics of the main characteristics of patients in both groups undergoing cycles that used the patients' own oocytes are shown in Table 1. Since the same patient could have ART cycles performed with and without MACS sperm selection, descriptive variables were expressed per ovarian stimulation cycle. Female patients' average age in the reference group (59,443 cycles) was 37.03 years (95% CI 37.00, 37.06) with a BMI of 23.22 (23.18, 23.25) kg/m^2, while patients in the MACS group (2627 cycles) were 36.76 years (36.62, 36.91) and 23.05 (22.88, 23.21) kg/m^2 on average. The average age for males was 38.88 (38.83, 38.92) years in the reference group and 38.77 (38.56, 38.97) years in the MACS group.

Table 1. Summary statistics for the main descriptive variables between the reference (semen samples processed according to routine clinical practice) and magnetic-activated cell sorting (MACS) groups in cycles using the patients' autologous oocytes. The data show the mean and the 95% CI, as well as the p-value obtained using a t-test for the quantitative variables. For the categorical variables (*), results are expressed in proportions with their corresponding 95% CI and the p-value was computed using the chi-squared test.

Variable	Reference	MACS
Patient's age (years)	37.03 (37.00, 37.06)	36.76 (36.62, 36.91)
Patient's BMI (kg/m^2)	23.22 (23.18, 23.25)	23.05 (22.88, 23.21)
Semen age (years)	38.88 (38.83, 38.92)	38.77 (38.56, 38.97)
Duration of sterility (years)	2.38 (2.36, 2.40)	2.53 (2.46, 2.60)
Number of oocytes retrieved	10.03 (9.97, 10.08)	11.43 (11.16, 11.69)
Number of MII oocytes	8.05 (8.01, 8.10)	9.03 (8.81, 9.26)
Number of available embryos per cycle	5.88 (5.85, 5.92)	6.50 (6.33, 6.68)

Table 1. *Cont.*

Variable	Reference	MACS
Number of viable embryos per cycle	2.37 (2.35, 2.39)	2.35 (2.25, 2.45)
Number of non-viable embryos per cycle	3.52 (3.49, 3.54)	4.16 (4.01, 4.30)
Days of ovarian stimulation (days)	10.66 (10.65, 10.68)	10.77 (10.70, 10.84)
Dose of gonadotropins (IU)	2247.82 (2240, 2256)	2159.57 (2124, 2195)
Estrogen level at day of ovulation induction (pg/mL)	1757.53 (1747, 1768)	1327.38 (1990, 2093)
Progesterone level at day of ovulation induction (ng/mL)	0.56 (0.44, 0.68)	0.76 (0.21, 1.31)
Days of endometrial preparation (days)	16.14 (16.11, 16.16)	16.83 (16.74, 16.93)
Last recorded endometrial lining (mm)	9.55 (9.54, 9.57)	9.53 (9.47, 9.6)
Last recorded estrogen level (pg/mL)	1438.09 (1428, 1448)	1526.18 (1479, 1573)
Last recorded progesterone level (ng/mL)	2.52 (2.09, 2.95)	1.54 (1.01, 2.07)
Male factor (%) *	15.03 (14.75, 15.32)	18.65 (17.16, 20.14)
Transfer over day 5 (%) *	36.74 (36.41, 37.07)	48.62 (47.05, 50.19)

3.2. Main Outcomes: Cumulative Live Birth Rates

There were 49,350 cycles considered for the assessment of cumulative rates and plotting of survival curves, 47,235 in the reference group and 2115 in the MACS group.

CLBR was first calculated per ET. For the MACS group, this rate was 43.0% (40.5%, 45.4%) for one transfer, 63.6% (60.1%, 66.8%) for two, 80.6% (75.5%, 84.6%) for three, and 88.3% (79.6%, 93.3%) for four, whereas the reference group presented a CLBR of 40.0% (39.5%, 40.5%), 59.6% (58.9%, 60.4%), 72.3% (71.2%, 73.4%), and 81.6% (80.0%, 83.4%), respectively. The plotted Kaplan–Meier curves are shown in Figure 1A. The Cox regression showed a statistically significant positive association between the processing of the semen sample through MACS (hazard ratio (HR) = 1.11, $p = 0.009$) and the CLBR per ET, which was consistent with the result obtained in the univariate analysis.

The CLBR per embryos transferred in the MACS group was 21.5% (19.4%, 23.6%) for one embryo, 55.5% (52.6%, 58.2%) for two, 65.4% (62.0%, 68.5%) for three, and 83.3% (78.9%, 86.7%) for four, while the reference group's CLBR was 15.0% (14.7%, 15.4%), 49.1% (48.6%, 49.7%), 58.0% (57.3%, 58.7%), and 73.3% (72.5%, 74.1%), respectively. Kaplan–Meier curves are shown in Figure 1B. The difference between both curves was statistically significant. Consistent with the results of the univariate analysis, the Cox regression showed a statistically significant positive association between CLBR per embryo transferred and the use of MACS in semen processing (HR = 1.26, $p < 0.001$).

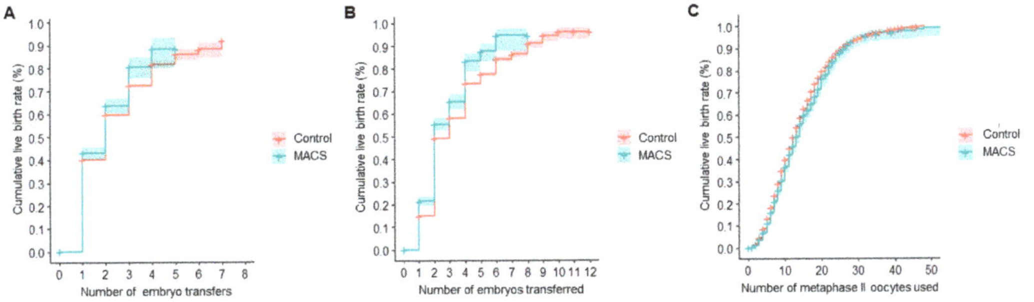

Figure 1. Cumulative live birth rates (CLBR) resulting from the unadjusted analysis of reproductive outcomes in cycles using autologous oocytes. (**A**) CLBR per embryo transfer. (**B**) CLBR per embryo replaced. (**C**) CLBR per patient's own metaphase II oocytes used.

If computed per MII oocytes used, the reference group showed a CLBR of 13.1% (12.8%, 13.5%) for five MII, 39.8% (39.2%, 40.4%) for 10, 62.7% (62.0%, 63.4%) for 15, and 79.55% (78.74%, 80.32%) for 20 MII oocytes, while the MACS group had a CLBR of

11.0% (9.6%, 12.4%), 36.6% (33.9%, 39.1%), 59.8% (56.3%, 63.0%), and 76.25% (72.33%, 79.62%) for the same number of MII oocytes consumed. The Kaplan–Meier curves shown in Figure 1C were statistically significantly different. The Cox regression exhibited no significant relationship between the outcome and sperm selection through MACS.

3.3. Secondary Outcomes: Gestational Outcomes

When computed per ET, the MACS group had a 46.9% (45.2%, 48.7%) biochemical pregnancy rate, a 39.7% (39.4%, 40.0%) clinical pregnancy rate, and a 32.4% (30.7%, 34.1%) ongoing pregnancy rate, while the reference group showed 45.4% (45.0%, 45.8%), 38.5% (38.1%, 38.9%), and 31.8% (31.4%, 32.2%), respectively. None of these differences was statistically significant.

In terms of LBR, the MACS group showed a 29.3% (27.6%, 31.0%) per ET and a 38.8% (36.7%, 40.9%) per cycle. The reference group exhibited a 29.2% (28.8%, 29.6%) LBR per ET and 37.4% (37.0%, 37.8%) per cycle. Neither of the comparisons was statistically significant.

The MACS groups exhibited an 8.2% (7.1%, 9.3%) miscarriage rate per ET, whereas the reference group had a 7.5% (7.2%, 7.7%). The difference between groups was not statistically significant.

All gestational outcomes measured per transfer and, in the case of the LBR also per cycle, are displayed in Table 2.

Table 2. Results from the unadjusted analysis of gestational outcomes in cycles using autologous oocytes. The proportions for each group and the odds ratio (OR) are displayed with their corresponding 95% CI. The proportions are also shown with the sample number, either transfers or initiated cycles, for each of the outcome measurements. The *p*-value was computed using the Fisher's exact test.

Per Transfer	Reference	MACS	OR	*p*-Value
Biochemical pregnancy rate (%)	45.42 (45.04, 45.81) *n* = 63,128	46.94 (45.15, 48.74) *n* = 2961	1.06 (0.99, 1.15)	0.1085
Clinical pregnancy rate (%)	38.48 (38.10, 38.86) *n* = 63,128	39.68 (37.92, 41.44) *n* = 2961	1.05 (0.97, 1.13)	0.1956
Ongoing pregnancy rate (%)	31.80 (31.43, 32.16) *n* = 62,807	32.41 (30.72, 34.11) *n* = 2931	1.03 (0.95, 1.11)	0.4904
Live birth rate (%)	29.20 (28.84, 29.56) *n* = 60,503	29.30 (27.62, 30.99) *n* = 2802	1.01 (0.92, 1.09)	0.9154
Clinical miscarriage rate (%)	7.45 (7.22, 7.67) *n* = 52,218	8.22 (7.11, 9.33) *n* = 2336	1.11 (0.95, 1.30)	0.1715
Per Cycle	**Reference**	**MACS**	**OR**	***p*-Value**
Live birth rate [1] (%)	37.40 (36.96, 37.84) *n* = 47,235	38.82 (36.74, 40.89) *n* = 2115	1.06 (0.97, 1.16)	0.1907

[1] Cycles with all oocytes consumed. Cycles in which the result was not a live birth rate but had still cryopreserved embryos to use in future transfer were not included.

When including selected clinical variables in the model, the adjusted OR for the association between the use of MACS and LBR per cycle was 1.02 (0.91, 1.14). The LBR per cycle in those where the embryos underwent PGT-A analysis showed an adjusted OR of 1.02 (0.84, 1.24), whereas in cycles in which embryos did not undergo PGT-A the adjusted OR was 1.02 (0.89, 1.17). None of these showed a statistically significant difference in LBR per cycle between the MACS group and the reference, as shown in Table 3.

Table 3. Results from the multivariate adjusted analysis, accounting for the relationship between the use of magnetic-activated cell sorting (MACS) and the live birth rate (LBR) per cycle, once adjusted for the female patient's age and BMI, the age of the male patient, the presence or absence of male factor infertility, the last recorded endometrial lining measurement, the transfer of the embryo at the blastocyst stage, and the fact that the embryos underwent pre-implantation genetic testing for aneuploidies (PGT-A). This table shows the adjusted odds ratio (OR) with its 95% CI, standard error, and *p*-value for each studied population, namely: all patients included in the database, those whose embryos underwent PGT-A and those who did not.

Population	*n*	Adjusted OR	Standard Error	*p*-Value
All	59,443 reference cycles / 2627 MACS cycles	1.018 (0.91, 1.14)	0.059	0.767
PGT-A	18,710 reference cycles / 974 MACS cycles	1.020 (084, 1.24)	0.102	0.846
No PGT-A	40,733 reference cycles / 1653 MACS cycles	1.017 (0.89, 1.17)	0.070	0.810

4. Discussion

MACS sperm selection is currently not a part of routine clinical practice in fertility clinics. It is offered to patients in very particular cases with no standard treatment, such as men with a high fDNA index or couples with several failed cycles with no apparent female cause. Patients and clinicians tend to be willing to try diverse *add-ons* after several attempts have failed. However, their added costs should not be dismissed if their use is not justified by a proven increase in possibility to achieve a successful pregnancy [31,32]. Bearing in mind the controversy around the introduction of *add-ons* into the clinical practice without proper security and regulatory reviews [33,34], it is of utmost importance that clinicians are provided with reliable information resulting from carefully designed research, both prospective (randomized clinical trials or RCTs) and retrospective, making use of powerful statistical tools, proper designs, and bias control [35], to ensure that patients receive treatments catered to their needs and situation. This study aimed to determine whether the use of MACS results in an improvement in reproductive outcomes, measured as the number of oocytes, embryos, and transfer procedures required to obtain a live birth. Thus, it evaluated the true clinical impact of the enrichment of semen samples from unselected males with non-apoptotic sperm.

Concerning CLBR, cycles in the MACS group required a lower number of embryos to be transferred until a live birth rate was reached compared to the non-MACS group. Despite the MACS group needing a higher number of MII oocytes than the non-MACS group to obtain the same result, this difference was around 3.3% when consuming 20 oocytes, which, clinically, is meaningless for the patients in terms of increasing or lowering their possibilities to achieve a pregnancy. Results of the Cox regressions were consistent with the conclusions obtained from the univariate analysis: Even though the Mantel–Cox test showed a significant difference between Kaplan–Meier curves of the reference and MACS groups, the small difference observed (a 10.0% increase in CLBR when four embryos were transferred) was more likely due to covariates such as the female patients' age and BMI rather than by the use of MACS during semen sample processing.

The use of MACS offered no improvement in pregnancy or live birth rates per transfer when compared to the reference group, as observed by the non-statistically significant differences between both groups of patients and cycles in their outcomes. Covariates such as the female patient's age and BMI, the presence of male factor infertility, the last recorded endometrial lining, the transfer of the embryo after day 5 (at blastocyst stage), and the assessment of embryo ploidy via PGT-A could be influencing the correlation between the application of MACS as an added sperm selection step in semen sample preparation and the live birth rates per cycle. However, the adjusted OR of these associations are very close to 1, meaning that the size of the effect of the covariates on the live birth rate per cycle is, although significant, truly quite low.

The main limitation of this study was the broad reference population: unselected males in infertile couples. This introduced considerable heterogeneity between patients' prognosis and etiologies, both between the two study groups (MACS and non-MACS) and within the reference group. As shown in Table 1, there was a statistically significant difference in terms of the proportion of patients labelled as 'with male factor infertility', which corresponds to a number of total count of progressive motile sperm lower than 5 million in the fresh ejaculate, and between the MACS (18.65%) and non-MACS (15.03%) groups. Even though the difference in absolute value was quite small, the authors acknowledge that the fact that MACS sperm selection is recommended in the clinic to patients with higher fDNA, several previously failed cycles and, overall, a worse prognosis introduce a bias in the comparison between these and cycles in which semen samples were processed following standard practice. The general scope that this study aimed to provide will progressively develop into separate analysis focused on different male and female etiologies and indications of the patients for undergoing ART, providing more information on the specific patient groups in which MACS could be more useful. As an example of a retrospective study focused on a distinct male population (>20% fDNA index) that could benefit from the use of MACS, Pacheco and colleagues recently reported a decrease in miscarriage rate and an increase in live birth rate when semen samples were processed via MACS versus the control group in cycles using autologous oocytes [36].

Incidentally, statistically significant differences between both groups in variables such as age and BMI were observed. Due to the vastly large amount of similar data that the study handled, any small difference between the groups would be picked up as significant by the analysis, also known as overfitting. Clinically, a difference in 0.2 years and 0.24 kg/m^2 is not a meaningful disparity between these women. Nevertheless, these variables were controlled for during the statistical analysis.

Regarding the level of evidence provided by this reproductive study, it is worth noticing that RCT on this topic considered 29 [37], 138 [25], and 18 [38] patients in their MACS study groups and could not determine statistically significant relationships between the performance of MACS to the semen samples and their outcome variables (clinical pregnancy or miscarriage rates) [19]. In this study, which considered data from 389,212 embryos, 500,260 oocytes, and 62,070 cycles, differences between study groups as slight as a 6.4% in CLBR for two embryos transferred and 2.1% for five MII oocytes used were detected as statistically significant. This level of evidence, even if targeting a heterogeneous population, cannot be understated when drawing conclusions about infertile patients overall.

There is discrepant evidence on the effect of MACS sperm selection on clinical outcomes. One RCT showed increased pregnancy rates per cycle from 24.2% in the reference (DGC) group to 54.5% in the MACS-DGC, even though there was no significant improvement of fertilization or implantation rates, in couples with men factor infertility and at least two of the semen parameters below WHO 2010 normalcy criteria [18,36]. A similar RCT reported an increase of around 21% in LBR when using MACS compared to standard ICSI in normozoospermic men [38]. However, both RCTs had methodological issues, mainly incomplete outcome data or unclear randomization methods [19]. In another example focused on ICSI outcomes, even though the MACS group showed a 67.7% of good quality blastocysts while the standard ICSI group exhibited a 44.2%, there were no significant differences in LBR between groups [39].

Our results agree with the limited clinical impact of the use of MACS reported by previous studies: no improvement of ongoing pregnancy rates [13] and LBR in the MACS group in couples with idiopathic infertility using patients' own semen samples [40], in unselected males in an ovum donation program [25], or in patients with a high level of spermatic DNA fragmentation [41], all of them undergoing ICSI. A meta-analysis by Gil and colleagues suggests that MACS, when compared to standard sperm selection, offers a slight improvement in pregnancy rates, though this does not translate into higher implantation or lower miscarriage rates [18]. The Cochrane database was not able to emit a clear conclusion on the effectiveness of MACS [19]. This lack of consensus could be due to differences in patient inclusion and exclusion criteria, the reduced number of patients recruited and then followed until the end of each cycle and pregnancy, or differences in semen sample processing techniques in each clinic. This may prevent meta-analyses to reliably compare results between studies [18,39].

5. Conclusions

Considering the largest sample size for these types of studies to date, our findings suggest that the separation of non-apoptotic sperm by MACS prior to ICSI in cycles in which autologous oocytes were used reduces the number of embryos required to be transferred in order to obtain a live birth when compared to the control group, although this difference seemed not clinically meaningful. As shown by the pregnancy rates and LBR measured per ET, the clinical impact of the selection of non-apoptotic sperm from samples from unselected males through MACS before performing an ICSI had no clinical effect on reproductive success measured in both classical parameters and CLBR. Even if the method itself is economically, practically, and logistically feasible, its application in fertility clinics cannot be justified unless its use provides a clear improvement in ART outcome rates.

Author Contributions: Conceptualization, M.G.J., I.H., A.N.-G.L., and N.G.; methodology and performance of treatments F.Q., D.A., A.P., C.G.-R., and R.R.-E.; data curation, analysis, and interpretation M.G.J., I.H., and A.N.-G.L.; writing—original draft preparation, M.G.J. and I.H.; writing—review and editing, A.N.-G.L. and N.G.; supervision, N.G. All authors have read and agreed to the published version of the manuscript.

Funding: This research received no external funding.

Institutional Review Board Statement: The study was conducted according to the guidelines of the Declaration of Helsinki and approved by the Research Ethics Committee of IVI Valencia for the use of retrospective anonymized patient data (protocol code 1904-FIVI-049-NG and date of approval 24 December 2020).

Informed Consent Statement: Not applicable.

Data Availability Statement: No new data were created or analyzed in this study. Data sharing is not applicable to this article.

Acknowledgments: The authors would like to acknowledge the inestimable help of Víctor Lozoya as a biostatistician in this project. The authors appreciate the participating clinics of IVIRMA Alicante, Almería, Barcelona, Bilbao, Burgos, Las Palmas, Madrid, Málaga, Mallorca, Murcia, Pamplona, Sevilla, Valencia, Vigo, and Zaragoza for compiling and providing all the data used in this project. M.G.J. develops her research at IVI Foundation granted by a Contrato Predoctoral de Formación en Investigación en Salud (PFIS) from the Instituto de Salud Carlos III (REF:2019/0172). Co-author I.H. is granted by the Consejería de Educación, Investigación, Cultura y Deporte, Generalitat Valenciana (ACIF/2019/261) and the European Social Fund. A.N.-G.L. is granted by the Ministerio de Ciencia, Innovación y Universidades para la Formación de Profesorado Universitario (FPU19/06126).

Conflicts of Interest: The authors declare no conflict of interest.

Abbreviations

ART	assisted reproduction technologies
BMI	body-mass index
CLBR	cumulative live birth rate
DGC	density-gradient centrifugation
ET	embryo transfer

fDNA	DNA fragmentation
hCG	human chorionic gonadotropin
HR	hazard ratio
ICSI	intracytoplasmic sperm injection
LBR	live birth rate
MII	metaphase II
MACS	magnetic-activated cell sorting
OR	odds ratio
PGT-A	pre-implantation genetic testing for aneuploidies
RCT	randomized clinical trial
WHO	World Health Organization
95% CI	95% confidence interval

References

1. WHO. *WHO Laboratory Manual for the Examination and Processing of Human Semen*, 5th ed.; World Health Organization (WHO): Geneva, Switzerland, 2010.
2. Said, T.M.; Land, J.A. Effects of advanced selection methods on sperm quality and ART outcome: A systematic review. *Hum. Reprod. Update* **2011**, *17*, 719–733. [CrossRef]
3. Berteli, T.S.; Da Broi, M.G.; Martins, W.P.; Ferriani, R.A.; Navarro, P.A. Magnetic-activated cell sorting before density gradient centrifugation improves recovery of high-quality spermatozoa. *Andrology* **2017**, *5*, 776–782. [CrossRef]
4. Grunewald, S.; Fitzl, G.; Springsguth, C. Induction of ultra-morphological features of apoptosis in mature and immature sperm. *Asian J. Androl.* **2017**, *19*, 533–537. [CrossRef] [PubMed]
5. Said, T.M.; Gaglani, A.; Agarwal, A. Implication of apoptosis in sperm cryoinjury. *Reprod. Biomed. Online* **2010**, *21*, 456–462. [CrossRef]
6. Mahfouz, R.Z.; Sharma, R.K.; Said, T.M.; Erenpreiss, J.; Agarwal, A. Association of sperm apoptosis and DNA ploidy with sperm chromatin quality in human spermatozoa. *Fertil. Steril.* **2009**, *91*, 1110–1118. [CrossRef] [PubMed]
7. Weng, S.L.; Taylor, S.L.; Morshedi, M.; Schuffner, A.; Duran, E.H.; Beebe, S.; Oehninger, S. Caspase activity and apoptotic markers in ejaculated human sperm. *Mol. Hum. Reprod.* **2002**, *8*, 984–991. [CrossRef]
8. Chen, Z.; Hauser, R.; Trbovich, A.M.; Shifren, J.L.; Dorer, D.J.; Godfrey-Bailey, L.; Singh, N.P. The relationship between human semen characteristics and sperm apoptosis: A pilot study. *J. Androl.* **2006**, *27*, 112–120. [CrossRef]
9. Hichri, R.; Amor, H.; Khammari, M.; Harzallah, M.; El Fekih, S.; Saad, A.; Ajina, M.; Ben Ali, H. Apoptotic sperm biomarkers and the correlation between conventional sperm parameters and clinical characteristics. *Andrologia* **2018**, *50*. [CrossRef]
10. Engel, K.M.; Springsguth, C.H.; Grunewald, S. What happens to the unsuccessful spermatozoa? *Andrology* **2018**, *6*, 335–344. [CrossRef] [PubMed]
11. Simopoulou, M.; Gkoles, L.; Bakas, P.; Giannelou, P.; Kalampokas, T.; Pantos, K.; Koutsilieris, M. Improving ICSI: A review from the spermatozoon perspective. *Syst. Biol. Reprod. Med.* **2016**, *62*, 359–371. [CrossRef]
12. Sharma, R.; Kattoor, A.J.; Ghulmiyyah, J.; Agarwal, A. Effect of sperm storage and selection techniques on sperm parameters. *Syst. Biol. Reprod. Med.* **2015**, *61*, 1–12. [CrossRef] [PubMed]
13. Dirican, E.K.; Ozgun, O.D.; Akarsu, S.; Akin, K.O.; Ercan, O.; Ugurlu, M.; Camsari, C.; Kanyilmaz, O.; Kaya, A.; Unsal, A. Clinical outcome of magnetic activated cell sorting of non-apoptotic spermatozoa before density gradient centrifugation for assisted reproduction. *J. Assist. Reprod. Genet.* **2008**, *25*, 375–381. [CrossRef] [PubMed]
14. Delbes, G.; Herrero, M.B.; Troeung, E.T.; Chan, P.T. The use of complimentary assays to evaluate the enrichment of human sperm quality in asthenoteratozoospermic and teratozoospermic samples processed with Annexin-V magnetic activated cell sorting. *Andrology* **2013**, *1*, 698–706. [CrossRef]
15. Bucar, S.; Goncalves, A.; Rocha, E.; Barros, A.; Sousa, M.; Sa, R. DNA fragmentation in human sperm after magnetic-activated cell sorting. *J. Assist. Reprod. Genet.* **2015**, *32*, 147–154. [CrossRef]
16. Gonzalez-Martinez, M.; Sanchez-Martin, P.; Dorado-Silva, M.; Fernandez, J.L.; Girones, E.; Johnston, S.D.; Gosalvez, J. Magnetic-activated cell sorting is not completely effective at reducing sperm DNA fragmentation. *J. Assist. Reprod. Genet.* **2018**, *35*, 2215–2221. [CrossRef] [PubMed]
17. Cakar, Z.; Cetinkaya, B.; Aras, D.; Koca, B.; Ozkavukcu, S.; Kaplanoglu, I.; Can, A.; Cinar, O. Does combining magnetic-activated cell sorting with density gradient or swim-up improve sperm selection? *J. Assist. Reprod. Genet.* **2016**, *33*, 1059–1065. [CrossRef]
18. Gil, M.; Sar-Shalom, V.; Melendez Sivira, Y.; Carreras, R.; Checa, M.A. Sperm selection using magnetic activated cell sorting (MACS) in assisted reproduction: A systematic review and meta-analysis. *J. Assist. Reprod. Genet.* **2013**, *30*, 479–485. [CrossRef]
19. Lepine, S.; McDowell, S.; Searle, L.M.; Kroon, B.; Glujovsky, D.; Yazdani, A. Advanced sperm selection techniques for assisted reproduction. *Cochrane Database Syst. Rev.* **2019**, *7*, CD010461. [CrossRef] [PubMed]
20. Garrido, N.; Bellver, J.; Remohi, J.; Simon, C.; Pellicer, A. Cumulative live-birth rates per total number of embryos needed to reach newborn in consecutive in vitro fertilization (IVF) cycles: A new approach to measuring the likelihood of IVF success. *Fertil. Steril.* **2011**, *96*, 40–46. [CrossRef]

21. Garrido, N.; Bellver, J.; Remohi, J.; Alama, P.; Pellicer, A. Cumulative newborn rates increase with the total number of transferred embryos according to an analysis of 15,792 ovum donation cycles. *Fertil. Steril.* **2012**, *98*, 341–346.e2. [CrossRef] [PubMed]
22. Cobo, A.; Garrido, N.; Pellicer, A.; Remohi, J. Six years' experience in ovum donation using vitrified oocytes: Report of cumulative outcomes, impact of storage time, and development of a predictive model for oocyte survival rate. *Fertil. Steril.* **2015**, *104*, 1426–1434.e8. [CrossRef]
23. Rivera-Egea, R.; Garrido, N.; Sota, N.; Meseguer, M.; Remohi, J.; Dominguez, F. Sperm lipidic profiles differ significantly between ejaculates resulting in pregnancy or not following intracytoplasmic sperm injection. *J. Assist. Reprod. Genet.* **2018**, *35*, 1973–1985. [CrossRef] [PubMed]
24. Esbert, M.; Pacheco, A.; Soares, S.R.; Amoros, D.; Florensa, M.; Ballesteros, A.; Meseguer, M. High sperm DNA fragmentation delays human embryo kinetics when oocytes from young and healthy donors are microinjected. *Andrology* **2018**, *6*, 697–706. [CrossRef]
25. Romany, L.; Garrido, N.; Motato, Y.; Aparicio, B.; Remohi, J.; Meseguer, M. Removal of annexin V-positive sperm cells for intracytoplasmic sperm injection in ovum donation cycles does not improve reproductive outcome: A controlled and randomized trial in unselected males. *Fertil. Steril.* **2014**, *102*, 1567–1575.e1. [CrossRef]
26. Esbert, M.; Godo, A.; Soares, S.R.; Florensa, M.; Amoros, D.; Ballesteros, A.; Vidal, F. Spermatozoa with numerical chromosomal abnormalities are more prone to be retained by Annexin V-MACS columns. *Andrology* **2017**, *5*, 807–813. [CrossRef]
27. Cobo, A.; Garcia-Velasco, J.; Domingo, J.; Pellicer, A.; Remohi, J. Elective and Onco-fertility preservation: Factors related to IVF outcomes. *Hum. Reprod.* **2018**, *33*, 2222–2231. [CrossRef] [PubMed]
28. Cobo, A.; Garcia-Velasco, J.A.; Coello, A.; Domingo, J.; Pellicer, A.; Remohi, J. Oocyte vitrification as an efficient option for elective fertility preservation. *Fertil. Steril.* **2016**, *105*, 755–764. [CrossRef] [PubMed]
29. ASEBIR. *Criterios ASEBIR de Valoración Morfológica de Oocitos, Embriones Tempranos y Blastocistos Humanos*, 3rd ed.; Asociación para el Estudio de la Biología de la Reproducción (ASEBIR): Madrid, Spain, 2015.
30. de Los Santos, M.J.; Diez Juan, A.; Mifsud, A.; Mercader, A.; Meseguer, M.; Rubio, C.; Pellicer, A. Variables associated with mitochondrial copy number in human blastocysts: What can we learn from trophectoderm biopsies? *Fertil. Steril.* **2018**, *109*, 110–117. [CrossRef] [PubMed]
31. Datta, A.K.; Campbell, S.; Deval, B.; Nargund, G. Add-ons in IVF programme—Hype or Hope? *Facts Views Vis. Obgyn.* **2015**, *7*, 241–250. [PubMed]
32. van de Wiel, L.; Wilkinson, J.; Athanasiou, P.; Harper, J. The prevalence, promotion and pricing of three IVF add-ons on fertility clinic websites. *Reprod. Biomed. Online* **2020**, *41*, 801–806. [CrossRef]
33. Rutherford, A.J. Should the HFEA be regulating the add-on treatments for IVF/ICSI in the UK?: FOR: Regulation of the fertility add-on treatments for IVF. *BJOG* **2017**, *124*, 1848. [CrossRef]
34. Hendricks, S.; Pearson, S.D. A Proposed Framework for Strengthening Regulatory Review of Innovative Reproductive Techniques in the United States. Available online: https://www.fertstertdialog.com/posts/38973-hendriks-consider-this (accessed on 31 April 2021).
35. Macklon, N.S.; Ahuja, K.K.; Fauser, B. Building an evidence base for IVF 'add-ons'. *Reprod. Biomed. Online* **2019**, *38*, 853–856. [CrossRef] [PubMed]
36. Pacheco, A.; Blanco, A.; Bronet, F.; Cruz, M.; García-Fernández, J.; García-Velasco, J.A. Magnetic-Activated Cell Sorting (MACS): A Useful Sperm-Selection Technique in Cases of High Levels of Sperm DNA Fragmentation. *J. Clin. Med.* **2020**, *9*, 3976. [CrossRef]
37. Ziarati, N.; Tavalaee, M.; Bahadorani, M.; Nasr Esfahani, M.H. Clinical outcomes of magnetic activated sperm sorting in infertile men candidate for ICSI. *Hum. Fertil. (Camb.)* **2019**, *22*, 118–125. [CrossRef] [PubMed]
38. Troya, J.; Zorrilla, I. Annexin V-MACS in infertile couples as method for separation of sperm without DNA fragmentation. *JBRA Assist. Reprod.* **2015**, *19*, 66–69. [CrossRef] [PubMed]
39. Stimpfel, M.; Verdenik, I.; Zorn, B.; Virant-Klun, I. Magnetic-activated cell sorting of non-apoptotic spermatozoa improves the quality of embryos according to female age: A prospective sibling oocyte study. *J. Assist. Reprod. Genet.* **2018**, *35*, 1665–1674. [CrossRef] [PubMed]
40. Sheikhi, A.; Jalali, M.; Gholamian, M.; Jafarzadeh, A.; Jannati, S.; Mousavifar, N. Elimination of apoptotic spermatozoa by magnetic-activated cell sorting improves the fertilization rate of couples treated with ICSI procedure. *Andrology* **2013**, *1*, 845–849. [CrossRef]
41. Sanchez-Martin, P.; Dorado-Silva, M.; Sanchez-Martin, F.; Gonzalez Martinez, M.; Johnston, S.D.; Gosalvez, J. Magnetic cell sorting of semen containing spermatozoa with high DNA fragmentation in ICSI cycles decreases miscarriage rate. *Reprod. Biomed. Online* **2017**, *34*, 506–512. [CrossRef] [PubMed]

biology

MDPI

Article

High-Efficiency Bovine Sperm Sexing Used Magnetic-Activated Cell Sorting by Coupling scFv Antibodies Specific to Y-Chromosome-Bearing Sperm on Magnetic Microbeads

Korawan Sringarm [1,2,3], Marninphan Thongkham [1], Supamit Mekchay [1,2], Chompunut Lumsangkul [1], Wannaluk Thaworn [4], Wiwat Pattanawong [4], Ekaphot Rangabpit [5], Pornchai Rachtanapun [3,6], Kittisak Jantanasakulwong [3,6], Anucha Sathanawongs [7] and Surat Hongsibsong [2,8,*]

1 Department of Animal and Aquatic Sciences, Faculty of Agriculture, Chiang Mai University, Chiang Mai 50200, Thailand; korawan.s@cmu.ac.th (K.S.); marninphan_t@cmu.ac.th (M.T.); supamit.m@cmu.ac.th (S.M.); chompunut.lum@cmu.ac.th (C.L.)
2 Cluster of Research and Development of Pharmaceutical and Natural Products Innovation for Human or Animal, Chiang Mai University, Chiang Mai 50200, Thailand
3 Cluster of Agro Bio-Circular-Green Industry, Faculty of Agro-Industry, Chiang Mai University, Chiang Mai 50100, Thailand; pornchai.r@cmu.ac.th (P.R.); kittisak.jan@cmu.ac.th (K.J.)
4 Faculty of Animal Science and Technology, Maejo University, Chiang Mai 50290, Thailand; wannaluk.333@gmail.com (W.T.); wpattanawong@gmail.com (W.P.)
5 Livestock Semen Production Center, Inthanon Royal Project, Department of Livestock Development, Ministry of Agriculture and Cooperatives, Chiang Mai 50360, Thailand; spcm_cmi@dld.go.th
6 Faculty of Agro-Industry, Chiang Mai University, Chiang Mai 50200, Thailand
7 Department of Veterinary Biosciences and Veterinary Public Health, Faculty of Veterinary Medicine, Chiang Mai University, Chiang Mai 50100, Thailand; anucha.sa@cmu.ac.th
8 School of Health Sciences Research, Research Institute for Health Sciences, Chiang Mai University, Chiang Mai 50200, Thailand
* Correspondence: surat.hongsibsong@cmu.ac.th

Citation: Sringarm, K.; Thongkham, M.; Mekchay, S.; Lumsangkul, C.; Thaworn, W.; Pattanawong, W.; Rangabpit, E.; Rachtanapun, P.; Jantanasakulwong, K.; Sathanawongs, A.; et al. High-Efficiency Bovine Sperm Sexing Used Magnetic-Activated Cell Sorting by Coupling scFv Antibodies Specific to Y-Chromosome-Bearing Sperm on Magnetic Microbeads. *Biology* **2022**, *11*, 715. https://doi.org/10.3390/biology11050715

Academic Editors: Pilar Santolaria Blasco, Rosaura Pérez-Pé and Jessica Rickard

Received: 31 March 2022
Accepted: 4 May 2022
Published: 6 May 2022

Publisher's Note: MDPI stays neutral with regard to jurisdictional claims in published maps and institutional affiliations.

Simple Summary: Female calves are favored for milk production and genetic advancement in the dairy industry, and sex selection by using sexed semen has been long considered. A potential alternative sperm sexing technique is magnetic-activated cell sorting combined with an immunological method that uses scFv antibodies against male-specific sites on Y-chromosome-bearing sperm; however, the technique should be evaluated for validity and accuracy. This study focuses on how well bovine sperm are separated by the use of magnetic microbeads coupled with scFv antibodies against Y-chromosome-bearing sperm (PY-microbeads). The results showed that sexed bovine sperm using PY-microbeads was a highly effective technique for distinguishing X- and Y-chromosome-bearing sperm. It had no negative impact on the quality of X-chromosome-bearing sperm. The technique produced 82.65% of X-chromosome sperm in the X-enriched fraction semen and 81.43% of Y-chromosome sperm in the Y-enriched fraction semen, which was utilized to generate target sexed bovine semen.

Abstract: Sperm sexing technique is favored in the dairy industry. This research focuses on the efficiency of bovine sperm sexing using magnetic-activated cell sorting (MACS) by scFv antibody against Y-chromosome-bearing sperm (Y-scFv) coupled to magnetic microbeads and its effects on kinematic variables, sperm quality, and X/Y-sperm ratio. In this study, the optimal concentration of Y-scFv antibody coupling to the surface of magnetic microbeads was 2–4 mg/mL. PY-microbeads revealed significantly enriched Y-chromosome-bearing sperm (Y-sperm) in the eluted fraction (78.01–81.43%) and X-chromosome-bearing sperm (X-sperm) in the supernatant fraction (79.04–82.65%). The quality of frozen–thawed sexed sperm was analyzed by CASA and imaging flow cytometer, which showed that PY-microbeads did not have a negative effect on X-sperm motility, viability, or acrosome integrity. However, sexed Y-sperm had significantly decreased motility and viability. The X/Y-sperm ratio was determined using an imaging flow cytometer and real-time PCR. PY-microbeads produced sperm with up to 82.65% X-sperm in the X-enriched fraction and up to 81.43% Y-sperm in the Y-enriched fraction. Bovine sperm sexing by PY-microbeads showed high efficiency in separating Y-sperm from

X-sperm and acceptable sperm quality. This initial technique is feasible for bovine sperm sexing, which increases the number of heifers in dairy herds while lowering production expenses.

Keywords: bull semen; sexing semen; magnetic-activated cell sorting; scFv antibody; semen quality

1. Introduction

Livestock reproductive management aims to produce a high number of offspring that are good genetic breeders to expand production [1,2]. In the dairy industry, it is preferred to produce female calves for milk production and genetic improvement [3]. Sex preselection by using sexed semen has been considered. Bovine sperm sexing is a well-known technology. Various techniques have been applied to separate X-chromosome-bearing sperm (X-sperm) and Y-chromosome-bearing sperm (Y-sperm) based on the difference in their DNA content, which is 3.7–4.2% depending on the breed [1,2]. Originally, separating X and Y sperm was performed using flow cytometry combined with a cell sorter. Meanwhile, the so-called Beltsville sperm sexing technology has proceeded to the point of commercialization, with USDA licensee XY Inc. promoting it across the world [4]. This precise method is able to produce sex-sorted semen with an accuracy demonstrated by a female calve birth rate of 85–95% [2]. However, sperm sexing equipment is expensive to purchase and also to operate. Another major problem of cell sorting is that it increases the possibility of sperm deterioration due to the chemical and mechanical stress applied during the procedure, leading to low sperm quality and low fertility [5,6].

The immunological sexing approach is an alternate method of sexing sperm. Thongkham et al. (2021) [7] provided proof of principle for immunological sexing by using monoclonal antibodies (Mab) against male-specific sites on the plasma membrane of Y-sperm in order to identify between X- and Y-sperm, followed by a cytotoxic reaction that destroys Y-sperm, leading to a high X-sperm ratio in sexed semen. This sperm sexing technique has no damaging impact on acrosome integrity of the sperm or sexed sperm yields, and has a conception rate similar to that of conventional semen, with female calves born at a rate above 74%. The success of immunological sexing comes from the use of a highly specific antibody against an epitope on the plasma membrane of the sperm. A Mab generated against male-specific plasma membrane of Y-sperm produced by hybridoma cells [8] is a high-quality source of engineered recombinant antibodies, such as single-chain fragment variable antibodies (scFv antibodies). An scFv antibody structure is composed of variable regions of heavy (VH) and light (VL) chains of immunoglobulin, which are connected by flexible peptide linker such as $(Gly_4Ser)_3$ linker sequence. The protein engineering technique produces scFv antibodies by using recombinant bacteria. In addition, this technique has been utilized to improve the qualities of scFv antibodies, such as affinity and specificity [9,10]. Thaworn et al. (2020) [11] generated VH and VL genes and utilized them to generate a scFv gene (650 bp) that was expressed in *Escherichia coli* TG1 cells and produced a soluble scFv antibody specific for male-specific regions on the plasma membrane of Y-sperm. This soluble scFv antibody was advantageous for use in precisely sexing sperm to create a novel bovine semen sexing technique.

The immunological sexing technique was less expensive than flow cytometry cell sorting for sperm sexing and also produced a higher number of sperm per dose and better frozen-thaw sperm quality [7], but the accuracy (i.e., the female calf birth rate) was still lower than that of flow cytometry cell sorting [12]. Therefore, improving the accuracy of immunological sexing techniques remains a challenge.

The recently developed magnetic-activated cell sorting (MACS) technique uses specific antibody-conjugated magnetic microspheres to bind to a target antigen or protein so that different cell fractions can be separated by exposing the combined populations to a magnetic field [13]. A previous study showed that MACS effectively separated dead sperm and apoptotic sperm [14,15]. Said et al. (2006) [16] reported a 73.8% recovery rate after excluding spermatozoa with externalized phosphatidylserine utilizing paramagnetic annexin V-conjugated microbeads. Ybarra et al. (2016) [17] discovered that distinguishing membrane-damaged spermatozoa from viable sperm cells by biotin-labeled DNA aptamers with avidin-coated nanoparticles enhanced the selection procedure of sperm without impairing their potential to form embryos. Assumpção et al. (2021) [18] selected a process for high-quality sperm in bovine sperm utilizing MACS, that was quite effective in generating samples with a high percentage of viable cells, membrane integrity, and mitochondrial potential.

The use of MACS by coupling scFv antibodies against male-specific sites on magnetic beads may have the potential to serve as an alternative sperm sexing technique that improves upon the immunological sexing technique with more precision, less effect on sperm quality, and a lower cost of sexed sperm production. Therefore, the present study focuses on the efficiency of the sexing procedure in bovine sperm using MACS combined with scFv antibodies specific to the plasma membrane of Y-sperm coupled to the surface of the magnetic microbeads on sperm motility, kinematic variables, sperm quality, and the X/Y-sperm ratio.

2. Materials and Methods

2.1. Chemicals and Reagents

Magnetic microbeads coated with a poly (lactic acid) polymer film (PLA-M) with a surface carboxylic acid group (-COOH), a particle size of 30 μm, and a concentration of 10 mg/mL in suspension (micromod Partikeltechnologie, GmbH, Rostock, Germany) were used. The scFv antibody against male-specific sites on the plasma membrane of Y-sperm (Y-scFv antibody) was produced and prepared in our laboratory as described by Thaworn et al. (2020); this antibody presents a high binding capacity for Y-sperm and low cross-reactivity (4.25%) with X-sperm. *N*-hydroxysuccinimide (NHS), 1-ethyl-3-(3-dimethylaminopropyl)-carbodiimide hydrochloride (EDC), Dulbecco's phosphate-buffered saline (DPBS), and 3,3′, 5,5′-tetramethylbenzidine (TMB) were purchased from Sigma Aldrich (USA). SYBR-14, propidium iodide (PI), Hoechst 33342, and lectin PNA from Arachis hypogaea (peanut) conjugated with Alexa Fluor 488 (PNA-Alexa 488) were purchased from Invitrogen (Eugene, OR, USA). Anti-HA tag peroxidase (Abcam, Cambridge, MA, USA), and commercially available sex-sorted X-sperm by cell sorter (GENEX, Shawano, WI, USA) were also used.

2.2. Estimation of the Amounts of Y-scFv Antibody Coupled to Magnetic Microbeads

The amount of Y-scFv antibody on the surface of the magnetic microbeads was evaluated. PLA-M microbeads (10 mg/mL) were added to each vial (50 μL), and then 400 μL of 0.1 mM EDC and 400 μL of 0.1 mM NHS were added and incubated for 4 h at 4 °C. Then, the carboxylic acid group on the surface of the magnetic microbeads was activated. After that, ten samples containing Y-scFv antibody solution were placed in each microbead vial to final concentrations of 0, 0.5, 1, 2, 4, 6, 8, 10, 12, 14, and 16 mg/mL, and each reaction volume was mixed thoroughly in a shaker incubator (SHKE480HP, Thermo Fisher Scientific, Kansas, MO, USA) at 4 °C overnight. After incubation, the microbeads coupled with scFv antibody (PY-microbeads) were trapped by a strong neodymium magnet and kept at 4 °C until use for the coupling capacity evaluation process.

In this situation, the coupling capacity of the Y-scFv antibody to the magnetic microbeads was determined by using an anti-HA tag. The magnetic microbeads coupled with scFv antibody at each concentration were washed five times with 0.2 mL of DPBS. Then, 20 μL of goat polyclonal anti-HA tag peroxidase (1:100,000) was added and incubated at room temperature for 1 h. The sample was subsequently washed five times with 200 mL PBS. After that, 100 μL of TMB solution was added as the substrate and kept for 15 min at room temperature. Then, adding 2 M sulfuric acid, the reaction was stopped, and the absorbance at 450 nm was determined using a SpectraMax M3 microplate reader (Molecular

Devices, San Jose, CA, USA). The immunoreactivities were compared using optical density (OD) measurements.

2.3. Semen Sample Collection

Semen samples were taken from four tropical Holstein Friesian bulls aged 3–5 years and were kept separately at the Inthanon Royal Project Livestock Semen Production Center (Chiang Mai, Thailand). The Department of Livestock Development of the Thailand Ministry of Agriculture and Cooperatives has established legislation and regulations governing experimentation on live animals. Fresh sperm samples were obtained via an artificial vagina. The only semen samples that were used in the experiment had a sperm concentration of greater than 650 $\times$ 10^6 cells/mL and total motility of greater than 80%.

2.4. Optimization of Y-scFv Antibody Coupling on Magnetic Microbeads for Sorting Sperm

The 10 mg/mL of PLA-M magnetic microbeads were activated with 0.1 mM EDC and 0.1 mM NHS. Then, six mixture samples were mixed with Y-scFv antibody solution in each microbead vial to final concentrations of 0, 0.5, 1, 2, 3, and 4 mg/mL, and each mixture was shaken thoroughly in a shaker incubator at 4 °C overnight. After incubation, the microbeads coupling scFv antibody (PY-microbeads) were trapped by a strong neodymium magnet. Then, the supernatant was removed, and the PY microbeads were collected. After that, the unbound carboxylic acid groups on the microbeads were enclosed by Tris (50 mmol/L, pH 7.4) for 15 min at room temperature, and the beads were then washed three times with DPBS and suspended in 100 µL of DPBS. The prepared PY-microbeads were kept at 4 °C until use in the evaluated sperm sexing procedure.

Fresh semen was diluted in a Tris–citric acid-based extender to 4 $\times$ 10^6 cells/mL. Then, each individual PY-microbead sample was added, and incubated for 20 min at 37 °C with gentle shaking and placed on a strong neodymium magnet for 5 min at room temperature. Then, unbound PY-microbeads, or the "supernatant fraction" were removed into new tubes. The sperm entrapped on PY-microbeads were separated from the microbeads using an eluting buffer (Tris–citric acid extender buffer containing 0.05 mM imidazole, pH 7.4), incubated for 20 min at 37 °C, and then placed on a strong neodymium magnet to trap the magnetic microbeads. The fractions eluted from the magnetic microbeads were transferred to new tubes and called the "eluted fraction". The numbers of sperm in each "supernatant fraction" and "eluted fraction" were measured and used to indicate the PY microbead's sperm binding capacity and sex sperm ratio. A schematic of the sperm sorting protocol using PY-microbeads is shown in Figure 1.

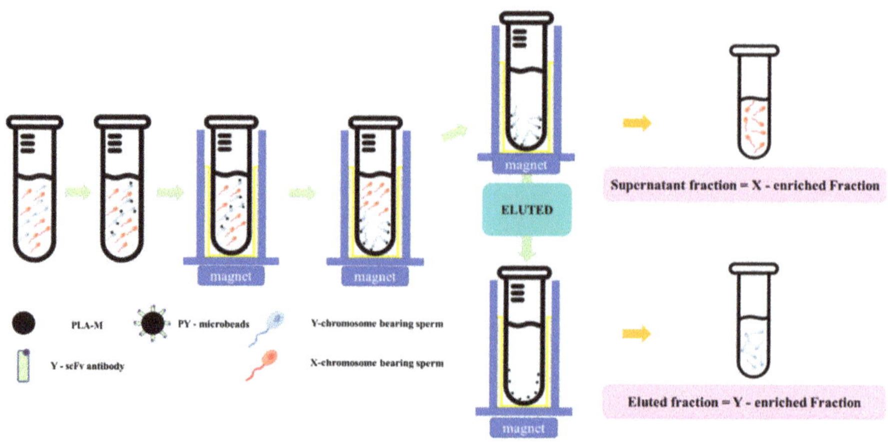

Figure 1. The schematic magnetic-activated cell sorting protocol for bovine sperm by using PY-microbeads.

2.5. Production of Frozen Sexed Bovine Semen by PY-Microbeads

Fresh semen samples from four tropical Holstein Friesian bulls were collected five times per bull using an artificial vagina. The first treatment was conventional semen treatment (T1; CON), in which fresh semen was diluted to a final concentration of 8×10^7 cells/mL with a Tris–egg yolk-based extender (glycerol 8% *v/v*, egg yolk 20% *w/v*, D-Fructose 0.2% *w/v*, penicillin 1000 IU, and gentamycin 0.3 mg/mL). In the second treatment, fresh semen was incubated with PLA-M magnetic microbeads (unbound with scFv antibody) at 37 °C for 20 min and then diluted to a final concentration of 8×10^7 cells/mL in Tris–egg yolk-based extender (T2; negative control; NC). The third treatment (T3) was sexed by using PY-microbeads (ratio 2 mg/mL Y-scFv: 10 mg/mL PLA-M beads), as follows: Fresh semen was diluted to a final concentration of 1×10^9 cells/mL and incubated with PY-microbeads at 37 °C for 20 min. Then, the PY-microbeads, which captured Y-sperm, were trapped by a strong neodymium magnet. After that, unbound sperm were removed to new tubes (T3: X-enriched fraction). The sperm entrapped on the PY-microbeads were separated from the beads by incubation in the eluting buffer for 20 min at 37 °C. After that, the released sperm were transferred to new tubes (T3: Y-enriched fraction). Meanwhile, the X- and Y-enriched fractions were diluted to a final concentration of 8×10^7 cells/mL with a Tris–egg yolk-based extender. In all treatments the sperm samples were frozen. A diagram of the protocol for bovine semen sexing and evaluation in the present study is shown in Figure 2.

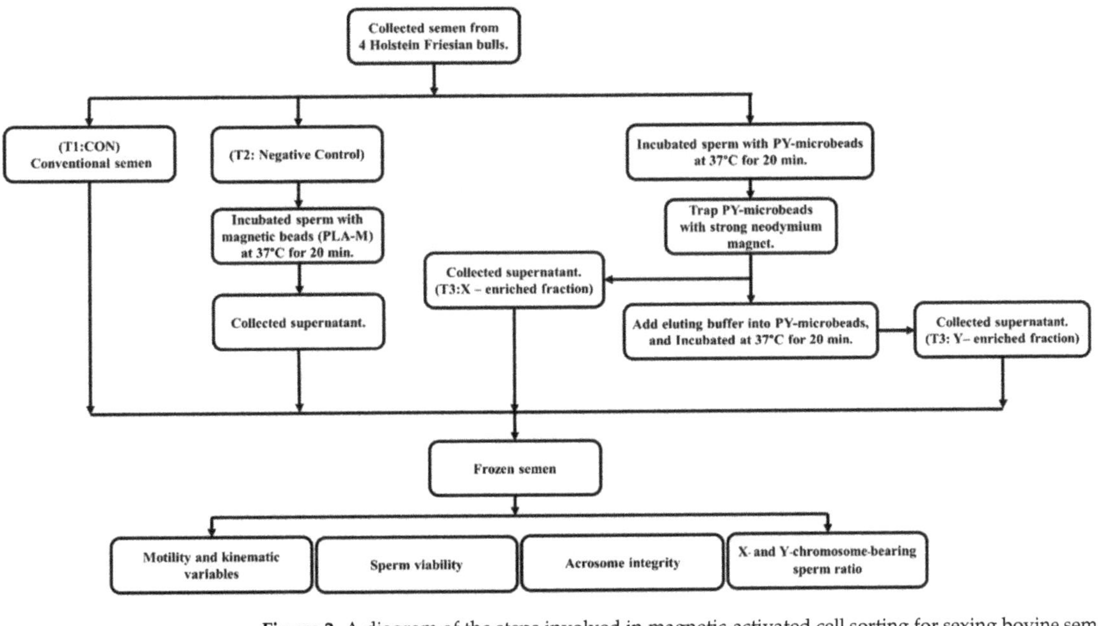

Figure 2. A diagram of the steps involved in magnetic-activated cell sorting for sexing bovine semen production and the evaluation of bovine semen in the present study.

2.6. Evaluation of Frozen–Thawed Sperm Motility and Kinematic Variables Using Computer-Aided Sperm Analysis (CASA)

The frozen semen samples from CON, NC, X-enriched, and Y-enriched fractions were thawed for 30 s in a water bath at 37 °C. AndroVision software (Minitube of America—MOFA®, Verona, WI, USA) linked to a Zeiss AxioScope with a heated stage at 37 °C (Carl Zeiss MicroImaging GmbH, Göttingen, Germany) was used to assess motility and kinematic characteristics. Aliquots of sperm (5.0 μL) were deposited on prewarmed slides, covered with a coverslip, and then subjected to rapid CASA analysis. The following parameters were measured and statistically analyzed in this study: total sperm motility (TM, %), progressive sperm motility (PM, %), distance average path (DAP, μm), distance curved line (DCL, μm), distance straight line (DSL, μm), velocity average path (VAP, μm/s), velocity curved line (VCL, μm/s), velocity straight line (VSL, μm/s), beat cross frequency (BCF, Hz), STR straightness VCL/VAP (STR, %), linearity VSL/VCL (LIN, %), and wobble VAP/VCL (WOB) [19].

2.7. Evaluation of Frozen–Thawed Sperm Viability Using Imaging Flow Cytometry

The viability of sperm in frozen semen straws from each treatment was assessed using two fluorescent probes: SYBR-14 for living sperm and PI for dead sperm, as described by Thongkham et al. (2021) [7]. After thawing, the sperm samples were centrifuged at a speed of 269× *g* for 8 min. Then, 300 mL of PBS containing 2×10^6 sperm was incubated for 10 min in the dark with 1.2 μL of diluted SYBR-14 and 3 μL of PI (2.4 mM) before imaging flow cytometry analysis (FlowSight, Seattle, WA, USA). SYBR-14 and PI were stimulated using a 60 mW 488-nm laser. IDEAS 6.2 was used to examine the data (Amnis, Seattle, WA, USA).

2.8. Evaluation of Frozen–Thawed Sperm Acrosome Integrity Using Imaging Flow Cytometry

The acrosome integrity status of sperm from each treatment was determined following Thongkham et al. (2021) [7]. After thawing, the sperm samples were centrifuged at 269× *g* for 8 min. Then, PNA-Alexa 488 (100 μg/mL) of 0.5 μL and PI (1 mg/mL) of 0.5 μL were added to each sample (final concentration of sperm 1.5×10^6 sperm/mL), and the mixture was incubated for 10 min at 37 °C. After that, the samples were centrifuged at 1075× *g* for 5 min, the supernatant was removed, and the pellets were resuspended in 50 μL of PBS before imaging flow cytometry analysis (FlowSight). The IDEAS version 6.2 (Amnis, Seattle, WA, USA) program was utilized for data analysis. A 60-mW 488-nm laser was used to stimulate PNA-Alexa 488 and PI.

2.9. Discrimination of X/Y-Sperm by Imaging Flow Cytometry

The X/Y sperm ratio in each treatment was determined following Thongkham et al. (2021) [7]. After thawing the semen sample, semen samples were diluted to 1×10^6 sperm/mL with PBS mixed with 1.2 μL of Hoechst 33,342 (50 μg/mL). Then, the sample was incubated for 10 min at 37 °C in the dark and the results were analyzed using image flow cytometry (FlowSight). A 15-mW 405-nm laser was used to stimulate the Hoechst 33,342 stain. Histograms were produced to evaluate the fluorescence of Hoechst 33,342.

2.10. Evaluation of Sexed Sperm by SYBR® Green RT–PCR

A pair of specific primers were constructed for the bovine Y and X chromosomal partial sequences using the NCBI website and the SYBR® Green Real-Time PCR settings. As a positive control, the internal housekeeping gene GAPDH was used. The pair of Y-specific primers was constructed to target a conserved region of the SRY gene connected to the bovine chromosome. For the bovine proteolipid protein gene, primers specific to the X chromosome were created (PLP). The forward and reverse primer sequences, Tm, and amplicon length are listed in Table 1 [20].

Table 1. Specific primers used to amplify bovine SRY, PLP and GAPDH.

Gene	Sequence (5′ → 3′)	Length (bp)	Accession No.
Y chromosome specific			
SRY-Forward	GAAAATAAGCACAAGAAAGTCCAGG	124	EU581861.1
SRY-Reverse	CAAAAGGAGCATCACAGCAGC		
X chromosome specific			
PLP-Forward	GGTGTGTTAGTTTCTGCTGTACAATAAATGG	96	AJ009913.1
PLP-Reverse	GATGGCAGGTGAGGGTAGGA		
Housekeeping gene			
GAPDH-Forward	GGCGCCAAGAGGGTCAT	120	NM_001034034.2
GAPDH—Reverse	GGTGGTGCAGGAGGCATT		

SRY: sex-determining region Y; PLP: proteolipid protein; GAPDH: glyceraldehyde-3-phosphate dehydrogenase.

2.10.1. DNA Sample Preparation

DNA was extracted from each semen sample by using the protocol of Chandler et al. (2002) [21] with some modifications. Sperm from 5 straws from each treatment were thawed and washed twice with D-PBS (composition). Proteinase K (Qiagen Inc., Valencia, CA, USA), 1 M dithiothreitol (DTT) in 0.01 M sodium acetate, and 5% Chelex (Sigma Aldrich (USA) were added to the samples. The samples were gently mixed and incubated at 37 °C. After incubation, the samples were extracted with DNA using the QIAamp® DNA mini kit (Qiagen, Valencia, CA, USA). The procedures for DNA extraction were carried out according to the manufacturer's instructions. Quantification of the samples was performed using a NanoDropTM 2000 (Thermo Scientific, Wilmington, DE, USA) at an absorbance ratio of 260–280 nm. The sequences of the forward and reverse primers and amplicon length are given in Table 1. GAPDH was used as an internal housekeeping gene.

2.10.2. Quantitative SYBR Green Real-Time PCR (qPCR) Analyses

Real-time PCR was performed on a CFX Connect™ Real-Time PCR System by using iTaq Universal SYBR Green Supermix (2X) (Bio–Rad Laboratories, CA). DNA (25 ng) was performed in triplicate along with a control for each test. Amplification was performed by an optimized protocol (5 min at 95 °C, 40 repeated cycles of two steps at 95 °C for 15 s, 55 °C X gene/56 °C Y gene for 15 s and 72 °C for 30 s).

2.11. Statistical Analysis

Motility and kinematic variables, acrosome integrity, percentages of X- and Y-sperm, and evaluation of sexed sperm by SYBR® Green RT–PCR were analyzed by one-way ANOVA using the statistical software program SPSS version 20.0 (SPSS Inc., Chicago, IL, USA). Duncan's new multiple range test was used to compare values between individual groups [7,22]. Data are presented as the mean ± standard deviation. A paired *t*-test was used to compare X- and Y-chromosome-bearing sperm on a histogram as a double Gaussian distribution.

3. Results

3.1. Efficiency of Y-scFv Antibody Coupling to Magnetic Microbeads

The amount of Y-scFv antibody coupled to the surface of the magnetic microbeads was determined by using goat polyclonal anti-HA tag peroxidase that reacted directly to the HA position on the scFv antibody. The concentration of Y-scFv antibody bound to 10 mg magnetic microbeads was increased stepwise until saturation at 3–16 mg/mL. The effectiveness of Y-scFv antibody coupling on the surface of the magnetic microbeads in the range of 2–4 mg/mL was tested for sperm sexing, as shown in Figure 3. According to this result, the Y-scFv antibody was fully bound to the surface of the PLA-M magnetic microbeads by using the EDC–NHS reaction.

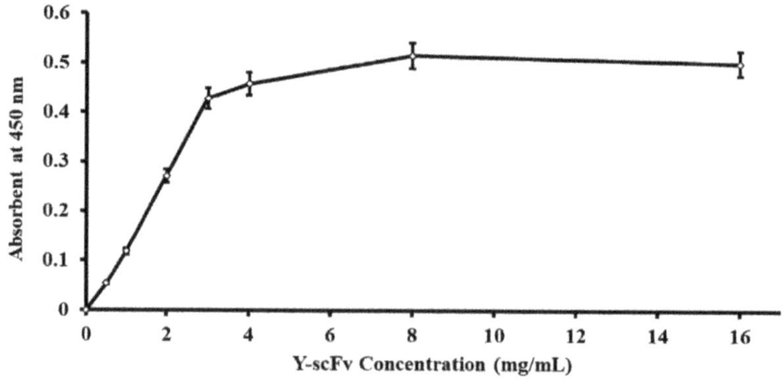

Figure 3. Correlative amount of Y-scFv antibody coupled to the surface of PLA-M magnetic beads.

3.2. Optimized Conditions for Sexed Sperm by PY-Magnetic Beads

The coupling of Y-scFv antibody on the surface of magnetic microbeads was performed via EDC–NHS. The subsequent process consisted of the specific targeting and removal of Y-sperm by the Y-scFv antibody coupled to the magnetic microbeads. Selected concentrations of Y-scFv antibody were coupled to the surface of the magnetic microbeads, and as shown in Figure 4, increasing the concentration of Y-scFv antibody significantly increased the percentage of sperm binding, compared with that bound by uncoupled PLA-M microbeads (0 mg/mL Y–scFv). Furthermore, magnetic microbeads coupled with 2–4 mg/mL Y-scFv demonstrated a high sperm binding efficiency of up to 51.5–52.4% of total sperm samples with an X/Y-sperm ratio of 1:1 (Figure 4).

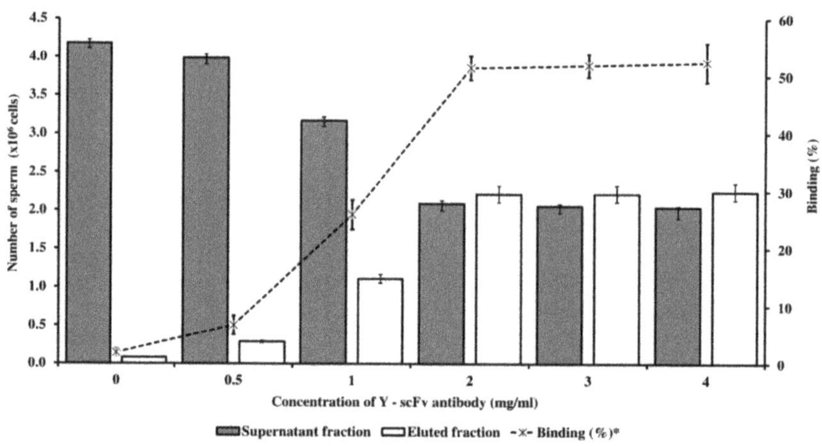

Figure 4. The concentration of Y-scFv antibody coupled to PLA-M microbeads bound with conventional bovine sperm with an X/Y-sperm ratio of 1:1 (expressed in the bar graph) and percentage of binding capacity (expressed in the linear graph) in fresh semen. * % Binding = (Eluted fraction sperm/total sperm) × 100.

The X/Y-sperm ratio in each fraction is shown in Table 2. According to these results, the high efficiency of PY-microbeads binding to Y-sperm was demonstrated when a high concentration of Y-scFv antibody was coupled to the PY-microbeads, leading to a high proportion of Y-sperm in the eluted fraction. In particular, a Y-scFv antibody concentration of 2–4 mg/mL produced significant enrichment of Y-sperm in the eluted fraction (78.01–81.43%) and of X-sperm in the supernatant fraction (79.04–82.65%). The optimal

Y-scFv antibody concentration on PY-microbeads appropriate for bovine sperm sexing was indicated as 2 mg/mL.

Table 2. Percentage of X- and Y-sperm expressed in each fraction produced by different concentrations of Y-scFv antibody coupled to PLA-M microbeads (mean ± SD).

Concentration of Y-scFv Antibody (mg/mL)	Supernatant Fraction		Eluted Fraction	
	X-Sperm	Y-Sperm	X-Sperm	Y-Sperm
0	50.86 ± 0.58 [a]	49.13 ± 0.58 [d]	49.90 ± 0.60 [b]	50.10 ± 0.60 [a]
0.5	52.26 ± 1.15 [ab]	47.73 ± 1.15 [cd]	49.13 ± 0.66 [b]	50.87 ± 0.66 [a]
1	54.93 ± 0.64 [b]	45.06 ± 0.64 [c]	46.50 ± 0.60 [b]	53.50 ± 0.60 [a]
2	79.04 ± 0.15 [c]	20.95 ± 0.15 [b]	21.90 ± 0.95 [a]	78.01 ± 0.95 [b]
3	80.54 ± 0.61 [cd]	19.46 ± 0.61 [ab]	21.37 ± 3.06 [a]	78.63 ± 3.06 [b]
4	82.65 ± 0.87 [d]	17.35 ± 0.87 [a]	18.57 ± 1.00 [a]	81.43 ± 1.00 [b]

Different superscript letters in the same column indicate significant differences ($p < 0.001$).

3.3. Frozen–Thaw Sperm Motility and Kinematic Variables Evaluated by CASA

The percentage motility and kinematic variables of frozen–thawed sperm in CON, NC, the X-enriched fraction, and the Y-enriched fraction are shown in Table 3. After sexing by PY-microbeads, the motility and kinematic variables of the X-enriched fraction did not differ from those of CON and NC ($p > 0.05$). However, the motility and kinematic variables of the Y-enriched fraction were significantly lower ($p < 0.05$) than those of CON, NC, and the X-enriched fraction.

Table 3. Evaluation of motility and kinematics of frozen–thawed sexed semen by PY-microbeads ($n = 20$; mean ± SD).

Parameter	Treatment				*p*-Value
	T1	T2	T3		
	CON	NC	(X-Enriched)	(Y-Enriched)	
TM (%)	68.79 ± 7.54 [b]	64.90 ± 14.30 [b]	62.90 ± 12.30 [b]	23.55 ± 6.82 [a]	0.030
PM (%)	57.11 ± 9.49 [b]	54.12 ± 16.13 [b]	52.14 ± 10.16 [b]	15.17 ± 6.61 [a]	0.021
VCL (μm/s)	89.52 ± 9.63 [b]	79.22 ± 12.85 [b]	74.36 ± 10.87 [b]	26.04 ± 6.77 [a]	0.040
VSL (μm/s)	35.69 ± 6.07 [b]	33.55 ± 10.11 [b]	31.25 ± 11.29 [b]	11.97 ± 5.66 [a]	0.001
VAP (μm/s)	45.25 ± 6.51 [b]	42.01 ± 10.67 [b]	40.24 ± 12.55 [b]	14.89 ± 5.41 [a]	0.001
DCL (μm)	30.03 ± 5.64 [b]	27.25 ± 9.39 [b]	24.11 ± 10.48 [b]	12.25 ± 5.30 [a]	0.010
DSL (μm)	10.53 ± 3.95 [b]	9.09 ± 3.12 [b]	8.02 ± 3.42 [b]	5.00 ± 2.89 [a]	0.024
DAP (μm)	12.37 ± 2.87 [b]	11.34 ± 4.12 [b]	10.42 ± 3.12 [b]	6.47 ± 3.12 [a]	0.041
ALH (μm)	0.90 ± 0.25 [b]	0.81 ± 0.12 [b]	0.79 ± 0.41 [b]	0.43 ± 0.25 [a]	0.025
BCF (Hz)	9.55 ± 5.45 [b]	9.02 ± 0.44 [b]	8.34 ± 3.69 [b]	4.33 ± 2.27 [a]	0.024
HAC (rad)	0.24 ± 0.15 [b]	0.23 ± 0.11 [b]	0.21 ± 0.07 [b]	0.11 ± 0.08 [a]	0.010
WOB (%)	0.52 ± 0.11 [b]	0.53 ± 0.15 [b]	0.51 ± 0.06 [b]	0.56 ± 0.09 [a]	0.008

Con: conventional frozen–thawed semen; NC: conventional semen incubated with non-active PLA-M microbeads; X-enriched: unbound sperm with PY-microbeads; Y-enriched: entrapped sperm with PY- microbeads. Different superscript letters in the same row indicate significant differences ($p < 0.05$).

3.4. Viability of Frozen–Thawed Sexed Semen

The viability of frozen–thawed sperm in CON, NC, the X-enriched fraction, and the Y-enriched fraction, is shown in Figure 5. The percentages of live sperm were significantly higher in CON (36.7%) and NC (33.8%) than in the X-enriched fraction (29.5%) and Y-enriched fraction (10.5%) ($p < 0.05$). In addition, the percentage of live sperm in the X-enriched fraction was significantly higher than in the Y-enriched fraction ($p < 0.05$).

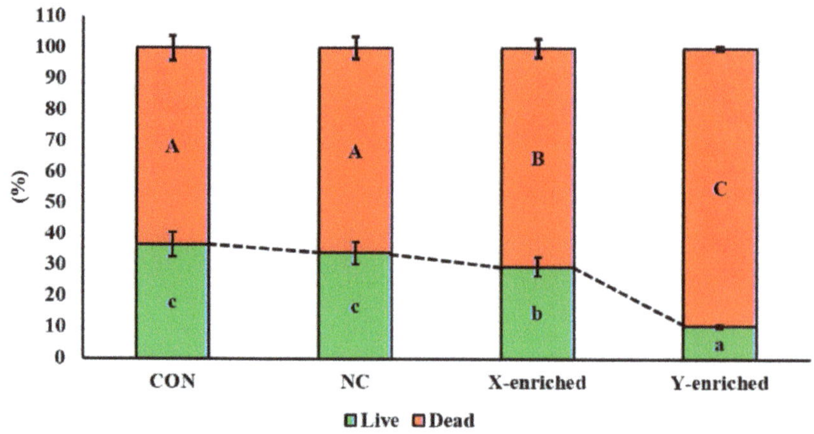

Figure 5. The percent viability of frozen–thawed sexed semen by PY-microbeads. Con: conventional frozen–thawed semen; NC: conventional semen incubated with nonactive PLA-M microbeads; X-enriched: sperm unbound by PY-microbeads; Y-enriched: sperm entrapped by PY-microbeads; A, B, C = comparison of dead sperm in each group; a, b, c = comparison of live sperm in each group.

3.5. Acrosome Integrity of Frozen–Thawed Sexed Semen

The percentage acrosome integrity of frozen–thawed sperm in CON, NC, X-enriched fraction and Y-enriched fraction is shown in Table 4. Live acrosome-intact sperm (LI) and dead acrosome-intact sperm (DI) in X-enriched sperm did not differ between CON and NC ($p < 0.05$). The proportion of live acrosome-reacted sperm (LR) in X-enriched fraction was significantly lower than in CON and NC ($p < 0.05$) but not significantly different from in the Y-enriched fraction ($p > 0.05$). Moreover, the proportions of live acrosome-intact sperm and live acrosome-reacted sperm were significantly lower in the Y-enriched group than in CON and NC ($p < 0.05$). The level of dead acrosome-intact sperm (DR) in the Y-enriched fraction was significantly higher than in CON, NC, and the X-enriched fraction. The proportions of dead acrosome-reacted sperm showed no differences among CON, NC, X-enriched fraction, and Y-enriched fraction.

Table 4. Acrosome integrity of frozen–thawed sexed semen by PY-microbeads ($n = 20$; mean ± SD).

Parameter	Treatment				*p*-Value
	T1	T2	T3		
	CON	NC	(X-Enriched)	(Y-Enriched)	
LI	42.40 ± 7.25^{b}	40.20 ± 6.25^{b}	38.90 ± 7.77^{b}	12.44 ± 3.25^{a}	0.001
DI	30.77 ± 4.42^{a}	35.72 ± 5.52^{a}	40.20 ± 7.25^{a}	60.72 ± 9.01^{b}	0.004
LR	0.77 ± 0.04^{b}	0.82 ± 0.24^{b}	0.44 ± 0.05^{a}	0.42 ± 0.12^{a}	0.032
DR	26.06 ± 2.94	23.26 ± 4.35	20.46 ± 8.88	26.42 ± 1.24	0.125

Con: conventional frozen–thaw semen; NC: conventional semen incubated with non-active PLA-M microbeads; X-enriched: unbound sperm with PY-microbeads; Y-enriched: entrapped sperm with PY- microbeads; LI: live-acrosome-intact sperm; LR: live acrosome-reacted sperm; DI: dead acrosome-intact sperm; DR: dead acrosome-reacted sperm. Different superscript letters in the same row indicate significant differences ($p < 0.05$).

3.6. X/Y-Sperm Ratio in Frozen-Thaw Semen Evaluated by Imaging Flow Cytometry

The sperm sex ratios calculated for frozen–thawed CON, NC, X-enriched fraction, and Y-enriched fraction are shown in Figure 6. The percentages of X- and Y-sperm did not differ between the CON and NC groups ($p > 0.05$). However, the percentage of X-sperm in X-enriched fraction (80.24 ± 3.85) was significantly higher than that of Y-sperm (19.76 ± 3.81) ($p < 0.05$). Conversely, the percentage of Y-sperm in the Y-enriched fraction (81.45 ± 4.95) was significantly higher than that of X-sperm (18.55 ± 4.91) ($p < 0.05$).

Y-scFv antibody concentration on PY-microbeads appropriate for bovine sperm sexing was indicated as 2 mg/mL.

Table 2. Percentage of X- and Y-sperm expressed in each fraction produced by different concentrations of Y-scFv antibody coupled to PLA-M microbeads (mean ± SD).

Concentration of Y-scFv Antibody (mg/mL)	Supernatant Fraction		Eluted Fraction	
	X-Sperm	Y-Sperm	X-Sperm	Y-Sperm
0	50.86 ± 0.58^{a}	49.13 ± 0.58^{d}	49.90 ± 0.60^{b}	50.10 ± 0.60^{a}
0.5	52.26 ± 1.15^{ab}	47.73 ± 1.15^{cd}	49.13 ± 0.66^{b}	50.87 ± 0.66^{a}
1	54.93 ± 0.64^{b}	45.06 ± 0.64^{c}	46.50 ± 0.60^{b}	53.50 ± 0.60^{a}
2	79.04 ± 0.15^{c}	20.95 ± 0.15^{b}	21.90 ± 0.95^{a}	78.01 ± 0.95^{b}
3	80.54 ± 0.61^{cd}	19.46 ± 0.61^{ab}	21.37 ± 3.06^{a}	78.63 ± 3.06^{b}
4	82.65 ± 0.87^{d}	17.35 ± 0.87^{a}	18.57 ± 1.00^{a}	81.43 ± 1.00^{b}

Different superscript letters in the same column indicate significant differences ($p < 0.001$).

3.3. Frozen–Thaw Sperm Motility and Kinematic Variables Evaluated by CASA

The percentage motility and kinematic variables of frozen–thawed sperm in CON, NC, the X-enriched fraction, and the Y-enriched fraction are shown in Table 3. After sexing by PY-microbeads, the motility and kinematic variables of the X-enriched fraction did not differ from those of CON and NC ($p > 0.05$). However, the motility and kinematic variables of the Y-enriched fraction were significantly lower ($p < 0.05$) than those of CON, NC, and the X-enriched fraction.

Table 3. Evaluation of motility and kinematics of frozen–thawed sexed semen by PY-microbeads ($n = 20$; mean ± SD).

Parameter	Treatment				*p*-Value
	T1	T2	T3		
	CON	NC	(X-Enriched)	(Y-Enriched)	
TM (%)	68.79 ± 7.54^{b}	64.90 ± 14.30^{b}	62.90 ± 12.30^{b}	23.55 ± 6.82^{a}	0.030
PM (%)	57.11 ± 9.49^{b}	54.12 ± 16.13^{b}	52.14 ± 10.16^{b}	15.17 ± 6.61^{a}	0.021
VCL (μm/s)	89.52 ± 9.63^{b}	79.22 ± 12.85^{b}	74.36 ± 10.87^{b}	26.04 ± 6.77^{a}	0.040
VSL (μm/s)	35.69 ± 6.07^{b}	33.55 ± 10.11^{b}	31.25 ± 11.29^{b}	11.97 ± 5.66^{a}	0.001
VAP (μm/s)	45.25 ± 6.51^{b}	42.01 ± 10.67^{b}	40.24 ± 12.55^{b}	14.89 ± 5.41^{a}	0.001
DCL (μm)	30.03 ± 5.64^{b}	27.25 ± 9.39^{b}	24.11 ± 10.48^{b}	12.25 ± 5.30^{a}	0.010
DSL (μm)	10.53 ± 3.95^{b}	9.09 ± 3.12^{b}	8.02 ± 3.42^{b}	5.00 ± 2.89^{a}	0.024
DAP (μm)	12.37 ± 2.87^{b}	11.34 ± 4.12^{b}	10.42 ± 3.12^{b}	6.47 ± 3.12^{a}	0.041
ALH (μm)	0.90 ± 0.25^{b}	0.81 ± 0.12^{b}	0.79 ± 0.41^{b}	0.43 ± 0.25^{a}	0.025
BCF (Hz)	9.55 ± 5.45^{b}	9.02 ± 0.44^{b}	8.34 ± 3.69^{b}	4.33 ± 2.27^{a}	0.024
HAC (rad)	0.24 ± 0.15^{b}	0.23 ± 0.11^{b}	0.21 ± 0.07^{b}	0.11 ± 0.08^{a}	0.010
WOB (%)	0.52 ± 0.11^{b}	0.53 ± 0.15^{b}	0.51 ± 0.06^{b}	0.56 ± 0.09^{a}	0.008

Con: conventional frozen–thawed semen; NC: conventional semen incubated with non-active PLA-M microbeads; X-enriched: unbound sperm with PY-microbeads; Y-enriched: entrapped sperm with PY- microbeads. Different superscript letters in the same row indicate significant differences ($p < 0.05$).

3.4. Viability of Frozen–Thawed Sexed Semen

The viability of frozen–thawed sperm in CON, NC, the X-enriched fraction, and the Y-enriched fraction, is shown in Figure 5. The percentages of live sperm were significantly higher in CON (36.7%) and NC (33.8%) than in the X-enriched fraction (29.5%) and Y-enriched fraction (10.5%) ($p < 0.05$). In addition, the percentage of live sperm in the X-enriched fraction was significantly higher than in the Y-enriched fraction ($p < 0.05$).

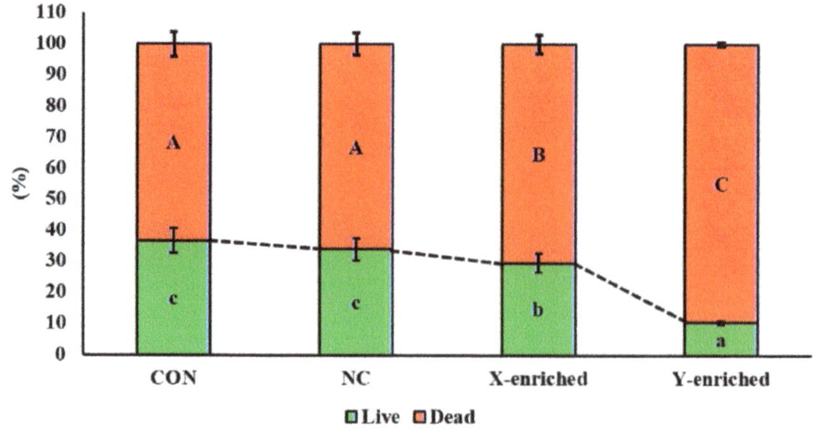

Figure 5. The percent viability of frozen–thawed sexed semen by PY-microbeads. Con: conventional frozen–thawed semen; NC: conventional semen incubated with nonactive PLA-M microbeads; X-enriched: sperm unbound by PY-microbeads; Y-enriched: sperm entrapped by PY-microbeads; A, B, C = comparison of dead sperm in each group; a, b, c = comparison of live sperm in each group.

3.5. Acrosome Integrity of Frozen–Thawed Sexed Semen

The percentage acrosome integrity of frozen–thawed sperm in CON, NC, X-enriched fraction and Y-enriched fraction is shown in Table 4. Live acrosome-intact sperm (LI) and dead acrosome-intact sperm (DI) in X-enriched sperm did not differ between CON and NC ($p < 0.05$). The proportion of live acrosome-reacted sperm (LR) in X-enriched fraction was significantly lower than in CON and NC ($p < 0.05$) but not significantly different from in the Y-enriched fraction ($p > 0.05$). Moreover, the proportions of live acrosome-intact sperm and live acrosome-reacted sperm were significantly lower in the Y-enriched group than in CON and NC ($p < 0.05$). The level of dead acrosome-intact sperm (DR) in the Y-enriched fraction was significantly higher than in CON, NC, and the X-enriched fraction. The proportions of dead acrosome-reacted sperm showed no differences among CON, NC, X-enriched fraction, and Y-enriched fraction.

Table 4. Acrosome integrity of frozen–thawed sexed semen by PY-microbeads ($n = 20$; mean ± SD).

Parameter	Treatment				*p*-Value
	T1	T2	T3		
	CON	NC	(X-Enriched)	(Y-Enriched)	
LI	42.40 ± 7.25^{b}	40.20 ± 6.25^{b}	38.90 ± 7.77^{b}	12.44 ± 3.25^{a}	0.001
DI	30.77 ± 4.42^{a}	35.72 ± 5.52^{a}	40.20 ± 7.25^{a}	60.72 ± 9.01^{b}	0.004
LR	0.77 ± 0.04^{b}	0.82 ± 0.24^{b}	0.44 ± 0.05^{a}	0.42 ± 0.12^{a}	0.032
DR	26.06 ± 2.94	23.26 ± 4.35	20.46 ± 8.88	26.42 ± 1.24	0.125

Con: conventional frozen–thaw semen; NC: conventional semen incubated with non-active PLA-M microbeads; X-enriched: unbound sperm with PY-microbeads; Y-enriched: entrapped sperm with PY- microbeads; LI: live-acrosome-intact sperm; LR: live acrosome-reacted sperm; DI: dead acrosome-intact sperm; DR: dead acrosome-reacted sperm. Different superscript letters in the same row indicate significant differences ($p < 0.05$).

3.6. X/Y-Sperm Ratio in Frozen-Thaw Semen Evaluated by Imaging Flow Cytometry

The sperm sex ratios calculated for frozen–thawed CON, NC, X-enriched fraction, and Y-enriched fraction are shown in Figure 6. The percentages of X- and Y-sperm did not differ between the CON and NC groups ($p > 0.05$). However, the percentage of X-sperm in X-enriched fraction (80.24 ± 3.85) was significantly higher than that of Y-sperm (19.76 ± 3.81) ($p < 0.05$). Conversely, the percentage of Y-sperm in the Y-enriched fraction (81.45 ± 4.95) was significantly higher than that of X-sperm (18.55 ± 4.91) ($p < 0.05$).

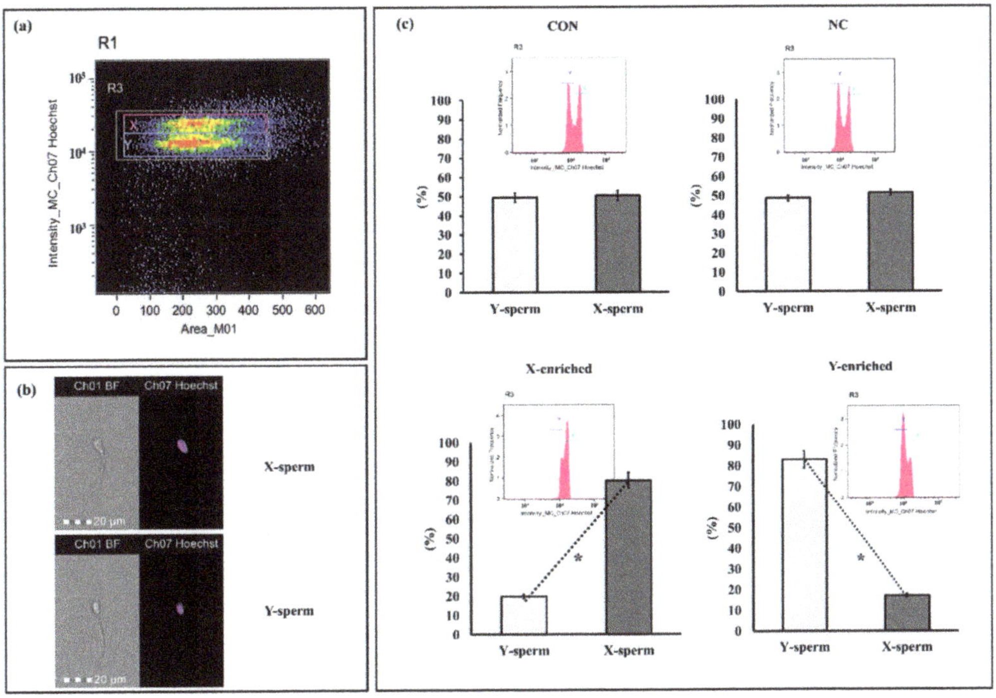

Figure 6. Discrimination between X- and Y-sperm after sexing by PY magnetic beads. (**a**) Regions are utilized to analyze sperm bearing the X and Y chromosomes. Separate populations of sperm bearing X and Y chromosomes were isolated, and the sex ratio in each sample was evaluated. (**b**) Patterns of sperm were observed with Hoechst 33,342. (**c**) Frequency histogram and percentage of sperm after staining with Hoechst 33,342. Con: conventional frozen–thawed semen; NC: conventional semen incubated with inactive magnetic PLA-M; X-enriched: fraction of sperm not bound by PY-magnetic beads; Y-enriched: fraction of sperm entrapped by PY-magnetic beads. * Paired *t*-tests were used to compare X- and Y-sperm from each group. *p*-value < 0.05 is statistically significant.

3.7. X/Y-Sperm Ratio in Frozen-Thaw Semen Evaluated by Real-Time PCR

The fold relative expression data for X- and Y-sperm content obtained using real-time PCR analysis are depicted in Figure 7. CON was taken as the baseline. The mean fold relative expression of Y-sperm in the Y-enriched fraction sexed by the PY-microbeads method was significantly higher ($p < 0.05$) for the Y primer (SRY) and the average 2 delta Ct value for the Y primers was significantly higher (4-fold) for the Y primers compared with the control. The mean fold relative expression of X-sperm in the Y-enriched group was lower (0.52-fold) ($p < 0.05$) for the Y primer and the results were comparable to those in the control. The mean fold relative expression of X-sperm in X-enriched sperm sexed by PY-microbeads was higher ($p < 0.05$) for the X primer (PLP), as was the average 2 delta Ct value for X primers (3–4-fold) compared with CON. In addition, the mean fold relative expression of X-sperm in X-enriched semen sexed by PY-microbeads did not differ from that of commercial cell sorter-sexed (SORT X) semen (4-fold, $p > 0.05$). However, the mean fold relative expression of Y-sperm in X-enriched fraction was significantly lower ($p < 0.05$), and the average 2 delta Ct value for the X primers was significantly lower than the control.

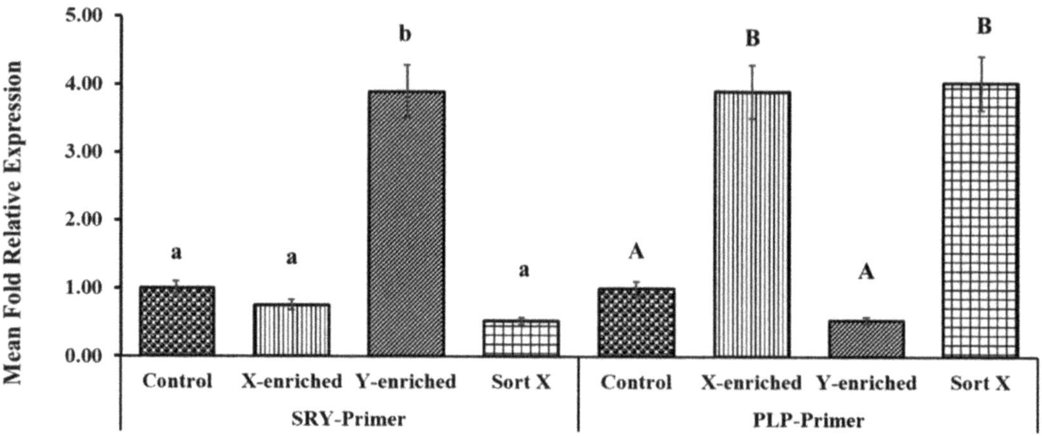

Figure 7. Expression levels of PLP and SRY genes in X- and Y-enriched semen samples were measured by RT–PCR to validate the sexing technique for bull semen. Con: conventional frozen–thawed semen; X-enriched: fraction of sperm not bound by PY-magnetic beads; Y-enriched: fraction of sperm entrapped by PY-magnetic beads; Sort X: commercial cell-sorter-sexed semen; A, B = comparison of PLP amplicons from each group; a, b = comparison of SRY amplicons from each group.

4. Discussion

Sexed semen has successfully been used in bovines and has revolutionized the dairy industry [23]. The most common use of sexed semen is for the sex preselection of females to achieve an adequate number of replacement heifers is [15]. Utilizing sexed sperm is a successful method of obtaining offspring of a specific gender [24]. Many previous studies have demonstrated that MACS is highly effective in separating sperm based on plasma membrane integrity [16,17]. The present work was the first MACS to use scFv antibodies specific to plasma membrane epitopes on Y-sperm coupled to magnetic microbeads to separate X- and Y-sperm in bovine semen. The advantage of producing scFv antibodies against Y-sperm is that it creates smaller protein molecules than monoclonal antibodies with a high affinity to capture Y-sperm [11]. Therefore, MACS, by using scFv antibodies specific to Y-sperm coupled to the surface of magnetic microbeads, is an interesting approach that might be utilized to develop bovine semen sexing technologies. The coupling of antibodies to the surface of PLA-magnetic microbeads via 1-(3-dimethylaminopropyl)-3-ethylcarbodiimide hydrochloride-N hydroxysuccinimide (EDC–NHS) activation involves a two-step procedure. First, alkane carboxylic acids containing thiols are bound to the particle surface to generate a carboxylic acid-terminated monolayer. Second, the terminal carboxylic acids are coupled with peptides or antibodies by EDC–NHS coupling to generate amide linkages [25]. Both the sperm binding efficiency of this method and its ability to produce high-quality sperm have been verified.

The approaches for estimating the amount of Y-scFv antibody coupled to PLA-M magnetic microbeads that rely on immobilization are simple and more easily evaluated. Anti-HA tag antibody peroxidase labels have been used to enhance the sensitivity of immunoassays [26]. In the present study, a concentration of Y-scFv antibody of 2–4 mg/mL was coupled to the surface of the magnetic microbeads. The concentration of Y-scFv antibody had a highly significant effect on the efficiency of immobilization.

This novel method of sexing bovine sperm with MACS by using Y-scFv antibody can effectively separate X- and Y-sperm. This study showed the optimal level of Y-scFv coupling to the surface of PLA-M magnetic microbeads to be 2, 3, or 4 mg/mL, which demonstrated high-efficiency binding with Y-sperm and high accuracy of Y-sperm in the eluted fraction, with up to 78.01%, 78.63%, and 81.43%, respectively. Furthermore, this approach yielded a high proportion of X-sperm in the supernatant fraction, which represented the X-enriched fraction, up to 82.65%. These results showed that PY-microbeads bound to Y-sperm, and left X-sperm in the supernatant fraction, making the supernatant fraction an enriched X-sperm fraction with a low concentration of Y-sperm. Therefore, PY-microbeads bound to Y-sperm were trapped by the powerful neodymium magnet. Eluting buffer was used to separate the Y-sperm from the entrapped PY-microbeads. The Y-sperm were liberated from the PY-microbeads and remained in the eluted fraction. However, this fraction still had a low concentration of X-sperm because of the cross-reactivity of the Y-scFv antibody. The eluted fraction represented an enriched Y-sperm fraction. The success of this technology can be utilized to produce sex-selected sperm, allowing farmers to enhance the production of replacement female dairy cows.

Sperm motility is one of the most critical aspects of sperm fertility, as it is important for sperm transportation and fertilization in the female reproductive tract [27]. CASA provides a wide range of sperm motility parameters that provide valuable information regarding the physiological status of sperm and thus has the potential to more accurately predict sperm fertility than the parameters assessed by routine microscopic semen evaluation [28,29]. After the production of sexed semen by PY-microbeads, the NC- and X-enriched fractions showed normal motility, with kinematics that did not differ from those of CON. This means that PY-microbeads do not damage sperm. The present results were in agreement with those obtained by Domínguez et al. (2018) [30], who found that magnetic bead separation had no effect on donkey sperm motility. However, sperm in the Y-enriched fraction showed significantly decreased motility after sexing with PY-microbeads. The evaluation of acrosome integrity is critical because acrosome enzymes are important for penetrating the zona pellucida and fertilization. The acrosome integrity indicators of frozen–thawed sexed semen generated with PY-microbeads did not differ from those of CON in the current study. Moreover, the viability of sperm in the Y-enriched fraction was lower than in the X-enriched fraction. The elution buffer and duration of the procedure likely affect the Y–sperm in the Y-enriched fraction. Therefore, the low motility and low viability of the Y-enriched fraction need to be improved in future research.

In accordance with Thongkham et al. (2021) [7] imaging flow cytometry was used to discriminate X- and Y-chromosome-bearing sperm. Hoechst 33,342 is more concentrated in X-bearing sperm than in Y-bearing sperm, allowing them to be distinguished using imaging flow cytometry. Sperm containing X and Y chromosomes has a double Gaussian distribution on a histogram. In this study, the X-enriched fraction had a higher proportion of X-sperm. In this way, the use of X-enriched fraction semen would be almost certain to increase the percentage of female calves born by artificial insemination. Additionally, the proportion of sperm carrying the Y-chromosome increased in the Y-enriched fraction after sex selection. However, sperm viability for artificial insemination continues to be a challenge. Real-time PCR is regarded as a practical, accurate, and reliable method for quantifying spermatozoa carrying the X and Y chromosomes in bovine semen [31,32]. The confirmation of bovine sperm sexed by PY-microbeads was performed in this work by utilizing SYBR® Green-based real-time quantitative PCR. During spermatogenesis, an equal number of sperm with both X and Y chromosomes are made, according to the meiotic model. The results of our real-time PCR validation of sexed bovine sperm are similar to previous investigations on bovine semen [33,34]. The 2 delta Ct value of X- and Y-chromosome-containing sperm differed significantly (3- to 4-fold) from the control. The SYBR® Green real-time PCR technology applied in this work accurately determined the sperm sex ratio in terms of the fold increase for both X- and Y-enriched sperm.

5. Conclusions

This initial MACS, which used a Y-scFv antibody linked to the surface magnetic microbeads, was demonstrated to be an efficient approach for sexing bovine sperm. The optimal concentration of Y-scFv antibody coupled to magnetic microbeads was 2 mg/mL. The bound sperm on PY-microbeads was highly Y-sperm, and the unbound sperm was highly X-sperm. This approach generated sperm with up to 82.65% X-sperm in the

X-enriched fraction and up to 81.43% Y-sperm in the Y-enriched fraction. Additionally, this method had no negative effect on X-sperm motility, kinetics, or quality; however, Y-sperm quality was low. PY-magnetic beads can be utilized to develop a technique for bovine X-sperm sexing.

Author Contributions: Conceptualization, K.S. and S.H.; methodology, M.T., W.T. and C.L.; validation, S.M. and S.H.; data curation, K.S., S.H. and M.T.; formal analysis M.T. and W.T.; investigation, K.S., C.L., W.P., A.S. and M.T.; project administration, K.S.; resources, E.R., P.R., K.J. and S.H.; writing—original draft, K.S., S.H. and M.T.; writing—review and editing, P.R., K.J., S.M., K.S., A.S. and S.H. All authors have read and agreed to the published version of the manuscript.

Funding: This research project was supported by the Fundamental Fund 2022, Chiang Mai University, and partially supported by Chiang Mai University.

Institutional Review Board Statement: All experiments in this study were approved by the Animal Care and Use Committee of Agricultural Animal Science at the Faculty of Agriculture, Chiang Mai University (protocol number: RAGIACUC005/2564). Following the animal welfare protocol, we minimized the pain and stress of the animals and used only the number of animals needed to obtain reliable scientific data.

Informed Consent Statement: Not applicable.

Data Availability Statement: Not applicable.

Conflicts of Interest: The authors declare no conflict of interest.

References

1. Xie, Y.; Xu, Z.; Wu, Z.; Hong, L. Sex Manipulation Technologies Progress in Livestock: A Review. *Front. Vet. Sci.* **2020**, *7*, 481. [CrossRef] [PubMed]
2. Garner, D.L. Flow cytometric sexing of mammalian sperm. *Theriogenology* **2006**, *65*, 943–957. [CrossRef] [PubMed]
3. Pahmeyer, C.; Britz, W. Economic opportunities of using crossbreeding and sexing in Holstein dairy herds. *J. Dairy Sci.* **2020**, *103*, 8218–8230. [CrossRef] [PubMed]
4. Rath, D.; Tiedemann, D.; Gamrad, L.; Johnson, L.; Klein, S.; Kues, W.; Mancini, R.; Rehbock, C.; Taylor, U.; Barcikowski, S. Sex Sorted Boar Sperm—An Update on Related Production Methods. *Reprod. Domest. Anim.* **2015**, *50*, 56–60. [CrossRef]
5. Li, X.X.; Yang, X.G.; Lu, Y.Q.; Lu, S.S.; Zhang, M.; Yao, H.L.; Meng, L.J.; Lu, K.H. Protective Effects of Melatonin against Oxidative Stress in Flow Cytometry-sorted Buffalo Sperm. *Reprod. Domest. Anim.* **2012**, *47*, 299–307. [CrossRef]
6. Wheeler, M.B.; Rutledge, J.J.; Fischer-Brown, A.; VanEtten, T.; Malusky, S.; Beebe, D.J. Application of sexed semen technology to in vitro embryo production in cattle. *Theriogenology* **2006**, *65*, 219–227. [CrossRef]
7. Thongkham, M.; Thaworn, W.; Pattanawong, W.; Teepatimakorn, S.; Mekchay, S.; Sringarm, K. Spermatological parameters of immunologically sexed bull semen assessed by imaging flow cytometry, and dairy farm trial. *Reprod. Biol.* **2021**, *21*, 100486. [CrossRef]
8. Patthanawong, W.; Pongpiachan, P.; Mekchay, S.; Sumretprasong, J. Production of Monoclonal Antibody Against Male Specific Anti-gen on Cell Membrane of Bovine Spermatozoa. *Indian J. Anim. Res.* **2010**, *44*, 22–27.
9. Ahmad, Z.A.; Yeap, S.K.; Ali, A.M.; Ho, W.Y.; Alitheen, N.B.M.; Hamid, M. scFv Antibody: Principles and Clinical Application. *Clin. Dev. Immunol.* **2012**, *2012*, 980250. [CrossRef]
10. Lamberski, J.A.; Thompson, N.E.; Burgess, R.R. Expression and purification of a single-chain variable fragment antibody derived from a polyol-responsive monoclonal antibody. *Protein Expr. Purif.* **2006**, *47*, 82–92. [CrossRef]
11. Thaworn, W.; Hongsibsong, S.; Thongkham, M.; Mekchay, S.; Pattanawong, W.; Sringarm, K. Production of single-chain fragment variable (scFv) antibodies specific to plasma membrane epitopes on bull Y-bearing sperm. *Anim. Biotechnol.* **2020**, 1–11. [CrossRef] [PubMed]
12. Yadav, S.K.; Gangwar, D.K.; Singh, J.; Tikadar, C.K.; Khanna, V.V.; Saini, S.; Dholpuria, S.; Palta, P.; Manik, R.S.; Singh, M.K.; et al. An immunological approach of sperm sexing and different methods for identification of X- and Y-chromosome bearing sperm. *Vet. World* **2017**, *10*, 498–504. [CrossRef] [PubMed]
13. Yadav, S.K.; Gangwar, D.K.; Singh, J.; Tikadar, C.K.; Khanna, V.V.; Saini, S.; Dholpuria, S.; Palta, P.; Manik, R.S.; Singh, M.K.; et al. The untapped potential of magnetic nano-particles for forensic investigations: A comprehensive review. *Talanta* **2021**, *230*, 122297.
14. Said, T.M.; Agarwal, A.; Zborowski, M.; Grunewald, S.; Glander, H.J.; Paasch, U. Utility of magnetic cell separation as a molecular sperm preparation technique. *J. Androl.* **2008**, *29*, 134–142. [CrossRef] [PubMed]
15. Gil Juliá, M.; Hervás, I.; Navarro-Gómez Lechón, A.; Quintana, F.; Amorós, D.; Pacheco, A.; González-Ravina, C.; Rivera-Egea, R.; Garrido, N. Sperm Selection by Magnetic-Activated Cell Sorting before Microinjection of Autologous Oocytes Increases Cumulative Live Birth Rates with Limited Clinical Impact: A Retrospective Study in Unselected Males. *Biology* **2021**, *10*, 430. [CrossRef]

16. Said, T.M.; Agarwal, A.; Grunewald, S.; Rasch, M.; Glander, H.; Paasch, U. Short communication Evaluation of sperm recovery following annexin V magnetic-activated cell sorting separation. *Reprod. Biomed. Online* **2006**, *13*, 336–339. [CrossRef]
17. Ybarra, G.; Viale, D.L.; Vichera, G.; Farini, V.L.; Cama, C.V.; Yakisich, J.S.; Radrizzani, M. Improvement of bovine semen quality by re-moval of membrane-damaged sperm cells with DNA aptamers and magnetic nanoparticles. *J. Biotechnol.* **2016**, *229*, 33–41.
18. Assumpção TI de Severo, N.C.; Zandonaide, J.P.B.; Macedo, G.G. Magnetic-activated cell sorting improves high-quality spermatozoa in bovine semen. *J. Anim. Reprod. Biotechnol.* **2021**, *36*, 91–98. [CrossRef]
19. Malama, E.; Zeron, Y.; Janett, F.; Siuda, M.; Roth, Z.; Bollwein, H. Use of computer-assisted sperm analysis and flow cytometry to detect seasonal variations of bovine semen quality. *Theriogenology* **2017**, *87*, 79–90. [CrossRef]
20. Parati, K.; Bongioni, G.; Aleandri, R.; Galli, A. Sex ratio determination in bovine semen: A new approach by quantitative real time PCR. *Theriogenology* **2006**, *66*, 2202–2209. [CrossRef]
21. E Chandler, J.; Canal, A.M.; Paul, J.; Moser, E. Collection frequency affects percent Y-chromosome bearing sperm, sperm head area and quality of bovine ejaculates. *Theriogenology* **2002**, *57*, 1327–1346. [CrossRef]
22. Ren, F.; Xi, H.; Ren, Y.; Li, Y.; Wen, F.; Xian, M.; Zhao, M.; Zhu, D.; Wang, L.; Lei, A.; et al. TLR7/8 signalling affects X-sperm motility via the GSK3 α/β-hexokinase pathway for the efficient production of sexed dairy goat embryos. *J. Anim. Sci. Biotechnol.* **2021**, *12*, 1–17. [CrossRef] [PubMed]
23. Seidel, G.E., Jr. Economics of selecting for sex: The most important genetic trait. *Theriogenology* **2003**, *59*, 585–598. [CrossRef]
24. Ashry, M.; Lee, K.B.; Folger, J.K.; Rajput, S.K.; Smith, G.W. Follistatin supplementation during in vitro embryo culture improves developmental competence of bovine embryos produced using sex-sorted semen. *Reprod. Biol.* **2018**, *18*, 267–273. [CrossRef] [PubMed]
25. Zhang, Q.; Li, R.-X.; Chen, X.; He, X.-X.; Han, A.-L.; Fang, G.-Z.; Liu, J.-F.; Wang, S. Study of Efficiency of Coupling Peptides with Gold Nanoparticles. *Chin. J. Anal. Chem.* **2017**, *45*, 662–667. [CrossRef]
26. Otieno, B.; Krause, C.; Rusling, J. Bioconjugation of Antibodies and Enzyme Labels onto Magnetic Beads. *Methods Enzymol.* **2016**, *571*, 135–150. [CrossRef]
27. Contri, A.; Valorz, C.; Faustini, M.; Wegher, L.; Carluccio, A. Effect of semen preparation on casa motility results in cryopreserved bull spermatozoa. *Theriogenology* **2010**, *74*, 424–435. [CrossRef]
28. Abaigar, T.; Holt, W.V.; Harrison, R.A.; del Barrio, G. Sperm subpopulations in boar (Sus scrofa) and gazelle (Gazella dama mhorr) semen as revealed by pattern analysis of computer-assisted motility assessments. *Biol. Reprod.* **1999**, *60*, 32–41. [CrossRef]
29. Alquézar-Baeta, C.; Gimeno-Martos, S.; Jiménez, S.M.; Santolaria, P.; Yániz, J.; Palacín, I.; Casao, A.; Cebrián-Pérez, J.; Muiño-Blanco, T.; Pérez-Pé, R. OpenCASA: A new open-source and scalable tool for sperm quality analysis. *PLoS Comput. Biol.* **2019**, *15*, e1006691. [CrossRef]
30. Domínguez, E.; Moreno-irusta, A.; Ramírez, H.; Flores, A.; Ugaz, C.; Clemente, H.; Giojalas, L.; Losnno, L. Sexing Mediated by Magnetic Nanoparticles in Donkeys, a Preliminary In Vitro Study. *J. Equine Vet. Sci.* **2018**, *65*, 123–127. [CrossRef]
31. Maleki, A.F.; Moussavi, A.R.H.; Nassiri, M.R.; Tahmoorespur, M.; Vakili, S.A. Introducing and validation of SYBR Green Real Time PCR method to determinate sex ratio in bovine semen. *Anim. Reprod. Sci.* **2013**, *140*, 1–6. [CrossRef] [PubMed]
32. Asma-ul-Husna Awan, M.A.; Mehmood, A.; Sultana, T.; Shahzad, Q.; Ansari, M.S.; Rakha, B.A.; Naqvi, S.M.S.; Akhter, S. Sperm sexing in Nili-Ravi buffalo through modified swim up: Validation using SYBR® green real-time PCR. *Anim. Reprod. Sci.* **2017**, *182*, 69–76. [CrossRef] [PubMed]
33. A Habermann, F.; Winter, A.; Olsaker, I.; Reichert, P.; Fries, R. Validation of sperm sexing in the cattle (Bos taurus) by dual colour fluorescence in situ hybridization. *J. Anim. Breed. Genet.* **2005**, *122*, 22–27. [CrossRef] [PubMed]
34. Reinsalu, O.; Scheler, O.; Mikelsaar, R.; Mikelsaar, A.-V.; Hallap, T.; Jaakma, Ü.; Padrik, P.; Kavak, A.; Salumets, A.; Kurg, A. A dual colour FISH method for routine validation of sexed Bos taurus semen. *BMC Veter. Res.* **2019**, *15*, 104. [CrossRef] [PubMed]

biology

MDPI

Article

The Effect of Semen Cryopreservation Process on Metabolomic Profiles of Turkey Sperm as Assessed by NMR Analysis

Gianluca Paventi [1], Michele Di Iorio [2], Giusy Rusco [2], Anatoly P. Sobolev [3], Silvia Cerolini [4], Emanuele Antenucci [2], Mattia Spano [5], Luisa Mannina [5] and Nicolaia Iaffaldano [2,*]

1 Department of Medicine and Health Sciences "Vincenzo Tiberio", University of Molise, 86100 Campobasso, Italy; paventi@unimol.it
2 Department of Agricultural, Environmental and Food Sciences, University of Molise, 86100 Campobasso, Italy; michele.diiorio@unimol.it (M.D.I.); giusy.rusco@unimol.it (G.R.); e.antenucci@studenti.unimol.it (E.A.)
3 Institute for Biological Systems, Magnetic Resonance Laboratory "Segre-Capitani", CNR, 00015 Monterotondo, Italy; anatoly.sobolev@cnr.it
4 Department of Veterinary Medicine, University of Milano, 26900 Lodi, Italy; silvia.cerolini@unimi.it
5 Department of Chemistry and Technology of Drugs, Sapienza University of Rome, 00185 Rome, Italy; mattia.spano@uniroma1.it (M.S.); luisa.mannina@uniroma1.it (L.M.)
* Correspondence: nicolaia@unimol.it

Citation: Paventi, G.; Di Iorio, M.; Rusco, G.; Sobolev, A.P.; Cerolini, S.; Antenucci, E.; Spano, M.; Mannina, L.; Iaffaldano, N. The Effect of Semen Cryopreservation Process on Metabolomic Profiles of Turkey Sperm as Assessed by NMR Analysis. *Biology* **2022**, *11*, 642. https://doi.org/10.3390/biology11050642

Academic Editors: Pilar Santolaria Blasco, Rosaura Pérez-Pé and Jessica Rickard

Received: 11 March 2022
Accepted: 20 April 2022
Published: 22 April 2022

Simple Summary: Semen cryobanking is a valuable tool for preserving the genetic resources of a wide range of species, providing the opportunituy to preserve representative samples and reconstruct the population or diversity. However, in avian species, the freezing–thawing process results in a sharp reduction in sperm quality and consequently fertility. This is mainly due to the lack of knowledge about the molecular basis of the cryopreservation process, especially in more sensitive species such as turkey. Thus, in this study, we took advantage of NMR technology to assess the changes in metabolic profile occurring in turkey sperm cryopreservation, which were correlated with sperm qualitative parameters measured in both fresh and frozen–thawed samples. Hence, the results reported here depict a clearer scenario about the changes in the levels of amino acids, other water-soluble compounds, and lipids resulting from the freezing–thawing process. Moreover, a wide discussion about the possible pathway affected by cryopreservation is provided. Therefore, this study allows us to: *(i)* identify biological markers related to the sperm freezability of male turkey donators; *(ii)* suggest a supplementation of specific metabolites in the diet or in the freezing medium in order to obtain spermatozoa abler to withstand the freezing process.

Abstract: Semen cryopreservation represents the main tool for preservation of biodiversity; however, in avian species, the freezing–thawing process results in a sharp reduction in sperm quality and consequently fertility. Thus, to gain a first insight into the molecular basis of the cryopreservation of turkey sperm, the NMR-assessed metabolite profiles of fresh and frozen–thawed samples were herein investigated and compared with sperm qualitative parameters. Cryopreservation decreased the sperm viability, mobility, and osmotic tolerance of frozen–thawed samples. This decrease in sperm quality was associated with the variation in the levels of some metabolites in both aqueous and lipid sperm extracts, as investigated by NMR analysis. Higher amounts of the amino acids Ala, Ile, Leu, Phe, Tyr, and Val were found in fresh than in frozen–thawed sperm; on the contrary, Gly content increased after cryopreservation. A positive correlation ($p < 0.01$) between the amino acid levels and all qualitative parameters was found, except in the case of Gly, the levels of which were negatively correlated ($p < 0.01$) with sperm quality. Other water-soluble compounds, namely formate, lactate, AMP, creatine, and carnitine, turned out to be present at higher concentrations in fresh sperm, whereas cryopreserved samples showed increased levels of citrate and acetyl-carnitine. Frozen–thawed sperm also showed decreases in cholesterol and polyunsaturated fatty acids, whereas saturated fatty acids were found to be higher in cryopreserved than in fresh sperm. Interestingly, lactate, carnitine ($p < 0.01$), AMP, creatine, cholesterol, and phosphatidylcholine ($p < 0.05$) levels were positively correlated with all sperm quality parameters, whereas citrate ($p < 0.01$), fumarate, acetyl-carnitine, and saturated fatty acids ($p < 0.05$) showed negative correlations. A detailed discussion aimed at explaining these correlations in the sperm cell context is provided, returning a clearer scenario of metabolic changes occurring in turkey sperm cryopreservation.

Keywords: sperm metabolic profile; turkey sperm cryopreservation; avian sperm; sperm NMR analysis

1. Introduction

Conservation of genetic variability in domestic animal species is a task for the sustainable production of human food resources, as well as for land management and, more importantly, the preservation of biodiversity [1,2]. Currently, despite the development of other innovative strategies, such as gonadal tissue allotransplantation [3] and diploid primordial germ cell methodologies [4], semen cryopreservation still remains the most effective method to store reproductive cells for the ex situ management of genetic diversity in birds [5–7]. Thus, the growth of this ex situ in vitro strategy, as a support to the in vivo strategy, is becoming more advanced. In this regard, the constitution of cryobanks for genetic resources would offer a crucial link between both strategies, leading to the improvement of conservation programs efficiency [8,9].

However, the main challenge for the creation of a poultry semen cryobank is the achievement of a successful freezing protocol [9]. In relation to this, it is known that the semen cryopreservation process causes a loss of sperm membrane integrity in poultry [2,5,10]. Thus, the improvement of sperm cryosurvival and increased fertility in artificial insemination (AI) with frozen–thawed sperm continues to be the focus in semen cryobanking in avian species. This is even more evident in turkey species, the semen of which has been proven to be more sensitive to cooling and freezing–thawing damages than chicken sperm [11–17]. Though chicken and turkey sperm share the same morphology (as a filiform shape, a long tail and a condensed nucleus [18]), turkey sperm presents a high cholesterol/phospholipid ratio, resulting in low membrane fluidity and permeability [18]; moreover, it has a low osmotic resistance at hypo-osmotic conditions [11].

For these reasons, over the last few decades, several studies have been performed, in which different factors involved in sperm cryosurvivability were taken into consideration, to identify an efficient freezing procedure for both chicken and turkey semen [14,19–25]. However, despite the encouraging results obtained so far, there is still a gap of knowledge about the biological bases involved in the cryopreservation process. This limits the development of a freezing procedure that results in fertilization rates closer to those obtained with fresh semen. In addition, more efficient semen cryopreservation, besides ensuring the conservation of genetic resources in a gene bank, could provide practical advantages to the turkey industry, since commercial farms are completely dependent on AI to obtain fertile eggs [14]. The avian sperm membrane contains more polyunsaturated fatty acids (PUFAs) than that of mammals, and it has lower protein content, a lower cholesterol/phospholipid ratio, and greater overall fluidity at physiological temperatures [26,27].

Specific biological and biophysical factors, such as membrane permeability, lipid composition, and membrane fluidity, can affect the ability of poultry spermatozoa to limit damages caused by the cryopreservation procedure [28,29]. In this regard, it was reported that the freezing–thawing procedure for avian spermatozoa induces a membrane rigidifying process that is accompanied by a dramatic and proportional decrease in the cholesterol/phospholipid ratio. Moreover, this effect is different from species to species; thus, it could behave as an indicator of between-species freezability [18,28]. Accordingly, it was also observed that the ratio of lipids in the sperm membrane determines the overall fluidity of the membrane and impacts the ability of sperm to remain viable during the cryopreservation process in chicken [18,30] as well as in mammals [31]. This is consistent with previous findings in turkey showing that lipids are involved in vital aspects of sperm metabolism and functions [32].

Besides lipids, other factors contribute to defining membrane fluidity in poultry [33], such as the nature and the level of insertion of the proteins in the membrane lipid bilayer [34]. In addition, amino acids could play a role in avian sperm function as shown for mammals: some amino acids participate in many metabolic processes involved in motility, acrosome reaction, and capacitation of human and other mammalian spermatozoa [35,36]. Amino acids also have antioxidant properties able to protect sperm cells from cold shock [37,38]; consistently, plasma amino acids seem to play a role in chicken sperm cryoresistance [39]. In mammals, it has been demonstrated that amino acids act at the extracellular level and improve sperm motility, acrosome integrity, and fertilizing potential after the freezing–thawing process [40–44]. It has also been reported that L-carnitine is involved in sperm energy metabolism, promoting sperm motility and maturation and the spermatogenic process [45,46]. Thus, its supplementation was proposed to increase both kinetics and morphological characteristics of sperm [47]. On the contrary, less information is known about these aspects in avian species, especially in turkey.

The sperm metabolite profile, which includes lipids, amino acids, and other water-soluble compounds, appears to be the main factor that affects sperm resilience the cryopreservation process, which in turn determines the fertilizing ability of sperm [48]. Even now, there is a complete lack of knowledge about the turkey sperm metabolite profile. The only investigation on metabolite profile of fresh turkey sperm reported in literature was our previous study, in which changes in metabolite levels occurring in male reproductive ageing were measured by using nuclear magnetic resonance (NMR) [49]. Currently, there is still no scientific evidence in the literature about the metabolite profile changes during the semen cryopreservation process. Thus, by detecting and simultaneously quantifying a wide range of metabolites with a high analytical precision, NMR represents a valuable tool for better understanding the biological bases of the turkey semen cryopreservation process.

In this study, deeper insight into the metabolite profile changes occurring in turkey sperm cryopreservation was obtained by assessing via NMR a relevant number of metabolites in both fresh and frozen–thawed samples. The changes in metabolite levels were further correlated to sperm quality variations after thawing and are herein discussed from a metabolic point of view.

2. Materials and Methods

2.1. Chemicals

The fluorescent dyes SYBR-14 and propidium iodide (PI) used were those provided in the LIVE/DEAD Sperm Viability kit (Invitrogen^{TM} by Thermo Fisher Scientific, Waltham, MA, USA). All the other chemicals used in this study were purchased from Sigma Chemical Co. (St. Louis, MO, USA).

2.2. Animals and Semen Treatment

Hybrid Large White turkey males from a private breeding group (Agricola Santo Stefano of Amadori's group, Canzano, TE, Italy) were used. Animals were housed there when they were 32 weeks old. They were maintained under standard management conditions and photostimulated on a daily basis with a 14L:10D photoperiod. The toms were kept in groups of 8–10 in floor pens. Feed and water were provided ad libitum. Toms were trained for semen collection by abdominal massage two times a week.

Semen was collected from 32 weeks of age males by abdominal massage. Ejaculates were pooled, with each pool originating from a minimum of 9 to a maximum of 12 males, and thoroughly mixed in order to reach at least 4 mL of semen/pool. In total, 5 pools of semen were used in this study.

2.3. Cryopreservation Process

Semen was cryopreserved by the pellet method [14]. In brief, semen samples were diluted (1:4) in Tselutin extender [50]. The diluted semen was cooled at 4 °C for 60 min, and then 8% (*v*/*v*; 0.860 M) of dimethylacetamide (DMA) was added as cryoprotectant. The semen was gently inverted and equilibrated for 5 min at 4 °C. Volumes of 80 μL of semen were plunged drop by drop directly into liquid nitrogen to form spheres of frozen semen (pellets). The pellets were rapidly placed in 2 mL polypropylene cryovials (Cryo.s™; Greiner Bio-One, Monroe, NC, USA) previously cooled by immersion in liquid nitrogen (3–4 pellets/cryovial) and then stored in a liquid nitrogen tank until analysis. After two weeks, the pellets were warmed by immersing the cryovials in a water bath at 75 °C for 12 s.

2.4. Semen Quality Evaluation

Mobility, viability and sperm osmotic tolerance were assessed on both freshly diluted and frozen–thawed samples. To this end, each sperm sample was divided in two aliquots: one to be immediately assessed (fresh), and the other to be subjected to freezing–thawing process.

Sperm mobility was evaluated using the Sperm Motility Test (SMT) according to the Accudenz® procedure (Accurate Chemical & Scientific Corp., Westbury, NY, USA) following the procedure described by Iaffaldano et al. [14,49]. This procedure is based on the ability of the spermatozoa with a forward progressive motility to penetrate a 4% Accudenz® layer. Semen was diluted to 1.0×10^9 as previously described [49]. A drop of 60 μL from each sperm suspension was superimposed onto 600 μL of 4% (*w*/*v*) Accudenz® solution in a semimicro polystyrene disposable cuvette. Cuvettes were incubated for 5 min in a 41 °C water bath, and absorbance was measured in a spectrophotometer at 550 nm after 60 s. The sperm motility was expressed by values of optical density (O.D.).

Sperm viability was measured using the Invitrogen™ LIVE/DEAD sperm viability kit according to the procedure set up by Iaffaldano et al [14]. Aliquots of 5 μL semen were diluted in 39 μL of Tselutin diluent containing 1 μL of SYBR-14 (diluted 1:100 into dimethylsulfoxide). Samples were incubated for 10 min at 38 °C. Then, 5 μL of propidium iodide (PI; dissolved 1:100 in PBS) was added, and the samples were further incubated at 38 °C for 5 min. The assessment of viable/nonviable spermatozoa was performed using fluorescence microscopy (blue excitation filter λ = 488 nm; ×100 oil immersion objective; magnification ×400). Viable sperm cells were stained green by SYBR-14, whereas dead cells were stained in red by PI. A minimum of 200 spermatozoa for each sample were counted. Percentages of viable spermatozoa were determined as the ratio: green cells/(green cells + red cells) × 100.

Sperm osmotic tolerance (SOT) was assessed using a hypo-osmotic swelling test (HOST) [14,15]. Five microliters of semen were added to 80 μL of distilled H_2O and then stained with SYBR-14/PI and read as described above for sperm viability.

This test is effective for assessing the percentage of viable spermatozoa that are capable of withstanding hypo-osmotic stress in vitro. Under hypo-osmotic conditions, viable thawed spermatozoa with intact membranes fluoresce green (SYBR) and exclude PI. Conversely, damaged membranes permit the passage of PI, staining spermatozoa that have lost their functional integrity red.

2.5. NMR Measurements

2.5.1. Sample Preparation

The Bligh–Dyer [51] method was used to extract and separate water-soluble and liposoluble metabolites from semen samples following the procedure previously reported [52].

Before NMR analysis, fresh and frozen semen was diluted at the same concentration and then centrifuged at 1500 rpm for 15 min to remove both diluent and seminal plasma. A chloroform/methanol (2:1, *v*/*v*) mixture was added to the sperm pellets, and the samples were homogenized with a vortex mixer for 60 s before adding distilled water in the proportion of 1:18:4 (spermatozoa–chloroform/methanol–water). The homogenate was centrifuged at a speed of 4000 rpm for 20 min at 5 °C. The liquid chloroform and water/methanol phases were separated and dried under vacuum in a rotary evaporator. The dried residues were dissolved into 0.75 mL of $CDCl_3/CD_3OD$ (2:3 *v*/*v*) or 0.75 mL of D_2O phosphate buffer (400 mM, pD = 7).

2.5.2. NMR Spectra

The NMR spectra of aqueous and organic extracts were recorded at 27 °C on a Bruker AVANCE 600 NMR spectrometer operating at the proton frequency of 600.13 MHz and equipped with a Bruker multinuclear z-gradient inverse probe head capable of producing gradients in the z-direction with a strength of 55 G/cm. ^1H spectra were referenced to methyl group signals of 3-(trimethylsilyl)-propionic-2,2,3,3-d4 acid sodium salt (TSP, $\delta = 0.00$ ppm) in D_2O and to the residual CHD_2 signal of methanol (set to 3.31 ppm) in $CDCl_3/CD_3OD$ mixture [53]. ^1H spectra of aqueous extracts were acquired by coadding 512 transients with a recycle delay of 3 s. The residual HDO signal was suppressed using a standard Bruker presaturation sequence zgpr. The experiment was carried out by using a 45° pulse of 7.25 µs and 32,000 data points. ^1H spectra of $CDCl_3/CD_3OD$ extracts were obtained using the following parameters: 256 transients, 32,000 data points, a recycle delay of 3 s, and a 90° pulse of 10 µs. The ^1H spectra were Fourier transformed using an exponential multiplication function with a line broadening factor of 0.3 Hz, and manual phase correction and baseline correction were applied.

2.5.3. Measurement of the Metabolic Content in Aqueous Extract

The intensity of 21 ^1H resonances due to water-soluble assigned metabolites (see Table 1) was measured with respect to the intensity of a TSP signal used as internal standard and normalized to 100.

2.5.4. Measurement of the Metabolic Content in Organic Extracts

The integrals of 8 ^1H resonances due to assigned liposoluble metabolites were measured and used to obtain the normalized integrals, see Table 1. All the integrals were normalized with respect to the integrals of α-CH_2 groups of all fatty acid chains at 2.31 ppm set to 100%. The molar percentages of lipids were calculated taking into account the number of equivalent protons corresponding to a specific resonance. The resonances due to the CH_3 of cholesterol (0.74 ppm), all allylic protons (2.08 ppm), α-CH_2 groups of all fatty acid (2.31 ppm), CH_2 diallylic protons of DUFA, (2.81 ppm), CH_2 diallylic protons of PUFA (2.88 ppm), CH_2N of PE (3.21 ppm), $(CH_3)_3N^+$ of PC (3.28 ppm), and CH (double bond) proton of SMN (5.76 ppm) were integrated. The molar percentage of all saturated fatty chains (SFA) was calculated as 100 UFA, where UFA was calculated using the all-allylic-protons signal at 2.08 ppm.

2.6. Statistical Analysis

Sperm qualitative parameters and metabolite levels determined by NMR analysis measured in fresh and frozen–thawed sperm were compared by paired-samples *t*-test (threshold at $p < 0.05$). Correlations between sperm variables and NMR-identified metabolites were assessed through Pearson's correlation coefficients, setting significance thresholds at the $p < 0.05$ level (one-tailed) and $p < 0.01$ levels (two-tailed). All statistical tests were performed using the software package SPSS v23.0 (SPSS, Chicago, IL, USA).

Table 1. Metabolites identified and quantified by NMR in fresh and frozen sperm of turkey males at 32 weeks of age.

Metabolite, ^{1}H Chemical Shift (ppm)	Fresh	Frozen	*p*-Value
	Water Extract		
Amino acids	mol % (n = 5)	mol % (n = 5)	
Ala (1.48)	**0.178 ± 0.007 [a]**	**0.113 ± 0.007 [b]**	***0.001***
Asp (2.83)	0.239 ± 0.022 [a]	0.221 ± 0.005 [a]	*0.466*
Gln (2.45)	1.614 ± 0.314 [a]	1.468 ± 0.197 [a]	*0.412*
Glu (2.07)	66.547 ± 0.391 [a]	66.956 ± 0.761 [a]	*0.725*
Gly (3.57)	**4.648 ± 0.043 [a]**	**4.951 ± 0.073 [b]**	***0.001***
Ile (1.02)	**0.015 ± 0.001 [b]**	**0.009 ± 0.001 [b]**	***0.009***
Leu (0.96)	**0.054 ± 0.003 [a]**	**0.026 ± 0.001 [b]**	***0.003***
Phe (7.43)	**0.022 ± 0.001 [a]**	**0.017 ± 0.001 [b]**	***0.030***
Tyr (6.92)	**0.048 ± 0.002 [a]**	**0.020 ± 0.002 [b]**	***0.001***
Val (0.99)	**0.038 ± 0.002 [a]**	**0.019 ± 0.002 [b]**	***0.002***
Organic acids			
Acetate (1.93)	0.486 ± 0.064 [a]	0.353 ± 0.087 [a]	*0.403*
Citrate (2.57)	**0.089 ± 0.006 [b]**	**0.125 ± 0.006 [a]**	***0.039***
Formate (8.46)	**0.039 ± 0.003 [a]**	**0.019 ± 0.004 [b]**	***0.002***
Fumarate (6.53)	0.032 ± 0.003 [a]	0.040 ± 0.002 [a]	*0.072*
Lactate (1.33)	**1.362 ± 0.072 [a]**	**0.884 ± 0.059 [b]**	***0.001***
Other compounds			
Ac-carnitine (3.20)	**0.037 ± 0.002 [a]**	**0.045 ± 0.002 [b]**	***0.006***
AMP (8.28)	**0.146 ± 0.007 [a]**	**0.115 ± 0.008 [b]**	***0.005***
Carnitine (3.24)	**0.080 ± 0.005 [a]**	**0.037 ± 0.002 [b]**	***0.002***
Creatine (3.94)	**1.828 ± 0.136 [a]**	**1.431 ± 0.071 [b]**	***0.028***
Glucose (3.26 and 5.25)*	16.445 ± 0.437 [a]	17.085 ± 0.592 [a]	*0.504*
Myo-inositol (3.65)	6.054 ± 0.039 [a]	6.067 ± 0.205 [a]	*0.949*
	Lipid extract		
	mol % (n = 3)	mol % (n = 3)	
CHO (0.74)	**9.587 ± 0.348 [a]**	**6.944 ± 0.533 [b]**	***0.036***
SFA	**38.001 ± 1.430 [b]**	**43.088 ± 1.032 [a]**	***0.006***
DUFA (2.81)	4.200 ± 0.136 [a]	3.857 ± 0.356 [a]	*0.293*
UFA (2.08)	**62.000 ± 1.430 [a]**	**56.912 ± 1.032 [b]**	***0.006***
PUFA (2.86)	**36.793 ± 0.561 [a]**	**34.665 ± 0.633 [b]**	***0.027***
PC (3.28)	24.703 ± 0.760 [a]	19.667 ± 0.612 [a]	*0.081*
PE (3.21)	14.107 ± 0.152 [a]	11.989 ± 0.611 [a]	*0.072*
SMN (5.76)	6.990 ± 0.239 [a]	6.200 ± 0.577 [a]	*0.167*

[a,b] Different superscript letters within the same row indicate significant differences ($p < 0.05$). Abbreviations:Ac-carnitine: acylcarnitine; AMP: adenosine monophosphate; CHO: cholesterol; SFA: total content of saturated fatty acids; DUFA: diunsaturated fatty acids; UFA: total content of unsaturated fatty acids; PUFA: polyunsaturated fatty acid; PC: phosphatidylcholine; PE: phosphatidylethanolamine; SMN: sphingomyelin.

3. Results

3.1. Sperm Quality

The sperm quality parameters recorded in freshly collected and frozen–thawed sperm samples are provided in Figure 1. The cryopreservation process severely affected all of the measured qualitative parameters. A significant reduction was found in sperm viability, as assessed by fluorescence microscopy (inset Figure 1a); these values, in fact, were lower by about 50% in frozen–thawed than in fresh samples (Figure 1a). A similarly remarkable decrease was observed in the sperm mobility of cryopreserved sperm (Figure 1b). The sperm osmotic tolerance also suffered from a dramatic decrease, with the values measured in cryopreserved sperm 40% lower than those measured in fresh samples (Figure 1c).

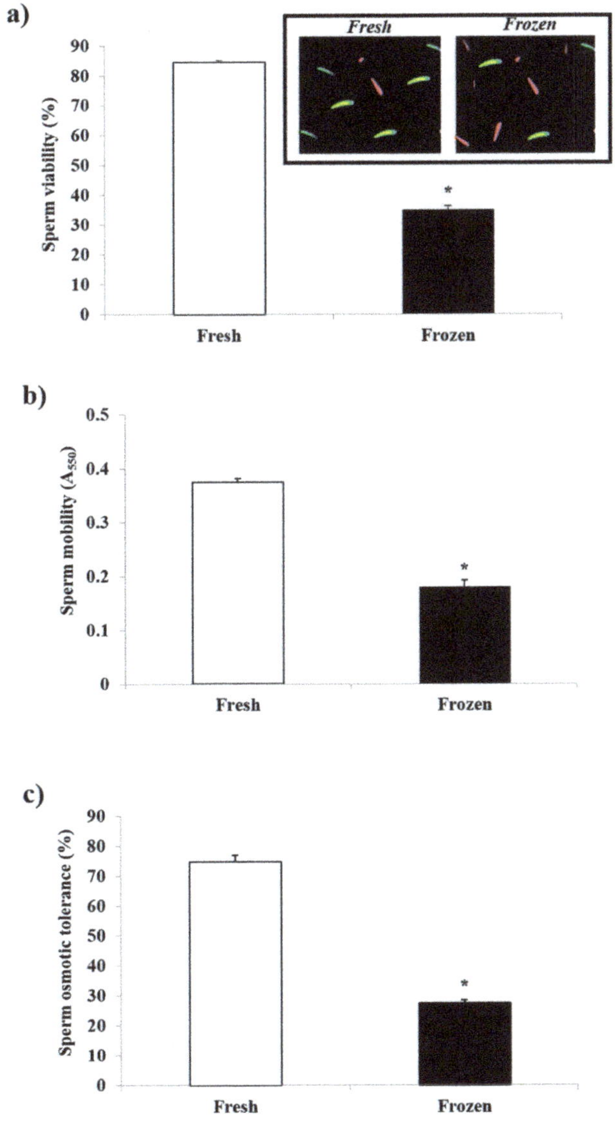

Figure 1. Effect of the cryopreservation process on the (**a**) viability, (**b**) mobility, and (**c**) osmotic tolerance of turkey sperm. Mean values ± SE ($n = 5$) of sperm qualitative parameters recorded for either fresh or frozen–thawed turkey sperm were reported. (**a**) Viability values, expressed as %, were measured by means of a dual staining technique (as shown in inset) using the stains SYBR-14 (green, viable cells) and PI (red, dead cells). (**b**) Mobility values, expressed as Abs 550 nm, were measured by the sperm motility test. (**c**) Sperm osmotic tolerance (SOT) was assessed by the hypo-osmotic H_2O test. For further details, see Methods section. * = $p < 0.05$.

3.2. NMR Analysis

To obtain a picture of the metabolite changes due to cryopreservation, water and lipid soluble components in turkey sperm were identified in NMR spectra using the same NMR experimental conditions and assignments from our previous study [49]. All identified metabolites quantified for both fresh and cryopreserved samples are reported in Table 1. The identified water-soluble metabolites were the amino acids alanine (Ala), isoleucine (Ile), phenylalanine (Phe), leucine (Leu), valine (Val), aspartic acid (Asp), glycine (Gly), tyrosine (Tyr), glutamate (Glu) and glutamine (Gln) and other water-soluble metabolites such as lactate, acetate, citrate, creatine, acetyl-carnitine (ac-carnitine), carnitine, glucose, *myo*-inositol, fumaric acid, formic acid, and adenosine monophosphate (AMP).

Compared with fresh samples, frozen–thawed sperm showed decreased content of Ala, Ile, Leu, Tyr, Val ($p < 0.01$), and Phe ($p < 0.05$), whereas Gly levels proved to increase in cryopreserved samples. No statistically significant differences were found for Asp, Glu, or Gln.

Moreover, significant decreases ($p < 0.01$) in formate, lactate, AMP, carnitine, and creatine levels were found in frozen–thawed sperm, which also showed increased values of citrate ($p < 0.05$) and ac-carnitine ($p < 0.01$). No significant differences were observed between fresh and frozen sperm for acetate, fumarate, glucose, or *myo*-inositol content.

In lipid extract, cholesterol (CHO), the total content of all unsaturated fatty acids (UFA), diunsaturated fatty acids (DUFA), phosphatidylcholine (PC), phosphatidylethanolamine (PE), sphingomyelin (SMN), polyunsaturated fatty acids (PUFA), and the total content of saturated fatty acids (SFA) were identified.

Significantly higher values of CHO, PUFA ($p < 0.05$), and UFA (0.01) were recorded in fresh than in frozen–thawed sperm, whereas a higher value of SFA was observed in the latter samples ($p < 0.01$). No significant differences between fresh and frozen–thawed samples were scored for other lipids identified.

3.3. Correlation

In order to study the correlations between sperm quality parameters and the amounts of the different metabolites, Pearson correlation coefficients were calculated (Table 2).

Table 2. Pearson correlations between sperm qualitative parameters and metabolites identified in fresh and frozen–thawed sperm of turkey males at 32 weeks of age.

Metabolite	Sperm Variables		
	Mobility	Viability	Osmotic Tolerance
Ala	0.867 **	0.930 **	0.902 **
Gly	−0.770 **	−0.771	−0.802 **
Ile	0.818 **	0.861 **	0.908 **
Leu	0.915 **	0.942 **	0.962 **
Phe	0.683 *	0.784 **	0.781 **
Tyr	0.915 **	0.969 **	0.972 **
Val	0.871 **	0.937 **	0.947 **
Citrate	−0.874 **	−0.833 **	−0.815 **
Formate	0.764 *	0.854 **	0.811 **
Fumarate	−0.723 *		−0.659 *
Lactate	0.806 **	0.887 **	0.830 **
Ac-carnitine	−0.740 *	−0.695 *	−0.689 *
AMP	0.653 *	0.728 *	0.708 *
Carnitine	0.923 **	0.925 **	0.957 **
Creatine		0.673 *	0.697 *
CHO	0.907 *	0.884 *	0.876 *
SFA	−0.865 *		−0.868 *
UFA	0.865 *		0.868 *
PC	0.969 **	0.882 *	0.884 *
PE	0.851 *	0.870 *	

Pearson correlation coefficients were calculated for the sperm qualitative parameters (Figure 1) versus metabolites detected by NMR (Table 1). Only significant correlation values * at the 0.05 level and ** at the 0.01 level are reported. Abbreviations:Ac-carnitine: acylcarnitine; AMP: adenosine monophosphate; CHO: cholesterol; SFA: total content of saturated fatty acids; UFA: total content of unsaturated fatty acids; PC: phosphatidylcholine; PE: phosphatidylethanolamine.

Sperm mobility, viability, and osmotic tolerance were positively correlated with Ala, Ile, Leu, Phe, Tyr, Val, formate, lactate, carnitine ($p < 0.01$), AMP, CHO, and PC ($p < 0.05$). The content of creatine was found to be positively correlated with sperm vitality and osmotic tolerance ($p < 0.05$). UFA levels correlated positively with sperm motility and osmotic tolerance ($p < 0.05$), and PE content showed positive correlations with sperm motility and sperm viability ($p < 0.05$). On the contrary, negative correlations with all qualitative parameters were found for Gly, citrate ($p < 0.01$), and ac-carnitine ($p < 0.05$), and both fumarate and SFA ($p < 0.05$) were negatively correlated with sperm motility and sperm osmotic tolerance.

4. Discussion

Results obtained by sperm quality analysis showed that the freezing–thawing process caused significant reductions in sperm motility, viability, and osmotic tolerance. These findings were in accordance with previous reports on avian sperm [14,15,18,54]. In particular, the post-thaw semen quality measured in this study was similar to that observed in a previous paper [14]: after thawing, in fact, the returned recovery rates (value found in cryopreserved semen/value found in the fresh semen $\times$ 100) were about 40%, 48%, and 37% for sperm viability, mobility, and osmotic tolerance, respectively.

It is well known that the semen cryopreservation process imposes numerous stresses not only on the physical features of sperm but on its chemical composition, which in turn is essential for sperm function, as in the case of energy metabolism, which is known to be absolutely crucial in supporting sperm motility. Thus, the high sensitivity of turkey sperm to the cryopreservation process is assumed to be a consequence of the sperm metabolic profile of this bird (also see [10]). In particular, the first pieces of evidence about changes in lipid content, cholesterol/phospholipid ratios, and glycoconjugate and ATP content as a result of semen cryopreservation process have been already reported [18,55,56]. However, a more exhaustive picture of the sperm metabolic profile before and after cryopreservation is still lacking. Therefore, in this study, by taking advantage of NMR technique, a relevant number of metabolites were assessed in both fresh and frozen–thawed spermatozoa in order to correlate their levels to sperm quality variations.

Data obtained by NMR analysis are discussed separately for each class of compounds.

Amino Acids. We observed a general decrease in amino acid levels in frozen–thawed sperm, with statistically significant differences for Ala, Ile, Leu, Phe, Tyr, and Val. Hence, we hypothesize that the reduction in the levels of these amino acids could play a key role in the reduced quality of post-thaw sperm. This notion is substantiated by the positive correlations detected here between sperm viability, mobility, and osmotic tolerance and these amino acids. Thus, it could be hypothesized that increasing the sperm content of these amino acids could improve the freezability of turkey semen and, consequently, the post-thaw sperm quality. This is also consistent with previous research involving both human [40] and other mammalian species [57–62] in which the supplementation of amino acids was successfully used to improve post-thawing sperm quality. More recently, amino acid supplementation was successfully checked in chicken [63–66].

To date, the mechanism by which amino acids could provide cryoprotection is not fully understood and remains unclear [60]. However, some authors have proposed that amino acids could form a layer over the sperm surface via the electrostatic interaction with the phosphate group of the sperm plasma membrane phospholipids, acting as a cushion for damage against ice crystal formation and therefore preventing thermal shock [57]. Thus, amino acids could also interact with phospholipid bilayers during freezing, allowing stabilization of the cell membrane [2]. In addition, the supplementation of amino acids to semen diluents can lead to a reduction in the concentration of toxic solutes to levels associated with lesser toxicity; moreover, some amino acids can protect sperm cells against the denaturing effects of hyperosmolality during cryopreservation process [2,67]. In accordance with our results, in a recent study, the supplementation of valine to chicken freezing extenders resulted in a decrease in DNA fragmentation and a positive effect on the fertilizing ability of frozen–thawed sperm, with a better response in a breed that is considered to have the lowest semen freezability [64].

In addition to the cryoprotective role played by exogenous amino acids as extender supplementation, putative effects of endogenous amino acids in specific sperm function should be taken in consideration. It was shown, in fact, that seminal plasma levels of Leu were higher in high- than in low-fertility bulls [68]; this was proposed to be due to the action of Leu in modulating active Ca^{2+} transport across sperm membrane, which would result in a delay in Ca^{2+} uptake in ejaculated sperm. Leu was also proposed to be one of the fertility biomarkers in bovine species [68]. Accordingly, free Leu content in chicken seminal plasma was positively correlated with sperm viability as well as DNA integrity [69].

Contrary to levels of the majority of detected amino acids, Gly levels increased in frozen–thawed samples, and a significant negative correlation between its levels and sperm osmotic tolerance (SOT) was found. This resembles what was already observed in a previous study, in which a decrease in SOT was associated with increased levels of Gly in sperm of ageing turkey males [49]. Surprisingly enough, Gly content was also found to be negatively correlated with sperm motility. This seems to be in contrast with other reports showing that the addition of Gly to the diluent prevented significant changes in chicken sperm motility during the freezing–thawing procedure [70]. However, this dissonant result could be due to differences in metabolism between the two avian species [71], as well as a different mechanism of action for Gly as a function of its concentration [70]. Nonetheless, further investigation into the role of amino acids in turkey sperm metabolism should be carried out in future studies.

Other water-soluble metabolites. Similarly to what was observed for most amino acids, significant reductions in carnitine, lactate, formate, creatine, and AMP content were found in cryopreserved sperm.

Furthermore, positive correlations between the aforementioned metabolites and sperm quality parameters were found, suggesting that these metabolites could somehow be involved in the decrease in sperm quality occurring in cryopreservation.

It is not surprising that a reduced post-thawed semen quality could be related to low content of carnitine. It is known, in fact, that carnitine plays a key role in sperm metabolism by providing readily available energy, thus affecting sperm motility and maturation and the spermatogenic process [46]. This should be more evident in turkey sperm, which presents a highly oxidative metabolism [71,72]. Moreover, carnitine has a protective action against reactive oxygen species (ROS) by exerting antioxidant properties [73,74]. The results therein were consistent with previous studies reporting that the addition of L-carnitine in the extender enhanced chicken sperm motility in vitro during liquid storage and frozen state [2,66,75] and that supplementation of L-carnitine in the diet could improve drake semen quality [76]. Accordingly, L-carnitine supplementation to freezing extender improved human sperm motility and vitality and reduced sperm DNA oxidation during cryopreservation [77].

The reduction in free carnitine levels observed in frozen–thawed sperm could also be due to the slight, but significant, increase in ac-carnitine levels observed in these samples, which was correlated with a sperm quality decrease. Since ac-carnitine seems to be involved in buffering or trapping the excessive production of acetyl-CoA [45], it could be speculated that as a result of cryopreservation process, overproduction of acetyl-CoA could occur, which in turn could increase ac-carnitine levels in sperm. In agreement with this hypothesis, cryopreserved samples also showed an increase in citrate, which derives from the condensation between oxaloacetate and acetyl-CoA. Thus, in frozen–thawed sperm, the increases in the amounts of both ac-carnitine and citrate, together with the unvaried levels of other Krebs cycle compounds such as fumarate, may suggest a reduced capability of mitochondria to utilize acetyl-CoA through this pathway and/or a reduced capability of the mitochondrial respiratory chain (linked to the Krebs cycle by its products NADH and $FADH_2$) to support sperm energy requirement. In this regard, it must be noted that among avian species, turkey presents a high oxidative metabolism [10]. The high aerobic metabolism for this species, in fact, was already reported [78] and further highlighted by a previous study in which it was found that the stimulation of the cytochrome c oxidase (complex IV of the respiratory chain) by He–Ne laser light increased post-thaw sperm motility in turkey, but not in chicken or pheasant [71].

In addition to mitochondrial oxidative phosphorylation, glycolysis constitutes the other energy source in sperm. In this regard, it was recently reported that water-soluble extract of quail cloacal gland secretion contained glucose as an energy source for the intrinsic sperm mobility after transportation to female vagina [79]. However, in partial agreement with the hypothesis of a cryopreservation-dependent reduction in mitochondrial activity, here we found no differences in the levels of glucose between fresh and cryopreserved samples. Thus, in our case, a reduced sperm cell capability of glucose uptake due to putative impairment of glucose carriers should be ruled out. Contrarily, in other species such as boars, it was found that the cryopreservation process impaired glucose uptake by affecting the distribution of glucose transporters, especially GLUT-3 [80], a GLUT family member that is also present in avians, as shown by proteomic and peptidomic analyses of chicken sperm [81]. On the other hand, the impairment of the glycolytic pathway as a result of cryopreservation also seems unlikely in light of the supposed increased acetyl-CoA production discussed above.

Despite unaltered glucose levels, in cryopreserved samples, a significant reduction in lactate content was found, which was correlated with the decreased sperm quality in these samples. This positive corelation is not surprising, since L-lactate was found to play an active role in sperm bioenergetics because of its mitochondrial metabolism [82,83]. In particular, because of the occurrence of a mitochondrial L-lactate dehydrogenase [84–86], mitochondria are able to actively metabolize this substrate for energy purposes [82]. For this reason, a reduction in sperm energy fuel such as L-lactate in cryopreserved sperm could be responsible for a decreased sperm quality. Therefore, further investigation of possible changes occurring in cryopreservation in cytosolic and mitochondrial L-lactate dehydrogenase (both protein levels and enzyme activities) represents a task to be addressed in future studies.

Myo-inositol levels were also investigated, since it has been reported that its addition to sperm could improve sperm motility [87] and mitochondrial membrane potential [88] (also see [89]). However, as for glucose, no significant variation between fresh and frozen–thawed sample was found. Thus, *myo*-inositol should not play a role in the sperm quality decrease during turkey semen cryopreservation, at least in our experimental conditions.

Another interesting result obtained by analyzing the water-soluble fraction of frozen–thawed sperm was the decreased AMP content in these samples. This low AMP content could also partially explain the observed increase in citrate levels, since the Krebs cycle rate is extremely sensitive to AMP concentration [90]. It is known, in fact, that AMP acts as a cell energy sensor via the AMP/ATP ratio; its increase is a signal of cell energy deficiency and leads to activation of the AMP-activated protein kinase (AMPK) [91], a mechanism well reported also in avian sperm [92]. Thus, the observed reduction in the levels of AMP in cryopreserved samples is quite surprising, especially in light of the 40-fold decrease in ATP levels that was already found as a result of the freezing–thawing process of turkey sperm [56]. In this regard, it must be noted that cell levels of AMP, ADP, and ATP give rise to the adenylate energy charge; moreover, these compounds are closely related to each other by the activity of the enzyme adenylate kinase (AK), which catalyzes the following equilibrium reaction: 2ADP = ATP + AMP. The activity of this enzyme was reported to allow ADP to partially support sperm motility [93], at least under energetic stress conditions [94]. Since this scenario seems to occur in turkey sperm cryopreservation, as evidenced by the dramatic ATP decrease, it could be speculated that a reduction in AK activity could play a role in the sperm quality decrease. Thus, further investigation into this aspect (i.e., the occurrence and activity of AK in turkey sperm and its variation during the freezing–thawing procedure), as well as the detection of ADP levels in sperm before and after cryopreservation, appears to be mandatory in further studies.

Lipids. It is widely accepted that lipids are the main component of sperm membrane and that they are responsible for the fluidity of membrane bilayers [95], which in turn influences the spermatozoa freezability of different animal species [48,96]. Accordingly, one study showed that the freezing–thawing process resulted in a rigidifying effect on the sperm membrane and suggested that sperm adaptability to freezing–thawing-induced stress could be dependent on its initial membrane fluidity [97]. The same authors argued that the initial membrane fluidity had practical implications for predicting the response of spermatozoa following freezing and thawing and for improving the recovery of viable spermatozoa.

Thus, by turning our attention to lipid extract, in this study, we observed significant reductions in CHO, UFA, PUFA and an increase in SFA in frozen compared with fresh sperm. In addition, cholesterol was positively correlated with sperm viability, motility, and osmotic tolerance. This outcome leads us to assume that the reduction in sperm quality after freezing–thawing could be due to increased spermatozoon membrane rigidity accompanied by the decrease in the cholesterol/phospholipid ratio. This would be consistent with previous studies in human [97] and in turkey [18,98] reporting that semen storage and, more importantly, the freezing–thawing procedure induced a rigidifying process in the sperm membrane followed by a dramatic decrease in the cholesterol/phospholipid ratio.

Moreover, besides the cholesterol content, membrane fluidity depends on the degree of saturation of fatty acids in membrane phospholipids [99], since saturated fatty acids are rigidifying components of the mammalian sperm membranes. The avian sperm membrane contains more polyunsaturated fatty acids (PUFAs) than mammal sperm and has a lower protein content, a lower cholesterol/phospholipid ratio, and greater overall fluidity at physiological temperatures [2]. Here, we found an increase in the SFA/UFA ratio occurring in cryopreservation, since it was 0.62 and 0.75 in fresh and frozen–thawed sperm, respectively. Thus, we can conclude that, distinctly from semen liquid storage, in which SFA/UFA ratio was not affected [100], more drastic storage conditions such as cryopreservation strongly affect the lipid composition, and consequently the fluidity, of turkey sperm membrane. In this regard, previous studies showed that the lipid composition of avian spermatozoa could be modified by the diet with subsequent effects on membrane fluidity [22,32,101]. Thus, a diet inducing a lower SFA/UFA ratio in turkey sperm could be tried out in the future as a putative way to increase the freezability of avian sperm. At the same time, it must be taken in consideration that sperm cryopreservation strongly increases ROS production [102,103], resulting in lipid peroxidation. Therefore, the reduction in sperm quality could also be due to a membrane destabilization deriving from PUFA peroxidation [98]. Thus, a higher UFA content in the sperm membrane may not necessarily be the best way to improve sperm freezability. However, proper knowledge of the relationship between metabolic profile and the freezability of the spermatozoa remains very interesting in light of the possibility of modifying the sperm metabolic profile via various factors such as diet manipulation, strain, and ageing [22,32,33,49].

5. Conclusions

We are confident that the findings reported here provide a valid contribution to the scientific community, since they returned a clearer scenario of metabolic changes occurring in turkey sperm cryopreservation. Semen cryopreservation is an important biotechnological strategy used to both preserve and protect genetic resources, which are subject to increasingly serious reductions in some species, as well as to enhance animal biodiversity in the case of inbreeding risks. The knowledge of metabolites responsible for the post-thawing sperm quality decrease allows *(i)* identifying several biological markers related to the sperm freezability of male turkey donators and *(ii)* suggesting a supplementation of specific metabolites in the diet or in the freezing medium in order to obtain spermatozoa abler to withstand the freezing process.

In addition, the analytic approach used here, which resorted to the NMR technique to determine the metabolites involved in semen cryopreservation, constitutes an important tool that can be applied also to other species, including humans.

Author Contributions: Conceptualization, N.I., G.P. and M.D.I.; methodology, M.D.I., N.I., G.P., A.P.S. and L.M.; validation, S.C., L.M. and N.I.; formal analysis, N.I., M.D.I., A.P.S., L.M. and G.P.; investigation, M.S., E.A. and G.R.; data curation, G.P., M.D.I. and A.P.S.; writing—original draft preparation, G.P., N.I. and M.D.I.; writing—review and editing, G.P., N.I., S.C., L.M. and M.D.I.; visualization, G.P. and M.D.I.; supervision, N.I.; funding acquisition, N.I. All authors have read and agreed to the published version of the manuscript.

Funding: This research was funded by The Italian Ministry of Agriculture, MIPAAFT-NRDP 2014-2020-Measure 10.2 Biodiversity—Poultry sector (DG DISR-DISR07-Prot. n. 0194727-28.04.2021) project name (Conservation of biodiversity in Italian poultry breeds: insights and monitoring—TuBAvI-2).

Institutional Review Board Statement: The handling of animals and semen collection was conducted in accordance with the Code of Ethics of the EU Directive 2010/63/EU.

Informed Consent Statement: Not applicable.

Data Availability Statement: The datasets used are available from the corresponding author on reasonable request.

Acknowledgments: The authors thank Carmine Marini for help with semen collection and the "Amadori" group for allowing to use their breeders.

Conflicts of Interest: The authors declare no conflict of interest.

References

1. Blesbois, E.; Seigneurin, F.; Grasseau, I.; Limouzin, C.; Besnard, J.; Gourichon, D.; Coquerelle, G.; Rault, P.; Tixier-Boichard, M. Semen Cryopreservation for Ex Situ Management of Genetic Diversity in Chicken: Creation of the French Avian Cryobank. *Poult. Sci.* **2007**, *86*, 555–564. [CrossRef] [PubMed]
2. Partyka, A.; Niżański, W. Advances in storage of poultry semen. *Anim. Reprod. Sci.* **2022**, 106921. [CrossRef] [PubMed]
3. Liptoi, K.; Buda, K.; Rohn, E.; Drobnyak, A.; Meleg, E.E.; Palinkas-Bodzsar, N.; Vegi, B.; Barna, J. Improvement of the application of gonadal tissue allotransplantation in the in vitro conservation of chicken genetic lines. *Anim. Reprod. Sci.* **2020**, *213*, 106280. [CrossRef] [PubMed]
4. Santiago-Moreno, J.; Blesbois, E. Animal board invited review: Germplasm technologies for use with poultry. *Animal* **2022**, *16*, 100475. [CrossRef]
5. Blesbois, E. Freezing avian semen. *Avian Biol. Res.* **2011**, *4*, 52–58. [CrossRef]
6. Kowalczyk, A.; Łukaszewicz, E. Simple and Effective Methods of Freezing Capercaillie (*Tetrao urogallus* L.) Semen. *PLoS ONE* **2015**, *10*, e0116797. [CrossRef]
7. Thélie, A.; Bailliard, A.; Seigneurin, F.; Zerjal, T.; Tixier-Boichard, M.; Blesbois, E. Chicken semen cryopreservation and use for the restoration of rare genetic resources. *Poult. Sci.* **2019**, *98*, 447–455. [CrossRef]
8. Prentice, J.R.; Anzar, M. Cryopreservation of Mammalian Oocyte for Conservation of Animal Genetics. *Vet. Med. Int.* **2011**, *2011*, 146405. [CrossRef]
9. Iaffaldano, N.; Di Iorio, M.; Rusco, G.; Antenucci, E.; Zaniboni, L.; Madeddu, M.; Marelli, S.; Schiavone, A.; Soglia, D.; Buccioni, A.; et al. Italian semen cryobank of autochthonous chicken and turkey breeds: A tool for preserving genetic biodiversity. *Ital. J. Anim. Sci.* **2021**, *20*, 2022–2033. [CrossRef]
10. Iaffaldano, N.; Di Iorio, M.; Cerolini, S.; Manchisi, A. Overview of Turkey Semen Storage: Focus on Cryopreservation—A Review. *Ann. Anim. Sci.* **2016**, *16*, 961–974. [CrossRef]
11. Blanco, J.M.; Gee, G.; Wildt, D.E.; Tselutin, K.; Donoghue, A.M. Semen cryopreservation in poultry and non-domestic species: A comparative approach to understanding the fundamentals of avian spermatozoa cryobiology. *Br. Poult. Sci.* **2000**, *41*, 3–5. [CrossRef]
12. Blanco, J.M.; Long, J.A.; Gee, G.; Donoghue, A.M.; Wildt, D.E. Osmotic tolerance of avian spermatozoa: Influence of time, temperature, cryoprotectant and membrane ion pump function on sperm viability. *Cryobiology* **2008**, *56*, 8–14. [CrossRef] [PubMed]
13. Blesbois, E. Current status in avian semen cryopreservation. *World's Poult. Sci. J.* **2007**, *63*, 213–222. [CrossRef]
14. Iaffaldano, N.; Romagnoli, L.; Manchisi, A.; Rosato, M.P. Cryopreservation of turkey semen by the pellet method: Effects of variables such as the extender, cryoprotectant concentration, cooling time and warming temperature on sperm quality determined through principal components analysis. *Theriogenology* **2011**, *76*, 794–801. [CrossRef]
15. Iaffaldano, N.; Di Iorio, M.; Miranda, M.; Zaniboni, L.; Manchisi, A.; Cerolini, S. Cryopreserving turkey semen in straws and nitrogen vapour using DMSO or DMA: Effects of cryoprotectant concentration, freezing rate and thawing rate on post-thaw semen quality. *Br. Poult. Sci.* **2016**, *57*, 264–270. [CrossRef]
16. Iaffaldano, N.; Rosato, M.; Manchisi, A.; Centoducati, G.; Meluzzi, A. Comparison of different extenders on the quality characteristics of turkey semen during storage. *Ital. J. Anim. Sci.* **2016**, *4*, 513–515. [CrossRef]

17. Izanloo, H.; Soleimanzadeh, A.; Bucak, M.N.; Imani, M.; Zhandi, M. The effects of varying concentrations of glutathione and trehalose in improving microscopic and oxidative stress parameters in Turkey semen during liquid storage at 5 °C. *Cryobiology* **2021**, *101*, 12–19. [CrossRef]
18. Blesbois, E.; Grasseau, I.; Seigneurin, F. Membrane fluidity and the ability of domestic bird spermatozoa to survive cryopreservation. *Reproduction* **2005**, *129*, 371–378. [CrossRef]
19. Di Iorio, M.; Rusco, G.; Iampietro, R.; Maiuro, L.; Schiavone, A.; Cerolini, S.; Iaffaldano, N. Validation of the Turkey Semen Cryopreservation by Evaluating the Effect of Two Diluents and the Inseminating Doses. *Animals* **2020**, *10*, 1329. [CrossRef]
20. Di Iorio, M.; Rusco, G.; Iampietro, R.; Colonna, M.A.; Zaniboni, L.; Cerolini, S.; Iaffaldano, N. Finding an Effective Freezing Protocol for Turkey Semen: Benefits of Ficoll as Non-Permeant Cryoprotectant and 1:4 as Dilution Rate. *Animals* **2020**, *10*, 421. [CrossRef]
21. Mosca, F.; Madeddu, M.; Abdel Sayed, A.; Zaniboni, L.; Iaffaldano, N.; Cerolini, S. Combined effect of permeant and non-permeant cryoprotectants on the quality of frozen/thawed chicken sperm. *Cryobiology* **2016**, *73*, 343–347. [CrossRef] [PubMed]
22. Zaniboni, L.; Rizzi, R.M.; Cerolini, S. Combined effect of DHA and α-tocopherol enrichment on sperm quality and fertility in the turkey. *Theriogenology* **2006**, *65*, 1813–1827. [CrossRef] [PubMed]
23. Iaffaldano, N.; Manchisi, A.; Gambacorta, M.; Di Iorio, M.; Rosato, M.P. Effect of different sperm concentrations on the post-thaw viability and motility of turkey spermatozoa cryopreserved by the pellet method. *Ital. J. Anim. Sci.* **2009**, *8*, 760–762. [CrossRef]
24. Mosca, F.; Zaniboni, L.; Abdel Sayed, A.; Madeddu, M.; Iaffaldano, N.; Cerolini, S. Effect of dimethylacetamide and N-methylacetamide on the quality and fertility of frozen/thawed chicken semen. *Poult. Sci.* **2019**, *98*, 6071–6077. [CrossRef]
25. Mosca, F.; Zaniboni, L.; Sayed, A.A.; Iaffaldano, N.; Soglia, D.; Schiavone, A.; Cerolini, S. Effect of N-Methylacetamide Concentration and Thawing Rate on Chicken Sperm Quality after Cryopreservation. *Animals* **2020**, *10*, 824. [CrossRef]
26. Parks, J.E.; Lynch, D.V. Lipid composition and thermotropic phase behavior of boar, bull, stallion, and rooster sperm membranes. *Cryobiology* **1992**, *29*, 255–266. [CrossRef]
27. Cerolini, S.; Zaniboni, L.; Maldjian, A.; Gliozzi, T. Effect of docosahexaenoic acid and α-tocopherol enrichment in chicken sperm on semen quality, sperm lipid composition and susceptibility to peroxidation. *Theriogenology* **2006**, *66*, 877–886. [CrossRef]
28. Blesbois, E.; Grasseau, I.; Seigneurin, F.; Mignon-Grasteau, S.; Saint Jalme, M.; Mialon-Richard, M.M. Predictors of success of semen cryopreservation in chickens. *Theriogenology* **2008**, *69*, 252–261. [CrossRef]
29. Long, J.A.; Purdy, P.H.; Zuidberg, K.; Hiemstra, S.-J.; Velleman, S.G.; Woelders, H. Cryopreservation of turkey semen: Effect of breeding line and freezing method on post-thaw sperm quality, fertilization, and hatching. *Cryobiology* **2014**, *68*, 371–378. [CrossRef]
30. Mussa, N.J.; Ratchamak, R.; Ratsiri, T.; Vongpralub, T.; Boonkum, W.; Semaming, Y.; Chankitisakul, V. Lipid profile of sperm cells in Thai native and commercial roosters and its impact on cryopreserved semen quality. *Trop. Anim. Health Prod.* **2021**, *53*, 321. [CrossRef]
31. Mandal, R.; Badyakar, D.; Chakrabarty, J. Role of Membrane Lipid Fatty Acids in Sperm Cryopreservation. *Adv. Androl.* **2014**, *2014*, 190542. [CrossRef]
32. Zaniboni, L.; Cerolini, S. Liquid storage of turkey semen: Changes in quality parameters, lipid composition and susceptibility to induced in vitro peroxidation in control, n-3 fatty acids and alpha-tocopherol rich spermatozoa. *Anim. Reprod. Sci.* **2009**, *112*, 51–65. [CrossRef] [PubMed]
33. Long, J.A. Avian semen cryopreservation: What are the biological challenges? *Poult. Sci.* **2006**, *85*, 232–236. [CrossRef] [PubMed]
34. Shinitzky, M.; Yuli, I. Lipid fluidity at the submacroscopic level: Determination by fluorescence polarization. *Chem. Phys. Lipids* **1982**, *30*, 261–282. [CrossRef]
35. Vickram, A.S.; Ramesh Pathy, M.; Sridharan, T.B. Effect of various biomolecules for normal functioning of human sperm for fertilization: A review. *Int. J. Pharm. Pharm. Sci.* **2012**, *4*, 18–24.
36. Cheah, Y.; Yang, W. Functions of essential nutrition for high quality spermatogenesis. *Adv. Biosci. Biotechnol.* **2011**, *02*, 182–197. [CrossRef]
37. Atessahin, A.; Bucak, M.N.; Tuncer, P.B.; Kızıl, M. Effects of anti-oxidant additives on microscopic and oxidative parameters of Angora goat semen following the freeze–thawing process. *Small Rumin. Res.* **2008**, *77*, 38–44. [CrossRef]
38. Ugur, M.R.; Dinh, T.; Hitit, M.; Kaya, A.; Topper, E.; Didion, B.; Memili, E. Amino Acids of Seminal Plasma Associated with Freezability of Bull Sperm. *Front. Cell Dev. Biol.* **2020**, *7*, 347. [CrossRef]
39. Santiago-Moreno, J.; Bernal, B.; Perez-Cerezales, S.; Castaño, C.; Toledano-Díaz, A.; Esteso, M.C.; Gutierrez-Adan, A.; López-Sebastián, A.; Garcia-Gil, M.; Woelders, H.; et al. Seminal plasma amino acid profile in different breeds of chicken: Role of seminal plasma on sperm cryoresistance. *PLoS ONE* **2019**, *14*, e0209910. [CrossRef]
40. Renard, P.; Grizard, G.; Francoisgriveaua, J.; Sion, B.; Boucher, D.; Le Lannou, D. Improvement of Motility and Fertilization Potential of Postthaw Human Sperm Using Glutamine. *Cryobiology* **1996**, *33*, 311–319. [CrossRef]
41. Li, Y.; Si, W.; Zhang, X.; Dinnyes, A.; Ji, W. Effect of amino acids on cryopreservation of cynomolgus monkey (*macaca fascicularis*) sperm. *Am. J. Primatol.* **2003**, *59*, 159–165. [CrossRef] [PubMed]
42. Khlifaoui, M.; Battut, I.; Bruyas, J.F.; Chatagnon, G.; Trimeche, A.; Tainturier, D. Effects of glutamine on post-thaw motility of stallion spermatozoa: An approach of the mechanism of action at spermatozoa level. *Theriogenology* **2005**, *63*, 138–149. [CrossRef] [PubMed]

43. Bucak, M.N.; Tuncer, P.B.; Sarıözkan, S.; Ulutaş, P.A. Comparison of the effects of glutamine and an amino acid solution on post-thawed ram sperm parameters, lipid peroxidation and anti-oxidant activities. *Small Rumin. Res.* **2009**, *81*, 13–17. [CrossRef]
44. Moradi, B.; Faramarzi, A.; Ghasemi-Esmailabad, S.; Aghaz, F.; Hashemian, A.H.; Khazaei, M. L-proline as a novel additive to cryopreservation media improved post-thaw quality of human spermatozoon via reducing oxidative stress. *Andrologia* **2022**, *54*, e14301. [CrossRef] [PubMed]
45. Jeulin, C.; Lewin, L.M. Role of free L-carnitine and acetyl-L-carnitine in post-gonadal maturation of mammalian spermatozoa. *Hum. Reprod. Update* **1996**, *2*, 87–102. [CrossRef]
46. Agarwal, A.; Said, T.M. Carnitines and male infertility. *Reprod. Biomed. Online* **2004**, *8*, 376–384. [CrossRef]
47. Stradaioli, G.; Sylla, L.; Zelli, R.; Chiodi, P.; Monaci, M. Effect of L-carnitine administration on the seminal characteristics of oligoasthenospermic stallions. *Theriogenology* **2004**, *62*, 761–777. [CrossRef]
48. Yu, J.; Wang, Z.; An, X.; Liu, H.; Wang, F.; Ji, C.; Liu, G.; Wang, Y.; He, F.; Dang, R.; et al. Metabolomic profiling of Dezhou donkey sperm associated with freezability. *Theriogenology* **2022**, *181*, 131–139. [CrossRef]
49. Iaffaldano, N.; Di Iorio, M.; Mannina, L.; Paventi, G.; Rosato, M.P.; Cerolini, S.; Sobolev, A.P. Age-dependent changes in metabolic profile of turkey spermatozoa as assessed by NMR analysis. *PLoS ONE* **2018**, *13*, e0194219. [CrossRef]
50. Tselutin, K.; Narubina, L.; Mavrodina, T.; Tur, B. Cryopreservation of poultry semen. *Br. Poult. Sci.* **1995**, *36*, 805–811. [CrossRef]
51. Bligh, E.G.; Dyer, W.J. A rapid method of total lipid extraction and purification. *Can. J. Biochem. Physiol.* **1959**, *37*, 911–917. [CrossRef] [PubMed]
52. Ingallina, C.; Maccelli, A.; Spano, M.; Di Matteo, G.; Di Sotto, A.; Giusti, A.M.; Vinci, G.; Di Giacomo, S.; Rapa, M.; Ciano, S.; et al. Chemico-Biological Characterization of Torpedino Di Fondi® Tomato Fruits: A Comparison with San Marzano Cultivar at Two Ripeness Stages. *Antioxidants* **2020**, *9*, 1027. [CrossRef] [PubMed]
53. Ingallina, C.; Spano, M.; Sobolev, A.P.; Esposito, C.; Santarcangelo, C.; Baldi, A.; Daglia, M.; Mannina, L. Characterization of Local Products for Their Industrial Use: The Case of Italian Potato Cultivars Analyzed by Untargeted and Targeted Methodologies. *Foods* **2020**, *9*, 1216. [CrossRef] [PubMed]
54. Lemoine, M.; Mignon-Grasteau, S.; Grasseau, I.; Magistrini, M.; Blesbois, E. Ability of chicken spermatozoa to undergo acrosome reaction after liquid storage or cryopreservation. *Theriogenology* **2011**, *75*, 122–130. [CrossRef]
55. Long, J.A.; Guthrie, H.D. Validation of a rapid, large-scale assay to quantify ATP concentration in spermatozoa. *Theriogenology* **2006**, *65*, 1620–1630. [CrossRef]
56. Słowińska, M.; Liszewska, E.; Judycka, S.; Konopka, M.; Ciereszko, A. Mitochondrial membrane potential and reactive oxygen species in liquid stored and cryopreserved turkey (*Meleagris gallopavo*) spermatozoa. *Poult. Sci.* **2018**, *97*, 3709–3717. [CrossRef]
57. Kundu, C.N.; Das, K.; Majumder, G.C. Effect of Amino Acids on Goat Cauda Epididymal Sperm Cryopreservation Using a Chemically Defined Model System. *Cryobiology* **2001**, *42*, 21–27. [CrossRef]
58. Sánchez-Partida, L.G.; Setchell, B.P.; Maxwell, W.M.C. Effect of compatible solutes and diluent composition on the post-thaw motility of ram sperm. *Reprod. Fertil. Dev.* **1998**, *10*, 347–358. [CrossRef]
59. Koskinen, E.; Junnila, M.; Katila, T.; Soini, H. A Preliminary Study on the Use of Betaine as a Cryoprotective Agent in Deep Freezing of Stallion Semen. *J. Veter. Med. Ser. A* **1989**, *36*, 110–114. [CrossRef]
60. Sangeeta, S.; Arangasamy, A.; Kulkarni, S.; Selvaraju, S. Role of amino acids as additives on sperm motility, plasma membrane integrity and lipid peroxidation levels at pre-freeze and post-thawed ram semen. *Anim. Reprod. Sci.* **2015**, *161*, 82–88. [CrossRef]
61. Ahmed, H.; Jahan, S.; Khan, A.; Khan, L.; Ullah, H.; Riaz, M.; Ullah, K.; Ullah, F. Supplementation of l-tryptophan (an aromatic amino acid) in tris citric acid extender enhances post-thaw progressive motility, plasmalemma, mitochondrial membrane potential, acrosome, and DNA integrities, and in vivo fertility rate of buffalo (*Bubalus bubalis*) bull spermatozoa. *Cryobiology* **2020**, *92*, 117–123. [CrossRef] [PubMed]
62. Beheshti, R.; Asadi, A.; Eshratkhah, B.; Kandi, J.G.G.; Ghorbani, A. The effect of cysteine on post-thawed buffalo bull (*Bubalus bubalis*) sperm parameters. *Adv. Environ. Biol.* **2011**, *5*, 1260–1263.
63. Thananurak, P.; Chuaychu-Noo, N.; Thélie, A.; Phasuk, Y.; Vongpralub, T.; Blesbois, E. Different concentrations of cysteamine, ergothioneine, and serine modulate quality and fertilizing ability of cryopreserved chicken sperm. *Poult. Sci.* **2020**, *99*, 1185–1198. [CrossRef] [PubMed]
64. Bernal, B.; Iglesias-Cabeza, N.; Sánchez-Rivera, U.; Toledano-Díaz, A.; Castaño, C.; Pérez-Cerezales, S.; Gutiérrez-Adán, A.; López-Sebastián, A.; García-Casado, P.; Gil, M.; et al. Effect of supplementation of valine to chicken extender on sperm cryoresistance and post-thaw fertilization capacity. *Poult. Sci.* **2020**, *99*, 7133–7141. [CrossRef] [PubMed]
65. Khiabani, A.B.; Moghaddam, G.; Kia, H.D. Effects of adding different levels of Glutamine to modified Beltsville extender on the survival of frozen rooster semen. *Anim. Reprod. Sci.* **2017**, *184*, 172–177. [CrossRef] [PubMed]
66. Partyka, A.; Rodak, O.; Bajzert, J.; Kochan, J.; Niżański, W. The Effect of L-Carnitine, Hypotaurine, and Taurine Supplementation on the Quality of Cryopreserved Chicken Semen. *BioMed Res. Int.* **2017**, *2017*, 7279341. [CrossRef]
67. Heber, U.; Tyankova, L.; Santarius, K.A. Stabilization and inactivation of biological membranes during freezing in the presence of amino acids. *Biochim. Biophys. Acta Biomembr.* **1971**, *241*, 578–592. [CrossRef]
68. Kumar, A.; Kroetsch, T.; Blondin, P.; Anzar, M. Fertility-associated metabolites in bull seminal plasma and blood serum: 1H nuclear magnetic resonance analysis. *Mol. Reprod. Dev.* **2015**, *82*, 123–131. [CrossRef]
69. Santiago-Moreno, J.; Blesbois, E. Functional Aspects of Seminal Plasma in Bird Reproduction. *Int. J. Mol. Sci.* **2020**, *21*, 5664. [CrossRef]

70. Cerolini, S.; Zaniboni, L.; Mangiagalli, M.G.; Gliozzi, T.M. Effect of glycine on cryopreservation of chicken spermatozoa. *Avian Poult. Biol. Rev.* **2007**, *18*, 65.
71. Iaffaldano, N.; Paventi, G.; Pizzuto, R.; Passarella, S.; Cerolini, S.; Zaniboni, L.; Marzoni, M.; Castillo, A.; Rosato, M.P. The post-thaw irradiation of avian spermatozoa with He-Ne laser differently affects chicken, pheasant and turkey sperm quality. *Anim. Reprod. Sci.* **2013**, *142*, 168–172. [CrossRef] [PubMed]
72. Iaffaldano, N.; Meluzzi, A.; Manchisi, A.; Passarella, S. Improvement of stored turkey semen quality as a result of He–Ne laser irradiation. *Anim. Reprod. Sci.* **2005**, *85*, 317–325. [CrossRef] [PubMed]
73. Vessey, W.; Saifi, S.; Sharma, A.; McDonald, C.; Almeida, P.; Figueiredo, M.; Minhas, S.; Virmani, A.; Dhillo, W.S.; Ramsay, J.W.; et al. Baseline levels of seminal reactive oxygen species predict improvements in sperm function following antioxidant therapy in men with infertility. *Clin. Endocrinol.* **2021**, *94*, 102–110. [CrossRef] [PubMed]
74. Vicari, E.S.D.; Calogero, A.E. Effects of treatment with carnitines in infertile patients with prostato-vesiculo-epididymitis. *Hum. Reprod.* **2001**, *16*, 2338–2342. [CrossRef] [PubMed]
75. Tabatabaei, S.; Aghaei, A. Effect of l-carnitine on sperm quality during liquid storage of chicken semen. *Comp. Clin. Pathol.* **2012**, *21*, 711–717. [CrossRef]
76. Aldaraji, H.J.; Tahir, A.O. Effect of L-carnitine supplementation on drake semen quality. *S. Afr. J. Anim. Sci.* **2014**, *44*, 18. [CrossRef]
77. Banihani, S.; Agarwal, A.; Sharma, R.; Bayachou, M. Cryoprotective effect of l-carnitine on motility, vitality and DNA oxidation of human spermatozoa. *Andrologia* **2014**, *46*, 637–641. [CrossRef]
78. Wishart, G.J. Maintenance of ATP concentrations in and of fertilizing ability of fowl and turkey spermatozoa in vitro. *Reproduction* **1982**, *66*, 457–462. [CrossRef]
79. Sasanami, T.; Izumi, S.; Sakurai, N.; Hirata, T.; Mizushima, S.; Matsuzaki, M.; Hiyama, G.; Yorinaga, E.; Yoshimura, T.; Ukena, K.; et al. A unique mechanism of successful fertilization in a domestic bird. *Sci. Rep.* **2015**, *5*, 7700. [CrossRef]
80. Sancho, S.; Casas, I.; Ekwall, H.; Saravia, F.; Rodriguez-Martinez, H.; Rodriguez-Gil, J.E.; Flores, E.; Pinart, E.; Briz, M.D.; Garcia-Gil, N.; et al. Effects of cryopreservation on semen quality and the expression of sperm membrane hexose transporters in the spermatozoa of Iberian pigs. *Reproduction* **2007**, *134*, 111–121. [CrossRef]
81. Labas, V.; Grasseau, I.; Cahier, K.; Gargaros, A.; Harichaux, G.; Teixeira-Gomes, A.-P.; Alves, S.; Bourin, M.; Gérard, N.; Blesbois, E. Qualitative and quantitative peptidomic and proteomic approaches to phenotyping chicken semen. *J. Proteom.* **2015**, *112*, 313–335. [CrossRef]
82. Paventi, G.; Lessard, C.; Bailey, J.L.; Passarella, S. In boar sperm capacitation l -lactate and succinate, but not pyruvate and citrate, contribute to the mitochondrial membrane potential increase as monitored via safranine O fluorescence. *Biochem. Biophys. Res. Commun.* **2015**, *462*, 257–262. [CrossRef] [PubMed]
83. Darr, C.R.; Varner, D.D.; Teague, S.; Cortopassi, G.A.; Datta, S.; Meyers, S.A. Lactate and Pyruvate Are Major Sources of Energy for Stallion Sperm with Dose Effects on Mitochondrial Function, Motility, and ROS Production. *Biol. Reprod.* **2016**, *95*, 34. [CrossRef] [PubMed]
84. Passarella, S.; de Bari, L.; Valenti, D.; Pizzuto, R.; Paventi, G.; Atlante, A. Mitochondria and L-lactate metabolism. *FEBS Lett.* **2008**, *582*, 3569–3576. [CrossRef] [PubMed]
85. Paventi, G.; Pizzuto, R.; Passarella, S. The occurrence of l-lactate dehydrogenase in the inner mitochondrial compartment of pig liver. *Biochem. Biophys. Res. Commun.* **2017**, *489*, 255–261. [CrossRef]
86. Passarella, S.; Paventi, G.; Pizzuto, R. The mitochondrial L-lactate dehydrogenase affair. *Front. Neurosci.* **2014**, *8*, 407. [CrossRef]
87. Condorelli, R.A.; La Vignera, S.; Bellanca, S.; Vicari, E.; Calogero, A.E. Myoinositol: Does it improve sperm mitochondrial function and sperm motility? *Urology* **2012**, *79*, 1290–1295. [CrossRef]
88. Condorelli, R.A.; La Vignera, S.; Di Bari, F.; Unfer, V.; Calogero, A.E. Effects of myoinositol on sperm mitochondrial function in-vitro. *Eur. Rev. Med. Pharmacol. Sci.* **2011**, *15*, 129–134.
89. Vazquez-Levin, M.H.; Verón, G.L. Myo-inositol in health and disease: Its impact on semen parameters and male fertility. *Andrology* **2020**, *8*, 277–298. [CrossRef]
90. Martínez-Reyes, I.; Chandel, N.S. Mitochondrial TCA cycle metabolites control physiology and disease. *Nat. Commun.* **2020**, *11*, 102. [CrossRef]
91. Hardie, D.G. AMP-activated/SNF1 protein kinases: Conserved guardians of cellular energy. *Nat. Rev. Mol. Cell Biol.* **2007**, *8*, 774–785. [CrossRef]
92. Nguyen, T.M.D.; Grasseau, I.; Blesbois, E. New insights in the AMPK regulation in chicken spermatozoa: Role of direct AMPK activator and relationship between AMPK and PKA pathways. *Theriogenology* **2019**, *140*, 1–7. [CrossRef] [PubMed]
93. Vadnais, M.L.; Cao, W.; Aghajanian, H.K.; Haig-Ladewig, L.; Lin, A.M.; AlAlao, O.; Gerton, G.L. Adenine Nucleotide Metabolism and a Role for AMP in Modulating Flagellar Waveforms in Mouse Sperm1. *Biol. Reprod.* **2014**, *90*, 128. [CrossRef] [PubMed]
94. Xie, M.; Zhang, G.; Zhang, H.; Chen, F.; Chen, Y.; Zhuang, Y.; Huang, Z.; Zou, F.; Liu, M.; An, G.; et al. Adenylate kinase 1 deficiency disrupts mouse sperm motility under conditions of energy stress. *Biol. Reprod.* **2020**, *103*, 1121–1131. [CrossRef]
95. Sanocka, D.; Kurpisz, M. Reactive oxygen species and sperm cells. *Reprod. Biol. Endocrinol.* **2004**, *2*, 12. [CrossRef] [PubMed]
96. Shabani, S.; Mehri, M.; Shirmohammad, F.; Sharafi, M. Enhancement of sperm quality and fertility-related parameters in Hubbard grandparent rooster fed diets supplemented with soybean lecithin and vitamin E. *Poult. Sci.* **2022**, *101*, 101635. [CrossRef] [PubMed]

97. Giraud, M.N.; Motta, C.; Boucher, D.; Grizard, G. Membrane fluidity predicts the outcome of cryopreservation of human spermatozoa. *Hum. Reprod.* **2000**, *15*, 2160–2164. [CrossRef] [PubMed]
98. Partyka, A.; Niżański, W. Supplementation of avian semen extenders with antioxidants to improve semen quality—Is it an effective strategy? *Antioxidants* **2021**, *10*, 1927. [CrossRef]
99. Ladha, S. Lipid heterogeneity and membrane fluidity in a highly polarized cell, the mammalian spermatozoon. *J. Membr. Biol.* **1998**, *165*, 1–10. [CrossRef]
100. Douard, V.; Hermier, D.; Blesbois, E. Changes in Turkey Semen Lipids During Liquid In Vitro Storage1. *Biol. Reprod.* **2000**, *63*, 1450–1456. [CrossRef]
101. Bongalhardo, D.C.; Leeson, S.; Buhr, M.M. Dietary lipids differentially affect membranes from different areas of rooster sperm. *Poult. Sci.* **2009**, *88*, 1060–1069. [CrossRef] [PubMed]
102. Ansari, M.S.; Rakha, B.A.; Akhter, S.; Akhter, A.; Blesbois, E.; Santiago-Moreno, J. Effect of glutathione on pre and post-freezing sperm quality of Indian red jungle fowl (*Gallus gallus murghi*). *Theriogenology* **2021**, *172*, 73–79. [CrossRef] [PubMed]
103. Zhang, B.; Wang, Y.; Wu, C.; Qiu, S.; Chen, X.; Cai, B.; Xie, H. Freeze-thawing impairs the motility, plasma membrane integrity and mitochondria function of boar spermatozoa through generating excessive ROS. *BMC Vet. Res.* **2021**, *17*, 127. [CrossRef] [PubMed]

biology

MDPI

Article

Should All Fractions of the Boar Ejaculate Be Prepared for Insemination Rather Than Using the Sperm Rich Only?

Chiara Luongo [1], Pedro José Llamas-López [2], Iván Hernández-Caravaca [3,4], Carmen Matás [1,4] and Francisco Alberto García-Vázquez [1,4,*]

[1] Departamento de Fisiología, Facultad de Veterinaria, Campus de Excelencia Mare Nostrum Universidad de Murcia, 30100 Murcia, Spain; chiara.luongo@um.es (C.L.); cmatas@um.es (C.M.)
[2] Departamento de Tecnología Agroalimentaria, Universidad Miguel Hernández, 03202 Elche, Spain; pedrojosellamas@outlook.es
[3] Departamento de Enfermería Comunitaria, Medicina Preventiva y Salud Pública e Historia de la Ciencia, Facultad Ciencias de la Salud, Campus de Sant Vicent del Raspeig, E-03080 Sant Vicent del Raspeig, Spain; ivan.hernandez@ua.es
[4] Institute for Biomedical Research of Murcia (IMIB-Arrixaca), 30120 Murcia, Spain
* Correspondence: fagarcia@um.es

Simple Summary: The swine industry is constantly looking for efficiency improvement, especially focusing on the artificial insemination (AI) process. One of the trends in AI centers is to maximize the number of doses obtained from one ejaculate. Seminal doses are usually prepared with the sperm-rich fraction or the whole ejaculate, but further studies are needed to understand how to prepare them properly. Thus, this study aims to analyze how accumulative ejaculate fractions may influence sperm storage, AI performance, and offspring. The results indicate that the presence of all ejaculate fractions within seminal doses does not affect either sperm quality or AI performance and offspring health. Therefore, this study highlights the possibility to use the bulk ejaculate for seminal dose preparation, leading to successful AI. Additionally, it results in a more time-efficient preparation of a greater number of seminal doses providing an economic advantage.

Abstract: Boar ejaculate is released in several well-characterized fractions, differing in terms of sperm concentration, seminal plasma volume, and composition. However, the inclusion of the last part of the ejaculate for artificial insemination (AI) purposes is still under debate due to its controversial effects. Thus, there is a need to study the potential synergistic impact of the different ejaculate fractions. We aimed to evaluate the effect of accumulative ejaculate fractions on sperm conservation, AI performance, and offspring health. Ejaculates (n = 51) were collected and distributed as follows: F1: sperm-rich fraction; F2: sperm-rich + intermediate fractions; F3: sperm-rich + intermediate + poor fractions. Each group was diluted in a commercial extender, packaged in seminal doses (2000×10^6 sperm/60 mL), and stored at ~16 °C. On day 3 of conservation, sperm were analyzed and used for AI (n = 174). High sperm quality was observed after storage without a significant difference between the groups ($p > 0.05$). Moreover, no differences were obtained for AI performance (pregnancy and farrowing rates, and litter size; $p > 0.05$) and offspring health (growth and blood analysis; $p > 0.05$). Conclusively, the presence of all ejaculate fractions within the seminal doses does not impair the reproductive performance, reporting important economic savings according to the economic model included here.

Keywords: porcine; sperm conservation; ejaculate portions; sperm function; reproduction; litter performance

Citation: Luongo, C.; Llamas-López, P.J.; Hernández-Caravaca, I.; Matás, C.; García-Vázquez, F.A. Should All Fractions of the Boar Ejaculate Be Prepared for Insemination Rather Than Using the Sperm Rich Only? *Biology* **2022**, *11*, 210. https://doi.org/10.3390/biology11020210

Academic Editors: Pilar Santolaria Blasco, Rosaura Pérez-Pé and Jessica Rickard

Received: 23 December 2021
Accepted: 21 January 2022
Published: 28 January 2022

1. Introduction

The porcine reproduction industry uses artificial insemination (AI) as the method to achieve fertilization. This fact implies the preparation of seminal doses from ejaculates of selected boars, which involves different steps, such as ejaculate collection, semen dilution, sperm quality control, packaging, distribution, and storage [1]. Today, modern breeding requires semen production in AI specialized centers, which improves the efficiency and accuracy of swine reproduction.

Boar ejaculate is characterized for its high volume (250–300 mL; [2]) and pulsatile ejaculation in well-differentiated fractions. The pre-sperm fraction is discarded because of a high degree of cell debris, urine, and smegma contamination. Then, the sperm-rich fraction is emitted, containing most of the spermatozoa of the ejaculates, which is well recognized by its creamy-white color. Then, the poor fraction is characterized by a lower number of sperm and high content of seminal plasma (watery aspect) [1,3]. Between the rich and poor fraction, there is a transition phase called the intermediate fraction in which the aspect is of a grayish color. Moreover, a gel fraction (tapioca) is expelled progressively during the intermediate fraction and is always discarded from ejaculate collection.

After obtaining the ejaculate, the process of preparing the seminal doses starts. There is a certain controversy in the way of preparing AI doses. Various ways of dilution rate/sperm concentration [4], semen management [5–7], semen conservation [8], or inclusion of seminal plasma [9] have been proposed. The rich fraction of the ejaculate is the base foundation of AI dose preparation. However, currently, boar studs are including semi-automatic ejaculate collectors [10–12] instead of the traditional manual gloved-hand method, where the entire ejaculate is collected. It is known that boar seminal plasma composition varies depending on the fraction [2,13,14], which influences sperm conservation. However, other factors in addition to seminal plasma composition could impact semen conservation, such as seminal plasma proportion [8,15] or sperm concentration [16,17]. These circumstances with the ongoing trend towards prolonged storage times and lower sperm numbers per semen dose [18,19] open, once again, the controversy of the detrimental effect of seminal plasma inclusion on semen conservation and further fertilization rates [8]. Moreover, the success of post-cervical AI under farm conditions triggered an increased number of insemination doses produced per male and, as a consequence, a reduction in the number of boars needed on the boar stud [20], which elevates the importance of maximizing the efficiency of each ejaculate.

When semen (comprising seminal plasma and spermatozoa) is deposited into the female genital tract, not only does a mere transport of the sperm towards the oocyte start, but an active response is also elicited, including interactions with cells (sperm, oocyte, epithelial cells, leukocytes, or dendritic cells), organs (female genital tract), and fluids (seminal plasma, uterine, and oviductal fluids) (reviewed by [21,22]). These responses and interactions induce changes in the female genital tract [21,23–26] but also have an impact on embryo development, gene expression [27,28], as well as offspring growth and metabolism [27]. Given the relevance of seminal plasma exposure on offspring in other species, it will be important to study the contributions of seminal plasma in the offspring of porcine species.

The present study aims to evaluate the effect of accumulative fractions of the ejaculate in seminal doses on in vitro sperm quality during conservation, in vivo reproductive performance after AI (fertility and prolificacy) in field conditions, and offspring analysis (growth and blood assay). Moreover, due to the importance of porcine production worldwide, an economic study has been included, taking into account the results of the study.

2. Materials and Methods

2.1. Ethics

All of the procedures for this study were approved by the Ethical Committee of the University of Murcia on 1 June 2020 (PID2019-106380RB-I00). Through the experiments, animals were handled carefully, avoiding any unnecessary stress. All the experiments were performed in accordance with relevant guidelines and regulations. The study was carried out in compliance with the ARRIVE guidelines (https://arriveguidelines.org/ (accessed on 1 March 2021)).

2.2. Boars and Semen Collection

The study was performed from March to September of 2021. A total of 6 fertility-proven boars (Pietrain German Genetics; 30.83 ± 2.63 months of age) showing well-differentiated ejaculate fractions were used for the experiment. Boars were housed in individual pens (according to the European Commission Directive for Pig Welfare) with sawdust in a commercial boar stud (Sergal Gestió Ramadera, Lleida, Spain). Temperature levels were controlled automatically by a climate control system, which maintained the temperature in the room between 18–22 °C. Boars had a restricted feeding regime according to their nutritional requirements. Water was available ad libitum.

A total of 51 ejaculates were collected in a pre-warmed thermal cup using the gloved-hand method. The type of ejaculate depending on the fraction/s included: (1) ejaculate containing only the rich fraction (one fraction—F1) (n = 17); (2) ejaculate containing rich fraction (F1) + intermediate fraction (two fractions—F2) (n = 17); (3) ejaculate containing rich fraction + intermediate fraction + poor fraction (three fractions—F3) (n = 17). The distribution of the ejaculate type and the number of males is shown in Table S1. For each extraction, different types of the ejaculate were collected, distinguished by sight: (F1) composed of the sperm-rich fraction of the ejaculate, characterized by a dense white color. The collection in this case ended with the transition to a less dense white color; (F2) included F1 and the transition fraction between rich and poor fractions, which consisted of a less dense white color of the ejaculate; (F3) included F2 and the poor fraction, characterized by a water-like liquid aspect. In any case, the pre-sperm phase of the ejaculate was discarded, and the gel fraction was removed using a filter. During the trial, semen collection was always carried out by the same technician. The following characteristics of the ejaculates were recorded: volume (mL), number of sperm ($\times 10^6$) per mL, number of sperm ($\times 10^9$) per ejaculate, and number of seminal doses per ejaculate (2000 $\times$ 10^6 sperm/60 mL). Sperm concentration was calculated using an automatic sperm analyzer (Androvision® Minitüb, Tiefenbach, Germany). Moreover, an estimation of the percentage of seminal plasma per dose was performed ((volume of ejaculate/n° of seminal doses per ejaculate) $\times$ 100/60 mL (volume of seminal dose)).

2.3. Seminal Dose Preparation and Conservation

Semen samples were diluted in AndroStar® Plus extender (Minitüb, Tiefenbach, Germany) until reaching a final concentration of ~33 $\times$ 10^6 sperm/mL. Semen was packaged in plastic bags (2000 $\times$ 10^6 sperm/60 mL) and color-labeled depending on the type of seminal doses for a better identification at the lab and farm (F1 was a white-colored label; F2 was a blue color; F3 was a pink color) (Figure 1A). Semen preparation was carried out by the same technician during the whole period of the trial. Finally, seminal doses were kept refrigerated until semen evaluation and AI (Figure 1A,B). The temperature inside the refrigerator was monitored (AKO group, Barcelona, Spain) every 15 min during the execution of the trial (from 1 March 2021 to 24 April 2021) (Figure 1C). The average temperature of the conservation was 16.72 ± 0.72 °C (mean ± SD).

2.4. Sperm Analysis

Sperm quality from seminal doses (the same doses used in the AI trial) was evaluated at day 3 of storage.

2.4.1. Motility Analysis by CASA

Spermatozoa motility and kinetic parameters were analyzed by the Computer Assisted Semen Analysis (CASA) by ISAS® software (PROiSER R+D S.L., Valencia, Spain) coupled to phase-contrast microscopy (negative-pH 10× objective; Leica DMR, Wetzlar, Germany) and a digital camera (Basler Vision, Ahrensburg, Germany). Each sample was warmed at 38 °C (heat block CH100, Biosan Laboratories, Inc., Warren, MI, USA) for 10 min before evaluation. Then, a 4 µL drop of the sample was placed in a prewarmed (38 °C) chamber (20 micron Spermtrack® chamber, Proiser R+D, SL; Paterna, Spain), and at least 3 fields per sample were recorded. CASA setting parameters used were 25 frames per second and particle size area between 10 and 80 mm^2. Spermatozoa were considered to be motile when there was an average path velocity (VAP) > 10 µm/s. Progressive motility was considered to exist when there was a straightness (STR) > 45%. The variables analyzed were total motility (%), progressive motility (%), curvilinear velocity (VCL, µm/s), average path velocity (VAP, µm/s), straight line velocity (VSL, µm/s), amplitude of lateral head displacement (ALH, µm), percentage linearity (LIN, ratio of VSL/VCL, %), percentage straightness (STR, ratio of VSL/VAP, %), percentage oscillation (WOB, %), and beat-cross frequency (BCF, Hz).

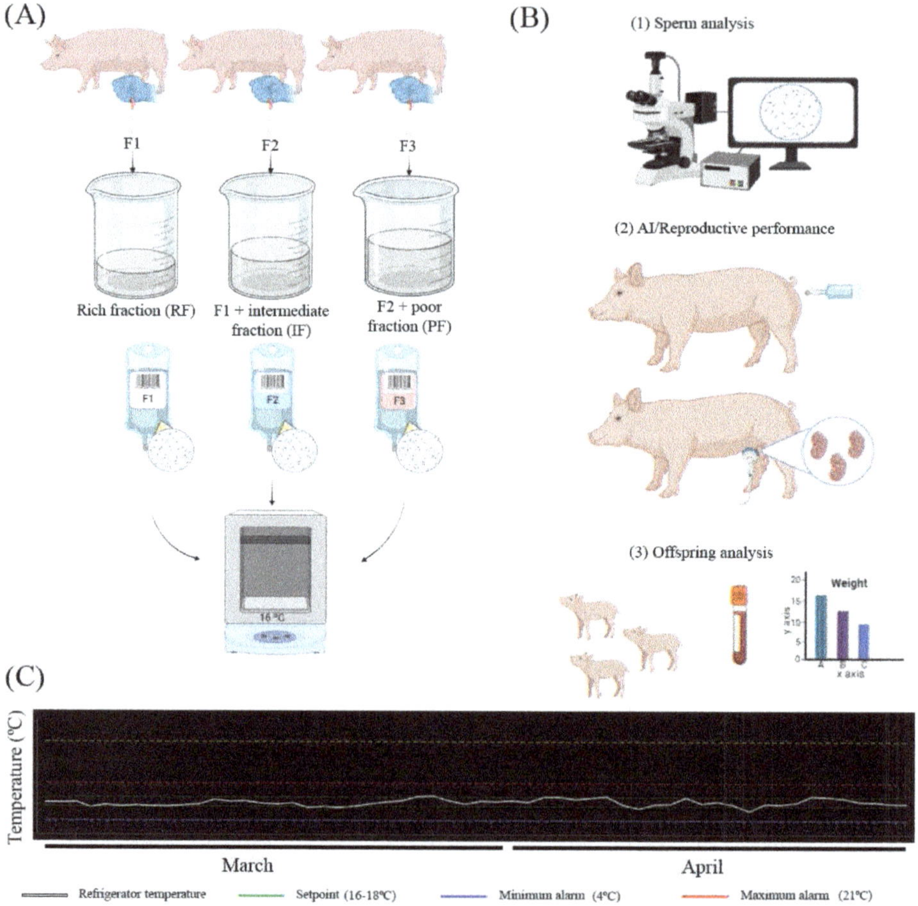

Figure 1. Scheme of the procedures carried out during the study. (**A**) The collection of the semen was performed according to the fractions of the ejaculate (F1 = rich fraction; F2 = F1 + intermediate fraction; F3 = F2 + poor fraction). The seminal doses were prepared for each type of ejaculate and kept at refrigeration until use. (**B**) On day 3 of conservation, the seminal doses were used for controlling the semen quality and for AI. The inseminated sows were diagnosed in pregnancy by ultrasound, and offspring was evaluated (growth and blood analysis). (**C**) Seminal doses were conserved in a temperature-controlled refrigerator during the whole period of the trial (March and April 2021). Images (**A**,**B**) were created on Biorender.com (accessed on 1 March 2021).

2.4.2. Viability Assay

For spermatozoa viability, the staining solution was prepared with 50 µL of propidium iodide 500 µg/mL (P-4170 Sigma-Aldrich®, Madrid, Spain) in 10 mL of PBS without calcium and magnesium (Sigma-Aldrich®, Madrid, Spain). Spermatozoa samples were incubated with propidium iodide solution for 10 min at room temperature in the dark. For the evaluation, spermatozoa were observed in transmitted light brightfield and fluorescence microscopy (Leica® DM4000 Led, Wetzlar, Germany, 495/520 nm) and were classified as live (without fluorescence) or dead (red fluorescence). At least 200 cells per sample were counted.

2.4.3. Acrosome Status

For acrosome status, the staining solution was prepared with 100 µL of Arachis hypogea lectin PNA-FITC 200 µg/mL (Sigma-Aldrich®, Madrid, Spain) in 10 mL of PBS without calcium and magnesium. Spermatozoa samples were incubated with PNA-FITC solution for 10 min at room temperature in the dark. For the evaluation, spermatozoa were observed in transmitted light brightfield and fluorescence microscopy (Leica® DM4000 Led, Wetzlar, Germany, 495/520 nm) and were classified as sperm with intact acrosome (without fluorescence) or with damaged acrosome (green fluorescence). At least 200 sperm per sample were counted.

2.4.4. Mitochondrial Activity

For mitochondrial activity, the staining solution was prepared with 10 µL of JC-1 0.017 µg/mL (5,5′,6,6′-tetrachloro-1,1′,3,3′-tetraethylbenzimidazolylcarbocyanine iodide; ThermoFisher Scientific Inc., Waltham, MA, USA) in 10 mL of PBS without calcium and magnesium. Spermatozoa samples were incubated with JC-1 solution for 30 min at 38 °C in the dark. For the evaluation, spermatozoa were observed under fluorescence microscopy (Leica® DM4000 Led, Wetzlar, Germany, 495/520 nm) and were classified as sperm with high mitochondrial membrane potential (orange fluorescence) or sperm with low mitochondrial membrane potential (green fluorescence). At least 200 sperm per sample were counted.

2.4.5. DNA Fragmentation

For DNA fragmentation, a Halomax kit for Sus scrofa (Halotech DNA, Madrid, Spain) was used following the manufacturer's instructions. Agarose was warmed at 100 °C for 5 min and then transferred to 37 °C for 5 min to equilibrate the temperature. Once agarose reached 37 °C, sperm samples were added to the vials containing agarose (1:2, *v*/*v*) and mixed thoroughly. Then, a 2 µL drop of this suspension was placed onto the slide, covered with a glass coverslip, and left at 4 °C for 10 min to solidify. The coverslip was removed, and the samples were treated with first lysis solution for 5 min and then distilled water for 5 min. Finally, slides were dehydrated with sequential 70 and 100% ethanol and stained with red fluorescent stain (HT-RF S100, Fluored®, Halotech DNA, Madrid, Spain). For the evaluation, spermatozoa were observed under fluorescence microscopy (Leica® DM4000 Led, Wetzlar, Germany, 495/520 nm) and were classified as sperm with unfragmented DNA (small and compact halo of chromatin dispersion) or with fragmented DNA (large and spotty halo of chromatin dispersion). At least 200 sperm per sample were counted.

2.5. Estrus Detection and Artificial Insemination (AI)

A total of 174 crossbred sows [Large-White X Landrace, Danbred genetic] from a commercial sow farm were used for the study (Genera S.L., Lorca, Spain). At weaning, sows were selected by parity (from 3 to 5; mean parity of 3.74) and body conditions before being randomly assigned to one of the treatment groups (F1, F2, or F3). The body conditions of the sows were assessed at the onset of the estrus visually (BC score 1 to 5; 1 was extremely thin, 5 was extremely fat) and by means of back-fat (BF) and loin depth (LD) measurements. The back-fat and loin thickness of the sows were measured at the P2 point (6.5 cm from the middle line of the last rib) using real-time ultrasound scan equipment with a linear probe (SF1 wireless BF and LD scanner, Sonivet, Beijing, China). A total of 4 measurements per animal were carried out (two on the right side and two on the left side) for the two parameters (BF and LD). After weaning, sows were housed in individual gestation crates with ad libitum access to water and 4.0 kg feed/day until the first AI.

Estrus detection was performed once a day in the presence of a sexually mature boar starting on the day of weaning. The weaning-to-estrus interval (WEI) was 4.28 days on average. AI was performed as previously described [20]. Briefly, sows were inseminated by a post-cervical AI method at estrus onset and every 24 h during the standing reflex period (average of 2.55 ± 0.53 inseminations per animal), using a combined catheter-cannula kit (Soft & Quick®, Tecno-Vet, S.L., Barcelona, Spain). Each sow was inseminated by the same boar (seminal doses were stored). The seminal doses were composed of 2000×10^6 sperm in 60 mL. The AI procedure was performed in individual crates. From the first AI until day 25 of gestation, sows were fed daily with 2.0–3.5 kg (depending on the initial BC score) of the same gestation feed. From day 25 until the entry to farrowing, all sows received 2.0 kg/day. From the day of farrowing, sows were fed with 1.0 kg/day, increasing the ration by 1 kg/day until maximum (8 kg/day).

2.6. Return to Estrus and Pregnancy Diagnosis

Fertility parameters were monitored with return-to-estrus starting at day 18 after the first insemination by boar exposure, while an experienced worker applied back pressure in search of a standing reflex. Sows showing estrus signs were considered non-pregnant. Moreover, pregnancy was detected by ultrasound 23–28 days after insemination by transabdominal ultrasonography (Echoscan T-300 S, Barcelona, Spain). Pregnant sows were then housed in pens with 8–10 sows/pen.

2.7. Farrowing and Litter Performance

Pregnant sows were moved from gestation facilities to the farrowing crates at 110 days of gestation. At farrowing, the following reproductive parameters were recorded: farrowing (%), gestation length (days), the total number of piglets born, and the number of piglets born alive. Moreover, the fecundity index (total number of live piglets born per 100 inseminations) was calculated (farrowing rate multiplied by the number of live piglets born per litter).

2.8. Offspring Growth Parameters

Body weight (kg) was evaluated at day 1 and 21 of life (weaning) using a precision scale (ZMISSIL F1-30). Moreover, daily weight gain (DWG, kg) was calculated as follows: weight at day 21—weight at day 1/days from first to second weight measured (21 days).

2.9. Blood Collection and Analysis

A total of 81 piglets (41 females and 40 males) from 3 experimental groups were randomly selected and used for blood analysis. Blood samples (about 4–5 mL) were collected (7 days after piglet born) by venipuncture of the jugular vein using a Vacutainer system (BD Vacutainer® 21G 0.8 × 25 mm needle; BD Vacutainer), including lithium heparin tubes. After collection, blood samples were transported to the lab within 1 h in a porexpan box and kept at refrigeration (4 °C) until analysis (within 18 h after collection). The hematological analysis was performed using analyzer equipment (Siemens ADVIA® 120, Holliston, MA, USA), while the biochemical serum parameters were evaluated by the Olympus AU600 and Mindray BS-200E analyzers.

The hematological parameters analyzed were hematocrit (HCT, %), concentration of erythrocytes (RBC, $\times 10^6$ cells/µL), hemoglobin (HB, g/dL), mean corpuscular volume (MCV, fL), mean corpuscular hemoglobin (MCH, pg), mean corpuscular hemoglobin concentration (MCHC, g/dL), cell hemoglobin concentration mean (CHCM, g/dL), red cell volume (RDW, %), cellular hemoglobin content (CH, pg), cellular hemoglobin distribution width (CHDW, g/dL), hemoglobin concentration distribution width (HDW, g/dL), white blood cells (WBC, $\times 10^3$ cells/µL), neutrophils (NEU, % and $\times 10^3$ cells/µL), lymphocytes (LYM, % and $\times 10^3$ cells/µL), monocytes (MON, % and $\times 10^3$ cells/µL), eosinophils (EOS, % and $\times 10^3$ cells/µL), basophils (BAS, % and $\times 10^3$ cells/µL), platelet indices (platelet (PLT, $\times 10^3$ cells/µL), platelet crit (PCT, %), mean platelet volume (MPV, fL), platelet distribution width (PDW, %), mean platelet volume component concentration (MPC, g/dL), platelet component distribution width (PCDW, g/dL), mean platelet mass (MPM, pg), platelet mass distribution width (PMDW, pg), and large platelets (large PLT, $\times 10^3$ cells/µL)) as well as reticulocyte indices (reticulocytes (RET, % and $\times 10^6$ cells/µL), average size of reticulocytes (MCVr, fL), and average hemoglobin content of reticulocytes (CHr, pg)).

The biochemical serum parameters analyzed were proteins (PRO, g/dL), albumin (ALB, g/dL), globulin (GLO, g/dL), creatinine (CR, mg/dL), urea (URE, mg/dL), glucose (GLU, mg/dL), cholesterol (CHOL, mg/dL), triglycerides (TRI, mg/dL), lipase (LIP, UI/L), creatine kinase (CK, UI/L), alkaline phosphatase (ALP, UI/L), gamma-glutamyl transferase (GGT, UI/L), aspartate aminotransferase (AST, UI/L), alanine aminotransferase (ALT, UI/L), total bilirubin (TBIL, mg/dL), calcium (Ca, mg/dL), potassium (K, mmol/L), sodium (Na, mmol/L), and chlorine (Cl, mmol/L).

2.10. Statistical Analysis

Statistical analyses were performed with the SPSS 24.0 software package (IBM SPSS Inc., Chicago, IL, USA). Ejaculate characteristics and semen quality parameters were analyzed for normality by a Kolmogorov–Smirnov test, which showed that all parameters had a normal distribution, except the percentage of seminal plasma per dose. A one-way ANOVA test followed by a post hoc Tukey test was applied. For the variable, which data were not normally distributed, the non-parametric Kruskal–Wallis test was used. Data are represented as the mean ± SD (standard deviation) for ejaculate parameters and means ± SEM (standard error of the mean) for sperm parameters. Differences were considered significant when $p < 0.05$.

For body condition, parity, and weaning-to-estrus interval, the non-parametric Kruskal–Wallis test was used. Regarding BF, LD, number of inseminations, gestation period, total and live-born piglets, fecundity index, weight at day 1 and day 21, and DWG, the assumption of normality was evaluated by a Kolmogorov–Smirnov test. All these variables were not normally distributed, and the non-parametric Kruskal–Wallis test was used. Concerning pregnancy and farrowing rates, a Chi-square test was used for comparison between experimental groups. The results obtained are presented as the mean ± SD, and differences were considered statistically significant when $p < 0.05$.

For blood parameters, the assumption of normality was evaluated by a Kolmogorov–Smirnov test. When normality was fulfilled (RBC, MCH, CHCM, CHDW, WBC, NEU%, LYM, MPC, PCDW, PMDW, RET, MCVr, CHr, PROT, ALB, CR, GLU, CHOL, TRI, AMI, GGT, Ca, K), a one-way ANOVA test followed by a post hoc Tukey test was applied. For those variables which data were not normally distributed, the non-parametric Kruskal–Wallis test was used. The results are represented as the mean ± SD. Values were considered significantly different when $p < 0.05$.

3. Results

The results of the ejaculate characteristics are depicted in Table 1, showing significant differences between groups in all of the parameters studied. The number of sperm per mL was significantly higher in the F1 group in comparison with F2 and F3, with F2 being statistically greater than F3. The total number of sperm per ejaculate and the number of seminal doses prepared per each type of ejaculate were statistically greater in F3 than F1, while F2 was similar to both. Having an estimation of six ejaculate collections per boar/month (data provided by a commercial boar stud), the use of F3 had an increase of seminal dose production of 24.16% per month compared to F1, while the use of F2 supposed an increase of 11.18%. After seminal dose preparation, semen was stored for

3 days before in vitro evaluation. Sperm quality did not show significant differences between the experimental groups (Table 2). All the parameters evaluated remained on a high level throughout 3 days of storage (e.g., total motility ranged from 89.11 ± 0.79% to 90.89 ± 0.92%; mitochondrial activity from 91.79 ± 0.34% to 92.37 ± 0.38%).

Table 1. Characteristics of the ejaculates from 6 boars used in the experiment. Data are represented as the mean ± SD (standard deviation).. The number of ejaculates evaluated per group (F1, F2, F3) is indicated between brackets.

	F1 (*n* = 17)	F2 (*n* = 17)	F3 (*n* = 17)	*p*-Value
Volume (mL)	102.35 ± 33.12 a	157.35 ± 31.87 b	308.23 ± 126.36 c	<0.0001
N° of Sperm/mL ($\times 10^6$)	540.94 ± 153.68 a	384.58 ± 119.29 b	241.05 ± 96.94 c	<0.0001
N° of Sperm/ejaculate ($\times 10^9$)	52.08 ± 11.20 a	57.90 ± 13.03 ab	64.67 ± 11.19 b	0.012
N° of seminal doses/ejaculate	26.04 ± 5.60 a	28.95 ± 6.51 ab	32.33 ± 5.59 b	0.012
Seminal plasma per dose (%) *	7.91 ± 3.76 a	9.15 ± 2.92 ab	15.19 ± 7.42 b	<0.0001

* The percentage of seminal plasma per dose was estimated as follows: (volume of ejaculate/n° of seminal doses per ejaculate) × 100/60 mL (volume of seminal dose). Different letters (a,b,c) in the same row indicate a significant difference between experimental groups ($p < 0.05$)

Table 2. Spermatozoa quality parameters from different accumulative ejaculate fractions (F1, F2, F3) analyzed after 3 days of storage at ~16 °C. Data are expressed as the mean ± SEM (standard error of the mean).

	F1 (*n* = 17)	F2 (*n* = 17)	F3 (*n* = 17)	*p*-Value
Total motility (%)	90.89 ± 0.92	90.32 ± 0.68	89.11 ± 0.79	0.65
Progressive motility (%)	37.53 ± 1.27	35.42 ± 1.89	39.16 ± 1.41	0.61
VCL (µm/s)	61.68 ± 2.72	59.26 ± 2.65	55.00 ± 1.54	0.50
VSL (µm/s)	22.32 ± 0.84	21.53 ± 0.85	23.00 ± 0.73	0.76
VAP (µm/s)	39.32 ± 1.50	38.32 ± 1.06	37.53 ± 0.93	0.82
ALH (µm)	1.95 ± 0.09	2.00 ± 0.06	1.89 ± 0.04	0.82
LIN (%)	38.26 ± 1.69	38.53 ± 1.75	42.53 ± 1.34	0.48
STR (%)	58.16 ± 1.62	57.00 ± 1.83	61.79 ± 1.51	0.47
WOB (%)	64.63 ± 1.25	66.32 ± 1.19	68.68 ± 0.97	0.35
BCF (Hz)	7.26 ± 0.15	6.79 ± 0.14	6.79 ± 0.11	0.24
Viability (%)	93.05 ± 0.25	92.89 ± 0.30	92.11 ± 0.27	0.32
Acrosome integrity (%)	95.16 ± 0.19	94.74 ± 0.23	94.32 ± 0.21	0.27
Mitochondrial activity (%)	92.37 ± 0.34	92.37 ± 0.38	91.79 ± 0.34	0.74
DNA fragmentation (%)	0.42 ± 0.09	1.11 ± 0.28	0.68 ± 0.15	0.33

Inseminated sows showed similar body parameters (BD, BF, LD) in all three experimental groups (Table 3). Moreover, parity and weaning-to-estrus interval (which ranged from 3.72 ± 0.74 to 3.76 ± 0.76 and from 4.21 ± 0.99 to 4.33 ± 0.98 days, respectively) showed no significant differences between the groups (Table 3). A total of 58 sows were inseminated per group, with no significant differences found between them, both in terms of gestation length (which ranged from 115.82 ± 1.09 to 116.02 ± 1.39 days) as well as pregnancy rate and farrowing rate (which ranged from 92.98% to 96.55% and from 82.46% to 89.66%, respectively) (Table 4). Additionally, in analyzing total and live-born piglets, no significant differences were observed, comparing the different types of seminal doses used (Table 4). When the fecundity index was calculated, it was similar between the groups, with a range from 1517.32 ± 466.68 to 1589.60 ± 320.15 piglets born alive. The results concerning the weight of piglets at days 1 and 21 after birth, and DWG are shown in Table 5. The weight at days 1 and 21 did not show significant differences between the groups, as well as the DWG, which ranged from 0.160 ± 0.059 kg to 0.167 ± 0.052 kg. Furthermore, the blood parameters evaluated in the offspring, including hematological and biochemical analyses, were statistically similar between the experimental groups (Tables S2–S5).

Table 3. Body condition score, back-fat thickness, loin depth, parity, and weaning-to-estrus interval (mean ± standard deviation) in sows from three experimental groups (F1, F2, F3).

Experimental Group	Sows (*n*)	Body Condition Score (1–5)	Back-Fat (mm)	Loin Depth (mm)	Parity	Weaning-to-Estrus Interval (Days)
F1	58	2.67 ± 0.66	11.33 ± 3.89	45.19 ± 7.12	3.76 ± 0.76	4.33 ± 0.98
F2	58	2.67 ± 0.66	11.24 ± 3.68	45.42 ± 7.78	3.74 ± 0.74	4.21 ± 0.99
F3	58	2.62 ± 0.59	11.57 ± 3.98	45.76 ± 7.04	3.72 ± 0.74	4.31 ± 0.90

Table 4. Pregnancy rate (%), farrowing rate (%), total litter size, live-born piglets, and fecundity index (mean ± standard deviation) in inseminated sows from three experimental groups (F1, F2, F3).

Experimental Group	Sows (*n*)	Number of Inseminations per Sow	Pregnancy Rate (%)	Farrowing Rate (%)	Total Born Piglets (*n*)	Live-Born Piglets (*n*)	Fecundity Index * (*n*)
F1	58	2.50 ± 0.50	92.98	82.46	22.57 ± 4.73	18.45 ± 4.81	1521.12 ± 396.89
F2	58	2.62 ± 0.52	96.55	89.66	20.50 ± 6.50	16.92 ± 5.20	1517.32 ± 466.68
F3	58	2.53 ± 0.57	96.55	84.48	21.86 ± 4.28	18.82 ± 3.79	1589.60 ± 320.15

* Fecundity index was calculated as follows: farrowing rate multiplied by the number of live-born Piglets per litter.

Table 5. Weight at day 1 (kg), weight at day 21 (kg), and daily weight gain (DWG, kg) of piglets derived from inseminated sows from three experimental groups (F1, F2, F3). Data are represented as the mean ± SD (standard deviation).

Experimental Group	Number of Piglets (Male/Female) (*n*)	Weight at Day 1 (kg)	Weight at Day 21 (kg)	DWG * (kg)
F1	611 (273/238)	1.363 ± 0.348	4.916 ± 1.200	0.167 ± 0.052
F2	672 (299/289)	1.392 ± 0.340	4.792 ± 1.279	0.160 ± 0.059
F3	626 (278/282)	1.418 ± 0.349	4.837 ± 1.275	0.162 ± 0.056

* DWG was calculated as follows: weight at day 21—weight at day 1/days from first to second weight measured (21 days).

Moreover, an economic study was performed, comparing the three experimental groups. This study showed lower costs for seminal doses prepared with F3 (3.18 €) compared to F1 and F2 (3.91 and 3.54 €, respectively) (Table 6). Additionally, F3 showed a seminal dose cost reduction per inseminated sow and piglet born alive compared to F1 and F2 (Table 7).

Table 6. Calculation of a seminal dose cost depending on the ejaculate fraction/s included (F1 vs. F2 vs. F3).

	F1	F2	F3
Fixed costs (€) [1]	3.172	2.853	2.555
Variable costs (€) [2]	0.315	0.284	0.254
Consumable costs (€) [3]	0.425	0.403	0.371
Dose packaging (€)	0.096	0.096	0.096
Extender (€)	0.076	0.075	0.067
Osmotized water (€)	0.050	0.049	0.044
PCR (€)	0.203	0.182	0.163
Seminal dose cost (60 mL) (€) [4]	3.91	3.54	3.18

Table 6. *Cont.*

	F1	F2	F3
Costs difference (%)	0.00	−9.53	−20.77

[1] Fixed costs include the workers'salary, energy, amortization facilities, boars, and others. Fixed costs of a boar per month are estimated at 495 €. Fixed cost per ejaculate fraction was calculated as: number of seminal doses per boar and month/cost of a boar per month (495 €). The number of seminal doses/boar/month was calculated as: n° of seminal doses per ejaculate (data included in Table 2) × 6 (number of collections per month). Data based on a real boar stud. [2] Variable costs: feed, medication, and sawdust. Variable costs of a boar per month are estimated at 49.26 €. Variable costs per ejaculate fraction were calculated as: number of seminal doses per boar and month/cost of a boar per month (49.26 €). The number of seminal doses/boar/month was calculated as: n° of seminal doses per ejaculate (data included in Table 2) × 6 (number of collections per month). Data based on a real boar stud. [3] Calculated as: dose packaging + extender + osmotized water + PCR (for detection of PRRS virus). [4] Calculated as: Fixed costs + variable costs + consumable cost.

Table 7. Economic comparison between ejaculate fraction/s included in the seminal dose (F1 vs. F2 vs. F3) in terms of cost reduction per AI and piglet born alive.

	F1	F2	F3
Number of sows inseminated	100	100	100
Number of AI/sow per estrus [1]	2.55	2.55	2.55
Seminal dose cost (€) [2]	3.91	3.54	3.18
Seminal dose cost/100 inseminated sows (€) [3]	997.05	902.70	810.90
Piglets born alive/100 inseminated sows (€) [4]	1521	1517	1589
Seminal dose cost reduction/100 inseminated sows (€) [#]	0	94.35 [†]	186.15 [††]
Cost reduction/inseminated sow (€) [#]	0	0.94 [¥]	1.86 [¥¥]
Cost reduction/piglet born alive (€) [#]	0	0.06 *	0.12 **

[1] Average of AIs per sow performed in our study. [2] Calculated in Table 6. [3] Calculated as: number of AIs (100) × number of AI/sow (2.55) × seminal dose cost. [4] Data collected from our study (Table 4). [#] Calculated based on F1 as the seminal dose type reference. [†] Calculated as: seminal dose cost per 100 inseminated sows using F1—seminal dose cost per 100 inseminated sows using F3. [††] Calculated as: seminal dose cost per 100 inseminated sows using F1—seminal dose cost per 100 inseminated sows using F2. [¥] Calculated as: seminal dose cost reduction per 100 inseminated sows using F2/100. [¥¥] Calculated as: seminal dose cost reduction per 100 inseminated sows using F3/100. * Calculated as: seminal dose cost reduction per 100 inseminated sows using F2/piglets born alive per 100 inseminated sows using F2. ** Calculated as: seminal dose cost reduction per 100 inseminated sows using F3/piglets born alive per 100 inseminated sows using F3.

4. Discussion

Nowadays, the preparation of seminal doses is a crucial step in the swine industry since AI is the dominant form of breeding. However, some aspects of its management are still controversial. Previous reports have been focused on the study of the boar ejaculate fractions per separate, claiming a negative impact of the poor fraction inclusion on sperm features. However, the synergy of different boar ejaculate fractions has yet to be elucidated. The present study indicates that under our conditions, the inclusion of the total fractions of the ejaculate in seminal doses does not impair sperm conservation, fertility, prolificacy, nor offspring performance.

Commonly, the rich fraction of the ejaculate, but not the poor fraction, is collected and processed for seminal doses. This process was established because the inclusion of the poor fraction to the seminal doses had a controversial effect during sperm conservation [2,20], suggesting negative impact upon sperm characteristics. However, the incorporation of the semi-automatic ejaculate collection, where the poor fraction is also collected, has again opened this debate. Our results have demonstrated that the inclusion of the total fractions of the ejaculate (rich, intermediate, and poor) did not affect the sperm quality during conservation at 16 °C. With the inclusion of the total ejaculate, the percentage of seminal plasma in the seminal doses increased in comparison with the single use of the rich fraction (~15% vs. ~8%; Table 1). Previous studies have reported a deleterious effect on sperm quality when a high proportion of seminal plasma is present during sperm conservation [9,15]. The inclusion of 10% of bulk seminal plasma reduced the total sperm motility in comparison with a lower level (0.5 and 5%) from the third day of conservation onwards [9]. This result contrasts ours, where a high sperm quality (~90% of total motility) was observed when a high proportion of seminal plasma was present (which ranged from ~8 in F1 to ~15% in F3), although no comparison has been established using lower levels of seminal plasma. Moreover, the sperm concentration used in both studies was different (18×10^6 vs. 30×10^6 sperm/mL), with the diluted semen being more susceptible to the putative adverse effect of seminal plasma [15]. The variations found between the studies are not surprising, considering that many factors can affect the outcome of semen processing and composition. The semen quality after medium–long term storage of sperm should not reduce the fraction of ejaculate or the proportion of seminal plasma used, as there are other factors that play an important role, such as semen processing [29,30] or the type of extender used [8,9,31]. The extender used in our study is recommended for medium-term conservation, which can aid in conserving the sperm quality when different seminal plasma proportions and fraction origin are used. Furthermore, the seminal plasma proteins not only vary between fractions of the ejaculates [13,14], but there are also inter-breed [32] and inter-male variations [33]. In our case, no male effect was observed. It could be that such variations may have been mitigated by the type of extender used, as previously reported by other authors [9], or there is also the chance that the six boars included in the study present a similar performance.

Although no effect on sperm features has been observed during storage using different accumulative ejaculate fractions, the situation in vivo is far from the in vitro analysis. It is known that seminal plasma influences the transition of the sperm through the female genital tract upon deposition [34] and interacts with the female reproductive fluids [22]. Particularly, seminal plasma proteins are involved during these interactions. For instance, AWN spermadhesins and PSP-I/PSP-II heterodimer, mostly abundant in the sperm-rich and poor fraction respectively, adhere to the sperm surface [13]. These proteins are involved during the interaction between sperm and uterine epithelium and fluid, as well as during sperm–zona pellucida binding [22,35]. However, PSPs seem to have a controversial effect, inducing a higher migration of leukocytes within the female reproductive tract [36]. For this reason, the poor fraction is supposed to exert a negative effect on the fertility outcome with respect to the sperm-rich fraction [37]. However, our data showed that the poor fraction combined with other ejaculate fractions (F3) leads to similar fertility outcomes compared with F1 and F2. Thus, seminal plasma volume and composition used seem to have neither a beneficial nor a harmful effect on pregnancy and farrowing rate, probably because this effect may be mitigated by the simultaneous presence of each ejaculate fraction. Nevertheless, there are controversial studies where the fertility rate and the number of embryos were higher in the presence or absence of seminal plasma [17,38]. Based on our findings, they do not mean that the seminal plasma is not acting in some way in sperm modulation during their transit towards the oocyte because the total absence of seminal plasma is not included in our experimental groups. However, they may indicate that seminal plasma present in F1, F2, and F3 is acting similarly for pregnancy and prolificacy. In fact, the infusion of seminal plasma from the entire ejaculate before insemination resulted in a higher number of collected embryos compared to the infusion of phosphate buffer saline as a control [39]. It is known that seminal plasma stimulation of the female tract is not essential for pregnancy success in a pig, but in animals showing poor reproductive performance or less optimal breeding, this fluid can improve reproductive outcomes (reviewed by [21]). Thus, further studies directed to compare F1, F2, and F3 within sows with low reproductive features could elucidate whether the seminal plasma fractions could impact the pregnancy and the number of piglets born.

Likewise, the seminal plasma according to the fractions present in the seminal doses did not influence the offspring's health. However, evidence suggesting that paternal factors (e.g., seminal plasma) may influence the offspring are emerging [21,27]. This is relevant in assisted reproduction techniques, such as porcine AI, where semen is diluted, reducing the components of this fluid; the addition of extra seminal plasma could improve the outcomes. Although studied in other species, the impact of female responses to seminal plasma in the offspring's health has not been evaluated in porcine yet. In our case, the type of seminal doses not only differed between them in protein and metabolite composition (as indicated by other studies; [13,14,40]), but also in the percentage of plasma. However, when the piglets born were evaluated for growth and blood analyses, no differences were found between the three experimental groups, indicating a possible lack of influence by the composition and/or percentage of seminal plasma present in the semen doses used. Therefore, we have tested the effect of the seminal plasma depending on the fractions collected from a practical point of view for the swine industry because ejaculates are not devoid of seminal plasma during processing.

Our results indicate that the inclusion of the entire ejaculate in the seminal doses has no detrimental effect in the different stages of swine production, such as sperm conservation, AI, and offspring health. This implies not only benefits from the collection method as mentioned, but also has positive implications on/in the environment and from an economic point of view. At the time of AI, seminal doses are deposited within the female reproductive tract, and a phenomenon consisting of semen loss through the vulva, named backflow, occurs [41,42]. During the semen backflow, part of the extender, which includes antibiotics, reaches the liquid manure [43]. Using the F3 for seminal dose production includes the addition of less extender and, in consequence, fewer antibiotics. Thus, fewer antibiotics will reach the purine through the backflow, helping to reduce the antimicrobial resistance. Antimicrobial resistance is among the most serious public health threats of the 21st century, with a great impact in terms of One Health (reviewed by [44]). In our seminal doses coming from F3, no bacterial growth was observed at day 3 of storage (data not shown). This fact opens the possibility of reducing the amount of antibiotic used because less extender is needed to process these seminal doses. Furthermore, the protein deoxyribonuclease-2-alpha, an acid endonuclease secreted by male accessory glands, is more highly expressed in the poor ejaculate fraction than in previous fractions [13], providing a bactericide activity protecting sperm in the transit along the female genital tract [45]. On the other hand, the economic impact of incorporating the bulk ejaculate (F3) versus other initial fractions (F1 or F2) on seminal doses has been evaluated, considering the results obtained in our study (Tables 6 and 7). The use of F3 in seminal doses would ensure important savings. Taking into account fixed, variable, and consumable costs (Table 6), the inclusion of F3 leads to a cost reduction compared to F1 and F2. Our analysis reveals that a dose's cost of F3 could save 0.73 € and 0.36 € in comparison with F1 and F2, respectively. Moreover, obtaining the bulk ejaculate increases the number of sperm collected and, in consequence, the number of doses and therefore the cost of doses processing per collection. This leads to a more efficient use of the boars, increasing the returns on investment, utilization efficiency, and profitability of the system by reducing the boar cost. Even more, the estimation savings per inseminated sow (considering 2.55 AIs per estrus) using F3 compared to F1 are 1.86 € (Table 7). Extrapolating to the piglet born alive (considering the fecundity index obtained for each group) involves a reduced cost of 0.06 € and 0.12 € per piglet for F2 and F3 compared with F1.

5. Conclusions

In conclusion, considering all the information reported here, our findings demonstrate that the inclusion of seminal plasma from the entire ejaculate in seminal doses has similar outcomes in terms of sperm conservation, reproductive performance, and piglet development than when only the rich fraction of the ejaculate is used. Moreover, an efficient use of the boar ejaculate increases the return investment, reduces the use of antibiotics, and reduces the cost per insemination and piglet production, which is of great importance for the swine industry.

Supplementary Materials: The following are available online at https://www.mdpi.com/article/10.3390/biology11020210/s1, Table S1: Number and type of ejaculates used per male in the study; Table S2: Red blood cells and reticulocyte parameters analyzed in three groups of piglets derived from AI (F1, F2, F3). Data are represented as the mean ± SD (standard deviation); Table S3: White blood cell parameters analyzed in the three groups of piglets derived from AI (F1, F2, F3). Data are represented as the mean ± SD (standard deviation); Table S4: Platelet parameters analyzed in the three groups of piglets derived from AI (F1, F2, F3). Data are represented as the mean ± SD (standard deviation); Table S5: Biochemical serum parameters analyzed in the three groups of piglets derived from AI (F1, F2, F3). Data are represented as the mean ± SD (standard deviation).

Author Contributions: Conceptualization, F.A.G.-V.; Data curation, C.L., P.J.L.-L., I.H.-C. and F.A.G.-V.; formal analysis, C.L. and F.A.G.-V.; funding acquisition, C.M. and F.A.G.-V.; investigation, C.L., P.J.L.-L., I.H.-C., C.M. and F.A.G.-V.; methodology, C.L., P.J.L.-L., I.H.-C. and F.A.G.-V.; supervision, F.A.G.-V.; validation, F.A.G.-V.; writing—original draft, C.L. and F.A.G.-V.; writing—review and editing, F.A.G.-V. All authors have read and agreed to the published version of the manuscript.

Funding: This research was supported by the Ministry of Science and Innovation (PID2019-106380RB-I00/AEI/10.13039/501100011033).

Institutional Review Board Statement: All the procedures carried out in this work were approved by the Ethical Committee of Animal Experimentation of the University of Murcia and by the Animal Production Service of the Agriculture Department of the Region of Murcia (Spain). Through the experiments, animals were handled carefully, avoiding any unnecessary stress. All experiments were performed in accordance with the relevant guidelines and regulations. The study was carried out in compliance with the ARRIVE guidelines (http://arriveguidelines.org (accessed on 1 March 2021)).

Informed Consent Statement: Not applicable.

Data Availability Statement: All data generated or analyzed during this study are included in the published paper.

Acknowledgments: We are grateful to the staff of Sergal Gestió Ramadera and Genera S.L. for their assistance in the successful development of the study. We would also like to thank Jordana S. Lopes and Paulina Grudzinska for English revision of the manuscript.

Conflicts of Interest: The authors declare no conflict of interest.

References

1. Lopez Rodriguez, A.; Van Soom, A.; Arsenakis, I.; Maes, D. Boar management and semen handling factors affect the quality of boar extended semen. *Porc. Health Manag.* **2017**, *3*, 15. [CrossRef] [PubMed]
2. Rodríguez-Martínez, H.; Kvist, U.; Saravia, F.; Wallgren, M.; Johannisson, A.; Sanz, L.; Peña, F.J.; Martínez, E.A.; Roca, J.; Vázquez, J.M.; et al. The physiological roles of the boar ejaculate. *Soc. Reprod. Fertil. Suppl.* **2009**, *66*, 1–21. [CrossRef] [PubMed]
3. Garner, D.L.; Hafez, E.S.E. Spermatozoa and seminal plasma. In *Reproduction in Farm Animals*; Hafez, B., Hafez, E.S.E., Eds.; Lippincott Williams & Wilkin: Baltimore, MD, USA, 2000; pp. 96–109. ISBN 9781119265306.
4. Schulze, M.; Ammon, C.; Schaefer, J.; Luther, A.M.; Jung, M.; Waberski, D. Impact of different dilution techniques on boar sperm quality and sperm distribution of the extended ejaculate. *Anim. Reprod. Sci.* **2017**, *182*, 138–145. [CrossRef] [PubMed]
5. Schulze, M.; Henning, H.; Rüdiger, K.; Wallner, U.; Waberski, D. Temperature management during semen processing: Impact on boar sperm quality under laboratory and field conditions. *Theriogenology* **2013**, *80*, 990–998. [CrossRef] [PubMed]
6. Schulze, M.; Rüdiger, K.; Waberski, D. Rotation of Boar Semen Doses During Storage Affects Sperm Quality. *Reprod. Domest. Anim.* **2015**, *50*, 684–687. [CrossRef] [PubMed]
7. Paschoal, A.F.; Luther, A.M.; Jakop, U.; Schulze, M.; Bortolozzo, F.P.; Waberski, D. Factors influencing the response of spermatozoa to agitation stress: Implications for transport of extended boar semen. *Theriogenology* **2021**, *175*, 54–60. [CrossRef]
8. Wiebke, M.; Hensel, B.; Nitsche-Melkus, E.; Jung, M.; Schulze, M. Cooled storage of semen from livestock animals (part I): Boar, bull, and stallion. *Anim. Reprod. Sci.* **2021**, 106822. [CrossRef]
9. Höfner, L.; Luther, A.M.; Waberski, D. The role of seminal plasma in the liquid storage of spermatozoa. *Anim. Reprod. Sci.* **2020**, *220*, 106290. [CrossRef]
10. Aneas, S.B.; Gary, B.G.; Bouvier, B.P. Collectis® automated boar collection technology. *Theriogenology* **2008**, *70*, 1368–1373. [CrossRef]
11. Lellbach, C.; Leiding, C.; Rath, D.; Staehr, B. Effects of automated collection methods on semen quality and economic efficiency of boar semen production. *Theriogenology* **2008**, *70*, 1389. [CrossRef]

12. Khalifa, T.; Rekkas, C.; Samartzi, F.; Lymberopoulos, A.; Kousenidis, K.; Dovenski, T. Highlights on artificial insemination (AI) technology in the pigs. *Maced. Vet. Rev.* **2014**, *37*, 5–34. [CrossRef]
13. Perez-Patiño, C.; Barranco, I.; Parrilla, I.; Valero, M.L.; Martinez, E.A.; Rodriguez-Martinez, H.; Roca, J. Characterization of the porcine seminal plasma proteome comparing ejaculate portions. *J. Proteom.* **2016**, *142*, 15–23. [CrossRef] [PubMed]
14. Fraser, L.; Wasilewska-Sakowska, K.; Zasiadczyk, Ł.; Piątkowska, E.; Karpiesiuk, K. Fractionated Seminal Plasma of Boar Ejaculates Analyzed by LC-MS/MS: Its Effects on Post-Thaw Semen Quality. *Genes* **2021**, *12*, 1574. [CrossRef] [PubMed]
15. Luther, A.M.; Waberski, D. In vitro aging of boar spermatozoa: Role of sperm proximity and seminal plasma. *Andrology* **2019**, *7*, 382–390. [CrossRef] [PubMed]
16. Leal, D.F.; Torres, M.A.; Ravagnani, G.M.; Martins, S.M.M.K.; Meirelles, F.V.; de Andrade, A.F.C. Absence of seminal plasma from sperm-rich fraction decreases boar sperm quality characteristics during the course of liquid storage. *Anim. Reprod. Sci.* **2018**, *198*, 20–26. [CrossRef] [PubMed]
17. Pavaneli, A.P.P.; da Silva Passarelli, M.; de Freitas, F.V.; Ravagnani, G.M.; Torres, M.A.; Martins, S.M.M.K.; Yeste, M.; de Andrade, A.F.C. Removal of seminal plasma prior to liquid storage of boar spermatozoa: A practice that can improve their fertilizing ability. *Theriogenology* **2019**, *125*, 79–86. [CrossRef]
18. García-Vázquez, F.A.; Llamas-López, P.J.; Jacome, M.A.; Sarrias-Gil, L.; López Albors, O. Morphological changes in the porcine cervix: A comparison between nulliparous and multiparous sows with regard to post-cervical artificial insemination. *Theriogenology* **2019**, *127*, 120–129. [CrossRef]
19. Waberski, D.; Luther, A.M.; Grünther, B.; Jäkel, H.; Henning, H.; Vogel, C.; Peralta, W.; Weitze, K.F. Sperm function in vitro and fertility after antibiotic-free, hypothermic storage of liquid preserved boar semen. *Sci. Rep.* **2019**, *9*, 14748. [CrossRef]
20. García-Vázquez, F.A.; Mellagi, A.P.G.; Ulguim, R.R.; Hernández-Caravaca, I.; Llamas-López, P.J.; Bortolozzo, F.P. Post-cervical artificial insemination in porcine: The technique that came to stay. *Theriogenology* **2019**, *129*, 37–45. [CrossRef]
21. Schjenken, J.E.; Robertson, S.A. The Female Response to Seminal Fluid. *Physiol. Rev.* **2020**, *100*, 1077–1117. [CrossRef]
22. Luongo, C.; González-Brusi, L.; Cots-Rodríguez, P.; Izquierdo-Rico, M.J.; Avilés, M.; García-Vázquez, F.A. Sperm Proteome after Interaction with Reproductive Fluids in Porcine: From the Ejaculation to the Fertilization Site. *Int. J. Mol. Sci.* **2020**, *21*, 6060. [CrossRef] [PubMed]
23. Álvarez-Rodríguez, M.; Martinez, C.A.; Wright, D.; Rodríguez-Martinez, H. The role of semen and seminal plasma in inducing large-scale genomic changes in the female porcine peri-ovulatory tract. *Sci. Rep.* **2020**, *10*, 5061. [CrossRef] [PubMed]
24. Alvarez-Rodriguez, M.; Atikuzzaman, M.; Venhoranta, H.; Wright, D.; Rodriguez-Martinez, H. Expression of immune regulatory genes in the porcine internal genital tract is differentially triggered by spermatozoa and seminal plasma. *Int. J. Mol. Sci.* **2019**, *20*, 513. [CrossRef]
25. Bogacki, M.; Jalali, B.M.; Wieckowska, A.; Kaczmarek, M.M. Prolonged Effect of Seminal Plasma on Global Gene Expression in Porcine Endometrium. *Genes* **2020**, *11*, 1302. [CrossRef] [PubMed]
26. Georgiou, A.S.; Snijders, A.P.L.; Sostaric, E.; Aflatoonian, R.; Vazquez, J.L.; Vazquez, J.M.; Roca, J.; Martinez, E.A.; Wright, P.C.; Fazeli, A. Modulation of the oviductal environment by gametes. *J. Proteome Res.* **2007**, *6*, 4656–4666. [CrossRef]
27. Bromfield, J.J.; Schjenken, J.E.; Chin, P.Y.; Care, A.S.; Jasper, M.J.; Robertson, S.A. Maternal tract factors contribute to paternal seminal fluid impact on metabolic phenotype in offspring. *Proc. Natl. Acad. Sci. USA* **2014**, *111*, 2200–2205. [CrossRef]
28. Martinez, C.A.; Cambra, J.M.; Gil, M.A.; Parrilla, I.; Alvarez-Rodriguez, M.; Rodriguez-Martinez, H.; Cuello, C.; Martinez, E.A. Seminal Plasma Induces Overexpression of Genes Associated with Embryo Development and Implantation in Day-6 Porcine Blastocysts. *Int. J. Mol. Sci.* **2020**, *21*, 3662. [CrossRef]
29. Luongo, C.; Garrappa, G.; Llamas-López, P.J.; Rodríguez-Tobón, E.; López-Úbeda, R.; Abril-Sánchez, S.; García-Vázquez, F.A. Effect of boar seminal dose type (cervical compared with post-cervical insemination) on cooling curve, sperm quality and storage time. *Anim. Reprod. Sci.* **2020**, *212*, 106236. [CrossRef]
30. Sebastián-Abad, B.; Llamas-López, P.J.; García-Vázquez, F.A. Relevance of the Ejaculate Fraction and Dilution Method on Boar Sperm Quality during Processing and Conservation of Seminal Doses. *Vet. Sci.* **2021**, *8*, 292. [CrossRef]
31. Dziekońska, A.; Świader, K.; Koziorowska-Gilun, M.; Mietelska, K.; Zasiadczyk, L.; Kordan, W. Effect of boar ejaculate fraction, extender type and time of storage on quality of spermatozoa. *Pol. J. Vet. Sci.* **2017**, *20*, 77–84. [CrossRef]
32. Xu, Y.; Han, Q.; Ma, C.; Wang, Y.; Zhang, P.; Li, C.; Cheng, X.; Xu, H. Comparative Proteomics and Phosphoproteomics Analysis Reveal the Possible Breed Difference in Yorkshire and Duroc Boar Spermatozoa. *Front. Cell Dev. Biol.* **2021**, *9*, 1845. [CrossRef] [PubMed]
33. Mills, K.M.; Aryal, U.K.; Sobreira, T.; Minton, A.M.; Casey, T.; Stewart, K.R. Shotgun proteome analysis of seminal plasma differentiate boars by reproductive performance. *Theriogenology* **2020**, *157*, 130–139. [CrossRef] [PubMed]
34. Kawano, N.; Araki, N.; Yoshida, K.; Hibino, T.; Ohnami, N.; Makino, M.; Kanai, S.; Hasuwa, H.; Yoshida, M.; Miyado, K.; et al. Seminal vesicle protein SVS2 is required for sperm survival in the uterus. *Proc. Natl. Acad. Sci. USA* **2014**, *111*, 4145–4150. [CrossRef] [PubMed]
35. Calvete, J.J.; Ensslin, M.; Mburu, J.; Lborra, A.; Martínez, P.; Adermann, K.; Waberski, D.; Sanz, L.; Töpfer-Petersen, E.; Weitze, K.F.; et al. Monoclonal antibodies against boar sperm zona pellucida-binding protein awn-1. Characterization of a continuous antigenic determinant and immunolocalization of AWN epitopes in inseminated sows. *Biol. Reprod.* **1997**, *57*, 735–742. [CrossRef]

36. Rodriguez-Martinez, H.; Saravia, F.; Wallgren, M.; Martinez, E.A.; Sanz, L.; Roca, J.; Vazquez, J.M.; Calvete, J.J. Spermadhesin PSP-I/PSP-II heterodimer induces migration of polymorphonuclear neutrophils into the uterine cavity of the sow. *J. Reprod. Immunol.* **2010**, *84*, 57–65. [CrossRef]
37. Rodriguez-Martinez, H.; Martinez, E.A.; Calvete, J.J.; Peña Vega, F.J.; Roca, J. Seminal Plasma: Relevant for Fertility? *Int. J. Mol. Sci.* **2021**, *22*, 4368. [CrossRef]
38. O'Leary, S.; Armstrong, D.T.; Robertson, S.A. Transforming growth factor-β (TGFβ) in porcine seminal plasma. *Reprod. Fertil. Dev.* **2011**, *23*, 748–758. [CrossRef]
39. O'Leary, S.; Jasper, M.J.; Warnes, G.M.; Armstrong, D.T.; Robertson, S.A. Seminal plasma regulates endometrial cytokine expression, leukocyte recruitment and embryo development in the pig. *Reproduction* **2004**, *128*, 237–247. [CrossRef]
40. Mateo-Otero, Y.; Fernández-López, P.; Gil-Caballero, S.; Fernandez-Fuertes, B.; Bonet, S.; Barranco, I.; Yeste, M. 1 H Nuclear Magnetic Resonance of Pig Seminal Plasma Reveals Intra-Ejaculate Variation in Metabolites. *Biomolecules* **2020**, *10*, 906. [CrossRef]
41. Hernández-Caravaca, I.; Izquierdo-Rico, M.J.; Matás, C.; Carvajal, J.A.; Vieira, L.; Abril, D.; Soriano-úbeda, C.; García-Vázquez, F.A. Reproductive performance and backflow study in cervical and post-cervical artificial insemination in sows. *Anim. Reprod. Sci.* **2012**, *136*, 14–22. [CrossRef]
42. García-Vázquez, F.A.; Hernández-Caravaca, I.; Yánez-Quintana, W.; Matás, C.; Soriano-Úbeda, C.; Izquierdo-Rico, M.J. Morphometry of boar sperm head and flagellum in semen backflow after insemination. *Theriogenology* **2015**, *84*, 566–574. [CrossRef] [PubMed]
43. Hensel, B.; Jakop, U.; Scheinpflug, K.; Mühldorfer, K.; Schröter, F.; Schäfer, J.; Greber, K.; Jung, M.; Schulze, M. Low temperature preservation of porcine semen: Influence of short antimicrobial lipopeptides on sperm quality and bacterial load. *Sci. Rep.* **2020**, *10*, 13225. [CrossRef] [PubMed]
44. Kim, D.W.; Cha, C.J. Antibiotic resistome from the One-Health perspective: Understanding and controlling antimicrobial resistance transmission. *Exp. Mol. Med.* **2021**, *53*, 301–309. [CrossRef]
45. Cheng, G.Z.; Li, J.Y.; Li, F.; Wang, H.Y.; Shi, G.X. Human ribonuclease 9, a member of ribonuclease A superfamily, specifically expressed in epididymis, is a novel sperm-binding protein. *Asian J. Androl.* **2009**, *11*, 240–251. [CrossRef] [PubMed]

biology MDPI

Article

Increasing the Yield and Cryosurvival of Spermatozoa from Rhinoceros Ejaculates Using the Enzyme Papain

Jessica P. Rickard [1,*], Kelsey Pool [1], Simon P. de Graaf [1], Timothy Portas [2], Natalie Rourke [3], Miriam Wiesner [4], Thomas B. Hildebrandt [5], Frank Göritz [5] and Robert Hermes [5]

1 School of Life and Environmental Sciences, Faculty of Science, The University of Sydney, Sydney, NSW 2006, Australia; kelsey.pool@uwa.edu.au (K.P.); simon.degraaf@sydney.edu.au (S.P.d.G.)
2 Zoo and Wildlife Veterinary Consultancy, Maleny, QLD 4552, Australia; timp@zoovet.com.au
3 Werribee Open Range Zoo, Werribee, VIC 3030, Australia; nrourke@zoo.org.au
4 Salzburg Zoo Hellbrunn, 5081 Anif, Austria; mwiesner@salzburg-zoo.at
5 Department of Reproductive Management, Leibniz Institute for Zoo and Wildlife Research, 10315 Berlin, Germany; hildebrandt@izw-berlin.de (T.B.H.); goeritz@izw-berlin.de (F.G.); hermes@izw-berlin.de (R.H.)
* Correspondence: jessica.rickard@sydney.edu.au; Tel.: +61-2-93518564

Simple Summary: Efficient collection and cryosurvival of semen from threatened wildlife species is key for the success of artificial reproductive technologies (ARTs). The viscous nature of ejaculates often collected from species such as rhinoceros, elephant, hippopotamus and primate, render the majority of spermatozoa collected useless and is therefore wasted. The enzyme papain has been used to reduce the viscosity of camelid semen but has yet to be tested in wildlife species. Therefore, the current study investigated the ability of papain to improve the yield and quality of spermatozoa collected from viscous fractions of the rhinoceros ejaculate during cryopreservation. Papain increased the quantity of useable spermatozoa collected from ejaculates, as well as the motility prior to freezing. It also improved the post-thaw motility, velocity, linearity and straightness of samples compared to spermatozoa frozen from the sperm-rich fraction of the ejaculate. There was no detrimental effect on membrane characteristics or DNA integrity. These results show that treating rhinoceros ejaculates with papain is able to salvage valuable spermatozoa and improve survival post-thaw, ultimately increasing the success of ARTs and creation of biobanks for the maintenance and survival of threatened species.

Abstract: The preservation of rhinoceros semen is vital for captive breeding programs. While successful collection and cryopreservation of rhinoceros semen has been reported, the volume and quality of semen produced is often low due to the high viscosity associated with ejaculates collected via electroejaculation. Reducing semen viscosity would enable access to previously unusable spermatozoa from viscous fractions and could improve quality post-thaw. The enzyme papain successfully reduced the viscosity of camelid semen but has yet to be tested in wildlife species. This study assessed the influence of papain on the in vitro quality of rhinoceros spermatozoa during cryopreservation using advanced semen assessment. In experiment 1, the motility of spermatozoa from the viscous fraction of an ejaculate, either untreated or treated with papain and its inhibitor E-64 prior to cryopreservation, was assessed post-thaw. In experiment 2, spermatozoa from papain-treated viscous fractions were compared to spermatozoa frozen from untreated sperm-rich fractions pre-freeze, as well as after 0, 1.5 and 3 h of incubation post-thaw (37 °C). Papain significantly increased the quantity of spermatozoa collected from ejaculates, as well as the motility prior to freezing. Papain also improved the post-thaw motility, velocity, linearity and straightness of samples compared to sperm-rich samples, with no detriment to sperm viability, lipid membrane disorder, production of ROS or DNA integrity ($p < 0.05$). Results show the benefit of supplementing rhinoceros spermatozoa with papain prior to cryopreservation on sperm cryosurvival and demonstrates the potential of using papain to improve the success of cryopreservation protocols, not only for the rhinoceros, but also for other wildlife species.

Citation: Rickard, J.P.; Pool, K.; de Graaf, S.P.; Portas, T.; Rourke, N.; Wiesner, M.; Hildebrandt, T.B.; Göritz, F.; Hermes, R. Increasing the Yield and Cryosurvival of Spermatozoa from Rhinoceros Ejaculates Using the Enzyme Papain. *Biology* **2022**, *11*, 154. https://doi.org/10.3390/biology11020154

Academic Editor: Marc Yeste

Received: 25 November 2021
Accepted: 15 January 2022
Published: 18 January 2022

Keywords: rhino; viscous; freezing; conservation; flow cytometry; CASA

1. Introduction

Current numbers of white, black, greater one-horned, Sumatran and Javan rhinoceros in situ of 10000, 3100, 2200, 30 and 18 adult individuals, respectively [1], emphasise the importance of captive breeding programs for the maintenance of genetic diversity and survival of these species [2–4]. Furthermore, it highlights the need for continued development of advanced assisted reproductive technologies (ARTs) to increase the likelihood of success. In this context, efficient collection and cryopreservation of spermatozoa play an important role for artificial breeding programs, in vitro production of embryos and long term genetic biobanking for safeguarding the species survival.

The first successful report of semen cryopreservation in the rhinoceros was published in 1979, reporting a sperm motility of less than 30% post-thaw [5]. Following the adaptation of equine protocols, motilities post-thaw increased to 55% [6–9], siring the first calf with frozen sperm in 2008 [6]. Since this time, work has continued to try to improve the success of cryopreservation protocols for all rhino species, improving the quality and quantity of spermatozoa post-thaw [10]. Consequently, extensive development of freezing extenders with different cryoprotectants and freezing rates has determined that freezing rhinoceros spermatozoa in ButoCrio® supplemented with glycerol and methylformamide above liquid nitrogen vapour can yield a post-thaw sperm motility of 75.6 ± 3.9% [11]. The presence of seminal plasma has also been investigated during the freezing protocol, with no significant improvements in motility post-thaw noted, if seminal plasma was removed via centrifugation from sperm-rich fractions prior to cryopreservation [11]. Despite these major developments, the use and success of cryopreservation protocols in the rhinoceros is limited by the quality and quantity of the original ejaculate produced [12]. Semen collection in the rhinoceros is predominantly collected via electroejaculation (EJ) [12,13] under anaesthesia [14]. While the development of ultrasonography and species-specific stimulation probes for EJ have resulted in a greater understanding of rhinoceros reproductive anatomy [13,15,16], creating greater consistency in the number of successful collections, the variability in semen quality associated with these collections is often still high [12]. Depending on the duration of accessory sex gland stimulation, the volume and composition of seminal plasma contributing towards the ejaculate can vary [12,13,17]. The ejaculate is collected in 1–6 fractions, with the first 1–2 deemed "sperm-rich" containing the highest concentration of viable sperm. This is followed by subsequent fractions containing large amounts of seminal plasma but lower sperm concentration [12]. In fact, the increasing degree of viscosity in fractions collected after the first, low volume sperm-rich fraction, is a characteristic not only of rhinoceros ejaculates [15–17] but also of other wildlife species, including the elephant [18], hippopotamus [19], pangolin [20], primates [21,22] and camelids [23,24]. Even marsupial semen has been reported to coagulate following EJ collection [25,26]. Regardless, the viscosity of ejaculates collected from these species makes the application of ARTs extremely challenging.

High viscosity fractions do not allow for obligatory separation of seminal fluids from spermatozoa by centrifugation or efficient dilution of spermatozoa with cryoprotectants, preventing sperm cryopreservation and ultimately wasting a large proportion of the ejaculate. For example, after successfully collecting 19 ejaculates from 14 African elephant bulls and recording a mean volume of 104.3 ± 80.0 mL and sperm concentration of 781 ± 622.8 × 10^6 spm/mL, up to 41% was discarded due to the high viscosity associated with fractions [18]. In the white and greater one-horned rhinoceros species, the degree of sperm wastage is even higher. Despite collecting a mean ejaculate volume of 80 ± 15 mL and 158 ± 33 mL, respectively, only 6 ± 1 mL and 26 ± 4 mL of the sperm-rich fractions were cryopreserved, representing a wastage of 92.5% and 84.5% per species [8,11,17]. Given the logistical challenges, potential risk and cost associated with the general anaesthesia of valuable wildlife species, this represents a considerable loss of valuable gametes.

If the viscosity of these collected fractions could be treated and reduced, this could rescue available spermatozoa for ARTs, improving the efficiency of semen collection via electroejaculation. The viscosity of rhinoceros ejaculates has been extensively characterised using a combination of subjective assessment, gel electrophoresis and mass spectrometry techniques [15]. It is mainly caused by secretions of the bulbourethral glands [27], in particular the presence of a 250 kDa glycoprotein, PBU250 [15]. By targeting PBU250 with the enzymes, α-amylase and collagenase, viscosity was reduced by 28% and 21% respectively; however, there was no improvement in the quality of spermatozoa following incubation at room temperature for 1 h [15]. Treatment of camelid semen to reduce viscosity saw greater success [23,28–30], recording an improvement in post-thaw sperm quality after the addition of papain, the cysteine protease present in papaya (*Carica papaya*) [30]. When alpaca spermatozoa were treated with 0.1 mg/mL of papain, viscosity was significantly reduced within 30 min, without having any detrimental impact on sperm motility, viability, acrosome integrity and DNA integrity [23]. Furthermore, when samples were frozen, papain-treated spermatozoa exhibited superior total motility post-thaw compared to untreated samples [30,31]. To date, papain has not been examined for its ability to reduce the viscosity of viscous ejaculates in wildlife species, or specifically of rhinoceros ejaculates, and holds tremendous promise to improve cryopreservation protocols in the rhinoceros as well as other endangered wildlife species with viscous ejaculates.

If papain was found to be successful, the yield of viable spermatozoa produced from one semen collection would significantly increase that which is available for use in advanced ARTs. It would also allow the use of advanced semen assessment tools, such as fluorescent flow cytometry, to examine more advanced sperm-cell characteristics, currently applied to spermatozoa from other domestic species. This additional information would ultimately contribute towards a better understanding of the fertility status of preserved rhinoceros spermatozoa, aiding in the success of ARTs. Furthermore, using the southern white rhinoceros as a model, papain may be a viable tool to reduce the viscosity and increase the yield of spermatozoa collected in other rhinoceros species, or from other wildlife species, such as elephants, tapirs, hippopotamus and primates, where electroejaculation is the method of choice for semen collection.

As such, the following study will aim to: (1) examine the influence of papain on the post-thaw quality of spermatozoa frozen from viscous or low sperm ejaculate fractions, (2) compare the post-thaw quality of papain-treated spermatozoa to that of sperm-rich samples, and (3) objectively assess, for the first time, advanced membrane characteristics of post-thaw rhinoceros spermatozoa using flow cytometry. It was hypothesised that the benefit of papain would be two-fold. It would increase the quantity of rhinoceros spermatozoa available for ARTs post collection, as well as improve the quality of spermatozoa post thaw.

2. Materials and Methods

2.1. Chemicals and Reagents

Unless otherwise stated, all chemicals were supplied by Sigma-Aldrich (St. Louis, MO, USA).

Low sperm, viscous fractions were diluted in a modified Tyrodes solution (TALP), which consisted of 2 mM $CaCL_2$, 3.1 mM KCl, 0.4 mM $MgCL_2$, 95 mM NaCl, 0.3 mM NaH_2PO_4, 10 mM HEPES, 21.6 mM Na Lactate, 5 mM glucose, 1 mM Na pyruvate, 25 mM $NaHCO_3$, 0.03 mM phenol red and 3 mg/mL BSA (fatty acid free).

2.2. Ethics Statement

This study was conducted on captive southern white rhinoceros (*Ceratotherium simum*), a species listed on the IUCN Red List of Threatened species™ as Near Threatened [32]. Semen was collected from bulls as part of a clinical examination of their fertility or for use in artificial inseminations. The study was approved by the IACUC animal ethics committee of the Leibniz Institute for Zoo and Wildlife Research (permit number: 2017-08-02).

2.3. Experimental Design

Two experiments were conducted to assess the effectiveness of the enzyme papain on improving the quality of spermatozoa sourced from low sperm fractions. Experiment 1 was a proof-of-concept study, where half of the viscous or low sperm fractions of each ejaculate collected (n = 4) was treated with papain (Sigma, St. Louis, MO, USA) before being processed for cryopreservation. Sperm motility was immediately assessed post-thaw on a computer-assisted sperm analyser (CASA; Hamilton Thorne IVOS II and CEROS II, Animal Breeder software, Version 1.8; Beverly, CA, USA) and compared to the non-treated viscous sample. Experiment 2 treated the low sperm, viscous ejaculate fractions collected (n = 6) with papain prior to cryopreservation. Sperm quality post thaw was then compared to spermatozoa frozen from the sperm-rich fraction of the same bull, immediately after thawing, as well as 1.5 and 3 h after incubation (37 °C). Spermatozoa were assessed for motility and associated kinematics using CASA. Membrane viability, acrosome integrity, lipid membrane disorder, production of reactive oxygen species and DNA integrity were also assessed using flow cytometry (Cytoflex; Beckman Coulter, Lane Cove, Australia and C6 BD Accuri; Becton Dickinson, NJ, USA).

2.4. Animals and Semen Collection

Over a 4-year period, semen was collected via electroejaculation (Seager model 14, Dalzell USA Medical Systems, The Plains, VA, USA) from 10 adult, captive, white rhinoceros bulls (n = 10) located at 3 European (Exp 1; n = 4 bulls) and 4 Australian (Exp 2; n = 6 bulls) zoos. For semen collection by electrostimulation, general anaesthesia in lateral recumbency was required. For this, 25 mg detomidine hydrochloride (Domidine1 10 mg/mL, Eurovet Animal Health B.V., Bladel, The Netherlands) and 25 mg butorphanol (Torbugesic1 Vet 10 mg/mL, Zoetis B.V., Capelle a/d IJssel, The Netherlands) were injected intra-muscularly as a premedication. A combination of 150 mg ketamine hydrochloride (Ketamin 10% WDT, Henry Schein VET GmbH, Hamburg, Germany) and 1.8–2.7 mg etorphine (Captivon, Wildlife Pharmaceuticals South Africa, Karino, South Africa) was injected intravenous into the ear vein after 20 min. Anaesthesia was antagonised by administration of 250 mg naltrexone hydrochloride (TrexonilTM, Wildlife Pharmaceuticals (PTY) Ltd., White River, South Africa) and 40 mg atipamezole hydrochloride (Atipam 5 mg/mL, Eurovet Animal Health B.V.). Half the reversal was given i.m. and the other half was given i.v. Animals were normal and alert two to three minutes after the antagonist was given. For electroejaculation, a custom-made electric probe, specifically designed for rhinoceros, was used for stimulation [27]. The stimulation probe expanded the lumen of the rectum providing maximum electric coupling of electrodes. Multiple sets of 2–3 electrical stimuli were applied with increasing voltage and amperage (5–15 V/200–800 mA). Each set of stimulations was followed by manual massage of the pelvic and penile aspects of the urethra [7,17]. After each set of stimulations, up to 6 fractions were collected into foam-insulated 50 mL isotherm collection tubes, labelled and transferred to the onsite laboratory at body temperature, where they were immediately analysed.

2.5. Assessment of Initial Ejaculate and Sperm Membrane Characteristics

Each fraction collected per bull was assessed for volume, concentration and total sperm produced per fraction. Sperm concentration was determined using a haemocytometer, as described by Evans and Maxwell [33], and total sperm per fraction ($\times 10^9$) was calculated by multiplying fraction volume by concentration ($\times 10^6$ sperm/mL).

A small aliquot of each fraction was assessed for motility approximately 20 min after collection. The percentage of motile spermatozoa was estimated subjectively to the nearest 5% on heated slides (37 °C) after examining several fields of view on a phase contrast light microscope (Olympus C41, Olympus, Germany). In experiment 2, only sperm-rich fractions which recorded a motility score of 40% and above were deemed acceptable for cryopreservation (as per current standards) and were included in the study.

Assessment of sperm membrane characteristics were conducted as described below. Acrosome integrity and morphology were assessed by fixing 10 μL aliquots of sample in 40 μL of Hancocks fixative [34]. Acrosomes were classified as intact versus modified or reacted (including completely detached acrosomes). Sperm morphology was assessed by visually searching several fields of view for a wide range of abnormalities. Assessment of sperm viability prior to freezing was evaluated using a hypo-osmotic swelling test (HOS), as described by Reid, Hermes, Blottner, Goeritz, Wibbelt, Walzer, Bryant, Portas, Streich and Hildebrandt [9]. Briefly, an aliquot of 20 μL of each sample was diluted in 100 μL hypo-osmolar TALP (100 mOsm), incubated at 37 °C for 30 min and subsequently fixed with paraformaldehyde (18.5%, 100 μL). From each sample, 10 μL was prepared on a glass slide with coverslip, whereby 200 sperm were evaluated. Spermatozoa were categorised as HOS-negative or non-viable when they remained unchanged in the hypo-osmotic environment. Swollen sperm or sperm with curved tails were classified as HOS-positive or viable spermatozoa with functional membranes.

2.6. *Treatment of Viscous Fractions with Papain and E-64*

Prior to enzyme treatment, viscous fractions were diluted 1:1 with a modified Tyrode's medium warmed to 37 °C and supplemented with albumin, lactate and pyruvate (TALP) [35], and 3 mg/mL BSA was added fresh on the day of collection.

Half of the sample (Exp 1) or all of the sample (Exp 2) was then incubated with papain (0.1 mg/mL; Sigma 76216, St. Louis, MO, USA) at 37 °C for 30 min, (Papain-treated; Exp 1 and 2). In experiment 1 the remaining sample was left untreated (Viscous control; Exp 1) before being processed for cryopreservation.

Following treatment, the reaction was then suspended with the addition of an inhibitor, trans-epoxysuccinyl-L-leucylamido (4-guanidino)butane (E-64;10 μM; Sigma E3132, St. Louis, MO, USA) and incubated at 37 °C for 5 min [30].

To demonstrate the change in viscosity due to papain treatment, the time taken for aliquots (3 μL) of treated and non-treated viscous fractions to fill a standard 20 μm Leja CASA slide (Hamilton Thorne, Beverly, MA, USA) was recorded (n = 1 bull only), as previously described [36].

2.7. *Dilution and Cryopreservation of Rhinoceros Spermatozoa*

Prior to cryopreservation, all samples (Exp 1: papain-treated and viscous control treatments; Exp 2: papain-treated and sperm-rich treatments) were centrifuged (1000× *g*; 20 min; room temp), underlain with 1 mL of a high-density gradient (OptiPrepTM, Sigma–Aldrich, Taufkirchen, Germany) to remove enzymes, seminal plasma and concentrate spermatozoa. Following centrifugation, the supernatant was discarded, the sperm-rich layer above the density gradient aspirated and concentration determined using a haemocytometer.

Samples were then resuspended in ButoCrio® (Nida-con, Mölndal, Sweden) to an approximate concentration of 100–150 × 10^6 sperm/mL.

Sperm suspensions were loaded into 0.5 mL straws and chilled for 45 min to 4 °C (0.5 °C/min). Straws were then held 4 cm above liquid nitrogen vapour for 8 min, before being plunged into liquid nitrogen. Straws were stored in liquid nitrogen until assessment in the laboratory.

2.8. *Thawing and Advanced In Vitro Semen Assessment*

Two straws per sample were thawed in a 37 °C water bath for approximately 60 s with gentle agitation and incubated over 3 h (Exp 2 only). Aliquots were then taken at 0, 1.5 and 3 h and diluted to 20 × 10^6 spermatozoa/mL with TALP supplemented with 3 mg/mL BSA. Spermatozoa was assessed for motility and other kinematics via computer-assisted sperm analysis (CASA CEROSII (Exp 1) IVOS II (Exp 2); Hamilton Thorne, USA). Viability, acrosome integrity, production of reactive oxygen species, membrane fluidity and DNA integrity of spermatozoa (Exp 2 only) was performed using fluorescent staining and flow cytometry (Cytoflex; Beckman Coulter, Lane Cove, AUS; C6 BD Accuri, Becton Dickinson, NJ, USA) following consideration of the recommendations published by Lee, et al. [37].

2.8.1. Computer-Assisted Sperm Analysis (CASA)

Post-thaw motility was assessed using CASA. Semen samples (5.5 μL) were placed on prewarmed slides (Cell Vu, Millenium Sciences Corp., New York, NY, USA) immediately prior to assessment. Motility (TM) and other kinematic parameters, including progressive motility (PM), average path velocity (VAP), straight line velocity (VSL), curvilinear velocity (VCL), amplitude of lateral head displacement (ALH), beat cross frequency (BCF), straightness (STR) and linearity (LIN) of spermatozoa was determined by capturing 8 microscopic fields (recording ≥ 200 cells/sample) using the following settings established for rhinoceros: frame rate = 60 Hz; number of frames acquired = 60; minimum contrast = 65; head size minimum 10 μm^2; head size maximum 30 μm^2; medium average path velocity cutoff = 25 μm/s; medium threshold straightness = 75%; slow average path velocity cutoff = 20 μm/s; slow straight line velocity cut off = 6 μm/s; head brightness minimum = 157; tail brightness minimum = 75; magnification = 1.2; illumination intensity = 85; temperature = 37 °C.

2.8.2. Flow Cytometric Analysis

Flow cytometric analysis of viability, acrosome integrity, reactive oxygen species and lipid membrane disorder was performed using a Cytoflex flow cytometer calibrated prior to use with Cytoflex daily QC fluorospheres (Beckman Coulter, Lane Cove, Australia). Sperm cells were isolated from total events based on forward and side scatter profiles using 3 different lasers including 50 mW 488 nm, 50 mW 638 nm and 80 mW 405 nm. Further separation from debris was performed by staining samples with Hoechst 33342 at a final concentration of 1 ug/mL. For every tested parameter, 10,000 cells were analysed per sample using CytExpert 2.0 software (Beckman Coulter, Lane Cove Australia).

The viability and acrosome integrity of sperm cells was assessed using a combination of fluorescein isothiocyanate (FITC)-PNA (final concentration; 0.4 μg/mL) and propidium iodide (PI, final concentration; 6 μM). Samples were incubated for 10 min at 37 °C prior to assessment. Propidium iodide fluorescence was detected using a 610/20 band-pass filter, and FITC-PNA fluorescence was detected on a 525/40 nm band-pass filter. If sperm cells were classified as both FITC-PNA and PI negative, they were considered viable cells with intact acrosomes.

The proportion of viable cells detected with lipid membrane disorder was assessed using a combination of merocyanine 540 (M540, final concentration; 0.8 μM) and YO-PRO (final concentration; 25 nM). Samples were incubated for 10 min at 37 °C prior to assessment. M540 fluorescence was detected on a 585/40 nm band-pass filter and Yo-Pro fluorescence detected on a 525/40 nm band-pass filter. Cells which were classified as Yo-Pro-negative were viable, with higher median M540 fluorescence corresponding to high levels of lipid disorder and membrane fluidity.

The proportion of viable cells with high levels of intracellular reactive oxygen species (ROS) was assessed using a combination of dichlorodihydrofluorescein diacetate acetyl ester (H2DCFDA, final concentration; 5 μM) and PI (final concentration; 6 μM). Samples were incubated with H2DCFDA for 60 min at 37 °C before being centrifuged (10 min at 600× *g*), supernatant removed, and sperm pellets resuspended in fresh buffer (TALP +0.3% BSA). An aliquot of this new sperm suspension was then counterstained with PI for 10 min at 37 °C at each time point for assessment. H2DCFDA fluorescence was detected on a 525/40 band-pass filter. Cells which were PI negative were classified as viable with higher median H2DCFDA fluorescence levels corresponding to a high production of intracellular ROS.

Flow cytometric analysis of the DNA integrity of frozen-thawed rhinoceros spermatozoa was performed using a C6 BD Accuri (Becton Dickinson, NJ, USA). Samples were stained with acridine orange (final concentration; 6 μg/mL) as described by Pool, et al. [38]. Stained samples were incubated for 3 min before assessment, where green fluorescence

(FL1) was detected using a 533/30 band-pass filter, and red fluorescence (FL3) was detected using a 670 long pass filter. Flow rate was set to around 200 events per second, and a minimum of 5000 sperm cells were recorded per sample. DNA fragmentation was estimated by the relative amount of single-stranded and double-stranded DNA, indicated by the proportion of sperm demonstrating red fluorescence to total fluorescence or cells outside the main population.

2.9. Statistical Analysis

Initial ejaculate characteristics, motility (Exp 1), viability, acrosome integrity and percent normal morphology (Exp 2 only) assessed prior to freezing, were analysed using an ANOVA and means were compared using a two-sample t-test comparison of means assuming unequal variances. Post-thaw motility kinematics (Exp 1) were assessed using an ANOVA and means were compared using a paired two-sample *t*-test. *t*-tests were conducted in Genstat (Version 18, VSN International, Hemel Hempstead, UK).

In Experiment 2, sperm motility, viability, acrosome integrity, lipid membrane disorder, intracellular ROS production and DNA integrity were analysed using a restricted maximum likelihood model (REML) in Genstat. Treatment and timepoint (0, 1.5 or 3) were specified as fixed effects, while individual bull was specified as a random effect. REMLS were used in the current study to investigate whether there was any interaction between the influence of treatments on changes in sperm parameters over time. These interactions were dropped from the model if not significant ($p > 0.05$).

Normal distribution of data and homogeneity of variances were assessed within Genstat. For both experiments and variables, means were reported with ± S.E.M and $p < 0.05$ was considered statistically significant.

3. Results

3.1. Initial Ejaculate Characteristics Collected in Experiment 1 and 2 Prior to Freezing

Table 1 presents the average number of ejaculates collected, number of ejaculates utilised, average volume, concentration and total sperm number of both the sperm-rich and viscous, low sperm fractions collected via electroejaculation from white rhinoceros bulls (n = 10). Across both experiment 1 and 2, there was a total of 11 ejaculates collected from 9 bulls over the collection period. Of these 11 ejaculates, only 10 contained a sperm-rich fraction deemed acceptable for further use in the study. Figure 1 provides a visual example of the viscosity noted in viscous, low sperm fractions during the study. Further, when a standard 20 μm Leja CASA slide was loaded with 3 μL of the viscous sperm fraction, it took 17 s for the chamber to load. Aliquots of papain-treated fractions filled the 20 μm chamber within 5 sec (n = 1 bull). There was a significant difference in the volume and concentration of spermatozoa (spermatozoa/mL) collected between the sperm-rich and viscous, low sperm fractions. The viscous, low sperm fraction recorded a larger volume but smaller sperm concentration compared to the sperm-rich fraction ($p < 0.05$; Table 1). There was no significant difference in the total number of sperm collected in both fractions ($\times 10^9$ spermatozoa; $p > 0.05$; Table 1).

Table 1. Semen characteristics of the sperm-rich and viscous, low sperm fractions collected in Experiment 1 and 2, from southern white rhinoceros bulls (n = 9) via electroejaculation. Data is mean ± SEM. Columns which differ in superscript indicate significant differences between means ($p < 0.05$).

Parameter	Ejaculate Fraction	
	Sperm-Rich	Viscous, Low Sperm
Number of ejaculate fractions collected	10	11
Volume (ml)	2.6 ± 0.58^{a}	15.1 ± 2.4^{b}
Concentration ($\times 10^6$ spermatozoa/mL)	863.3 ± 116.99^{a}	236.0 ± 46.61^{b}
Total sperm number per ejaculate ($\times 10^9$ spermatozoa)	2.7 ± 0.93	3.4 ± 0.71

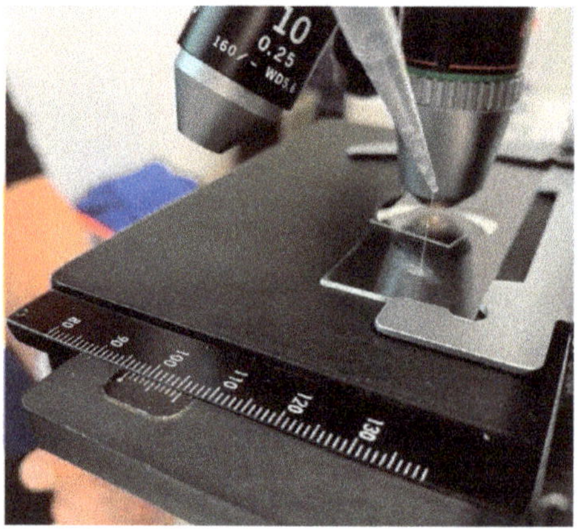

Figure 1. Visual example of viscosity witnessed in non-sperm-rich fraction of rhino ejaculate collected via electro-ejaculation.

3.2. Papain Improves the Quality of Spermatozoa Originating from Viscous, Low Sperm Fractions Both Prior to and after Cryopreservation

Table 2 presents the characteristics of samples collected and assessed in Experiment 1, prior to freezing as well as immediately post-thaw. Papain-treated samples recorded a significantly higher subjective motility prior to freezing compared to non-treated samples (88.3% ± 1.70% vs. 78.8% ± 2.39%, respectively; $p < 0.05$; Table 2). Post-thaw, papain-treated samples also recorded a significantly higher TM compared to untreated samples (84.9% ± 1.60% vs. 37.5% ± 10.56%, respectively; $p < 0.05$; Table 2), with almost no difference in the pre-freeze and post-thaw motility of treated samples.

Post-thaw, the PM, VAP, VCL, VSL and ALH of papain-treated samples in experiment 1 was significantly higher compared to the non-treated viscous samples immediately post-thaw ($p < 0.05$; Table 2). There was no significant difference in the post-thaw BCF, LIN or STR of treated and non-treated spermatozoa originating from the viscous fraction ($p > 0.05$; Table 2).

The viability of viscous spermatozoa was significantly higher post-thaw if treated with papain prior to freezing, compared to the untreated control (86.3 ± 3.33% vs. 71.5 ± 3.66, respectively; $p < 0.05$).

Table 2. Sperm characteristics of viscous, low sperm control and papain-treated samples collected from southern white rhinoceros bulls (n = 3 bulls, 4 ejaculates) assessed prior to and following cryopreservation (Exp 1). Percent viability, acrosome intact and normal morphology was assessed subjectively. Data are mean ± SEM. Columns which differ in superscript indicate significant differences between the viscous control and viscous treated samples ($p < 0.05$).

Time	Sperm Parameters	Viscous, Low Sperm Fraction	
		Non-Treated Control	Papain-Treated
Pre-freeze	Subjective motility (%)	78.8 ± 2.39 ^{a}	88.3 ± 1.7 ^{b}
Post-thaw	TM (%)	37.5 ± 10.56 ^{a}	84.9 ± 1.60 ^{b}
	PM (%)	25.4 ± 9.32 ^{a}	56.6 ± 5.72 ^{b}
	VAP	43.8 ± 1.92 ^{a}	74.4 ± 5.48 ^{b}
	VCL	77.1 ± 3.16 ^{a}	115.8 ± 8.30 ^{b}
	VSL	34.7 ± 4.58 ^{a}	57.6 ± 3.51 ^{b}
	ALH	3.7 ± 0.40 ^{a}	4.5 ± 0.18 ^{b}
	BCF	39.6 ± 2.81	41.2 ± 0.23
	LIN	50.8 ± 6.02	52.0 ± 1.86
	STR	80.9 ± 5.92	78.3 ± 2.87
	Percent viable (%)	71.5 ± 3.66 ^{a}	86.3 ± 3.33 ^{b}
	Percent acrosome intact (%)	68 ± 6.48	80.0 ± 3.70
	Percent normal morphology (%)	70.8 ± 4.37	80.5 ± 2.72

Table 3 displays the pre-freeze and post-thaw sperm characteristics of samples collected and analysed in Experiment 2, where viscous spermatozoa treated with papain were compared to spermatozoa from the sperm-rich fraction. There was no significant difference in the mean sperm motility of treatments prior to freezing ($p > 0.05$; Table 3).

Post-thaw, there was no significant interaction of treatment and time on the TM of frozen-thawed rhinoceros spermatozoa. In other words, there was no significant difference in the decline of treatments over time. However, when pooled over time, spermatozoa treated with papain prior to cryopreservation recorded a significantly higher motility than the sperm-rich control ($p < 0.05$; Table 3). When pooled over treatment, the TM of spermatozoa post-thaw declined significantly over time from 69.3% ± 3.10% at 0 h, 44.0% ± 6.91% to 1.5 h and 24.4% ± 4.37% at 3 h ($p < 0.001$).

There was no significant interaction of treatment and time on the PM of spermatozoa. However, when pooled over time, spermatozoa treated with papain recorded significantly higher PM compared to the sperm-rich sample ($p < 0.05$; Table 3). When pooled over treatment, the PM of spermatozoa post-thaw declined significantly over time from 45.8% ± 4.67% at 0 h to 32.9% ± 6.87% at 1.5 h and 15.6% ± 3.63% at 3 h ($p < 0.001$).

There was no significant interaction of treatment and time on the VAP, VCL or VSL of spermatozoa. However, when pooled over time, the VAP, VCL and VSL of spermatozoa treated with papain prior to cryopreservation was significantly higher compared to the sperm-rich control samples ($p < 0.05$; Table 3). When pooled over treatment, the VAP, VCL and VSL of spermatozoa post-thaw declined significantly over time from 75.1 ± 6.57, 118.0 ± 11.76 and 58.4 ± 4.67 µm/s at 0 h to 59.0 ± 6.58, 86.2 ± 8.47 and 51.0 ± 6.53 µm/s at 1.5 h and finally 42.0 ± 4.83, 64.5 ± 6.91 and 58.4 ± 4.67 µm/s at 3 h ($p < 0.001$).

Table 3. Pre-freeze and post-thaw characteristics of sperm-rich and viscous low sperm papain-treated spermatozoa collected from southern white rhinoceros bulls (n = 6) following treatment, centrifugation and dilution in ButoCrio (Exp 2). Data is pooled over bull and time (0, 1.5 and 3 h post-thaw) ± SEM. Columns which differ in superscript indicate significant differences between sperm-rich control and papain-treated spermatozoa ($p < 0.05$).

Time Point	In Vitro Sperm Parameters	Sperm-Rich Control	Viscous Low Sperm Papain-Treated
Pre-freeze	Subjective motility (%)	78.4 ± 6.66	76.6 ± 5.22
	Percent viable (%)	92.0 ± 1.10	93.2 ± 0.92
	Percent acrosome intact (%)	88.2 ± 2.48	89.2 ± 1.86
	Percent normal morphology (%)	82.2 ± 4.59	78.0 ± 5.10
Post-thaw	TM (%)	38.0 ± 5.50 ^{a}	54.2 ± 6.13 ^{b}
	PM (%)	21.8 ± 3.99 ^{a}	40.9 ± 5.20 ^{b}
	VAP (μm/s)	48.3 ± 4.41 ^{a}	68.8 ± 6.04 ^{b}
	VCL (μm/s)	78.0 ± 8.19 ^{a}	101.2 ± 9.45 ^{b}
	VSL (μm/s)	37.9 ± 3.38 ^{a}	58.9 ± 4.60 ^{b}
	LIN (%)	9.0 ± 2.33 ^{a}	8.15 ± 1.98 ^{b}
	STR (%)	81.0 ± 2.04 ^{a}	87.4 ± 1.79 ^{b}
	BCF (%)	39.0 ± 1.30	40.2 ± 0.96
	ALH (%)	3.66 ± 0.31	3.57 ± 0.29
	Viable acrosome intact (%)	40.2 ± 3.89	39.3 ± 1.90
	Membrane lipid disorder (median M540 fluor.)	2966.4 ± 357.26	13,993.6 ± 4901.47
	Level of ROS (median H2DCFDA fluor.)	17,278.9 ± 3923.57	9073.0 ± 2103.47
	DNA integrity (%)	11.0 ± 0.71	8.51 ± 0.69

There was no significant interaction of treatment and time on the STR or LIN of frozen-thawed rhinoceros spermatozoa. However, when pooled over time, the STR and LIN of papain-treated spermatozoa was significantly higher compared to the sperm-rich control samples ($p < 0.05$; Table 3). When pooled over treatment, there was a significant effect of time on the LIN but not the STR of spermatozoa post-thaw. The LIN of samples increased from 0 h (54.6% ± 3.03%) to 1.5 h (61.7% ± 2.18%; $p < 0.05$). There was no difference between 1.5 and 3 h (60.1% ± 3.08%; $p > 0.05$).

There was no significant interaction of treatment and time on the BCF or ALH of spermatozoa post-thaw. However, when pooled over treatments, there was a significant effect of time on the BCF and ALH of spermatozoa post-thaw. The BCF of spermatozoa was significantly lower at 3 h (36.6 ± 1.50%) post-thaw compared to 0 h (41.0% ± 0.96%) and 1.5 h (41.0% ± 0.96%; $p < 0.05$). The ALH of spermatozoa was significantly higher at 0 h (4.5% ± 0.36%) compared to 1.5 h (3.3% ± 0.23%) and 3 h (2.9% ± 0.31%; $p < 0.001$). There was no effect of treatment alone on the BCF and ALH of spermatozoa ($p > 0.05$).

3.3. Papain Does Not Influence the Viability, Acrosome Integrity, Membrane Lipid Disorder, Intracellular ROS Production or DNA Integrity of Frozen-Thawed Rhinoceros Spermatozoa Assessed Using Flow Cytometry

In experiment 2, prior to freezing, there was no significant difference between the percent viable, acrosome intact or normal morphology of sperm-rich and papain-treated samples ($p > 0.05$; Table 3). There was no significant interaction of time and treatment on the percentage of viable, acrosome intact spermatozoa post-thaw. Nor was there any significant effect of treatment when pooled over time ($p > 0.05$; Table 3). However, when pooled over treatments, the percentage of viable acrosome-intact spermatozoa was significantly lower at 3 h (30.1% ± 2.66%) compared to 0 h (48.4% ± 1.94%) and 1.5 h (41.3% ± 2.91%; $p < 0.001$).

There was no significant interaction of time and treatment on the median level of H2DCFDA or M540 fluorescence, otherwise known as level of ROS production and membrane lipid disorder, respectively ($p > 0.05$; Table 3). Nor was there any significant effect of treatment or time alone ($p > 0.05$).

There was no significant interaction of time and treatment on the percentage of DNA fragmentation ($p > 0.05$; Table 3). Nor was there any significant effect of treatment or time alone ($p > 0.05$).

4. Discussion

The near-threatened southern white rhinoceros serves as an excellent example of a species whose semen is characteristically viscous post-semen collection. This is caused in part by the presence of particular components within the seminal plasma, secreted in large amounts when the accessory sex glands are stimulated during electro-ejaculation [12,15]. This phenomenon is shared by other species, including camelids [23,28], elephants [18,34], hippopotamus' [19] and primates [21,22]. Spermatozoa trapped within the viscous (low sperm) fractions of the ejaculate equate to almost half of the total spermatozoa collected [11,17,20], but are often discarded. In the current study, treatment of viscous, low sperm rhinoceros spermatozoa, with papain prior to cryopreservation, not only enabled the utilisation of spermatozoa within these fractions, but resulted in improved motility and viability post-thaw. It also resulted in a similar motility to sperm-rich spermatozoa prior to freezing, however, it surprisingly exceeded the motility of sperm-rich treatments post-thaw. Furthermore, papain was not detrimental to the membrane or acrosome status of spermatozoa, nor found to have any impact on the level of intracellular reactive oxygen species (ROS) produced, lipid membrane disorder or degree of DNA fragmentation when compared to sperm-rich treatments, key characteristics for the timely maturation and survival of spermatozoa within the female. As such, treatment of viscous, low sperm rhinoceros fractions with papain increased sperm yield by almost two-fold post collection. Considering the ethical, logistical and financial challenges associated with the collection of semen from threatened or endangered wildlife species, this represents a substantial improvement in the efficiency of existing semen collection and cryopreservation protocols for the rhinoceros and other species which suffer from viscous ejaculates post collection.

Treatment of rhinoceros spermatozoa with papain enabled the assessment and successful cryopreservation of rhinoceros spermatozoa from the viscous, low sperm fraction of the ejaculate. In the current study, despite the sperm-rich fractions recording higher sperm concentration per mL compared to the viscous fractions (as shown previously [12,17]), given the larger volume produced, the total number of spermatozoa contained within this fraction ($3.5 \pm 0.79 \times 10^9$ sperm) was similar to that produced by the sperm-rich fraction ($2.0 \pm 0.98 \times 10^9$ sperm), spermatozoa which would have been discarded, if not for the addition of papain. Consequently, this enabled the cryopreservation of an additional 3.5 billion cells on average per male. To the best of our knowledge, when both fractions are combined, this amount exceeds what has previously been reported ($2.8 \pm 0.8 \times 10^9$) for high quality spermatozoa collected from the southern white rhinoceros via electroejaculation [10,17]. It is also higher than that recorded for the black (200×10^6 spm [13]) and Sumatran (2.5×10^9 [39]) rhinoceros. Although the greater one-horned rhinoceros recorded an impressive yield of 30.4×10^9 spm [8], the quality associated with this collection was not confirmed. The importance of improving the quantity of gamete cryopreservation in endangered species is best demonstrated in the Sumatran rhinoceros, where semen collection attempts have returned poor results. With the species listed as critically endangered, successful semen collection attempts are paramount. This would enable the rapid generation of a biobank containing genetically distinct gametes to help safeguard the species from extinction—a luxury not afforded to the northern white rhinoceros species, where in vitro fertilisation of oocytes harvested from the one last remaining female, depends on the quantity and quality of frozen spermatozoa collected from one deceased male.

The treatment of viscous fractions with papain not only resulted in an increase in sperm quantity collected per ejaculate, it also demonstrated the ability to improve motility prior to and following freezing. Following papain treatment and dilution in Butocrio, the motility of spermatozoa from viscous, low sperm fractions prior to freezing was improved by 10%. Post-thaw, papain-treated samples displayed superior freezing resilience (motility and viability) compared to untreated viscous samples, as well as samples frozen from the sperm-rich fraction. The post-thaw motility of the sperm-rich samples recorded in the current study was similar to that recorded previously in southern white rhinos by O'Brien, et al. [40] but lower than that reported by Hermes, Hildebrandt and Göritz [11]. We hypothesise this difference to be related to the age, sexual maturity and original semen quality of several bulls included in Exp 2. In any event, the low post-thaw motility of the sperm-rich samples in the current study highlights more so the improvements which could be made following treatment with papain, evidenced by a 16% increase in post-thaw motility when compared to the sperm-rich control (Table 2). This could also suggest a beneficial effect of papain on spermatozoa originating from the sperm-rich fraction, where removal or enzymatic treatment of any existing seminal plasma prior to freezing could improve survival rates. While seminal plasma has been shown to be beneficial to the freezing process for species such as the ram [41], its presence during stallion sperm cryopreservation has also been questioned [42,43]. This similarity in the contradictory nature of seminal plasma during freezing is unsurprising given that the horse is the closest domestic relative of the rhinoceros [44]. Regardless, the above results suggest that treatment of all rhinoceros ejaculate fractions with papain could be beneficial for improving the cryosurvival of spermatozoa for ARTs.

The mode of action by which papain improves the quality of rhinoceros semen collected from viscous fractions remains to be fully understood. We suspect that the issue of semen viscosity is not related to the volume of plasma produced but rather the presence or concentration of particular protein/s. We also know from previous studies in alpaca, that the proteinase papain targets glycosylated proteins within the seminal fluid, rather than glycosaminoglans [23]. This strongly supports the findings by Behr and colleagues in 2009, who identified the presence of a 250 kDa glycoprotein (P250) within the viscous seminal plasma of southern white rhino seminal plasma originating from the bulbourethral gland [15]. The occurrence and intensity of P250 also correlated strongly with increasing grades of fraction viscosity [15]. In other species, glycoproteins secreted by the bulbourethral glands have contributed to both the gelatinous plug produced at the end of the boars ejaculate [45,46] and the entrapment of spermatozoa within the uterus of camelids [47]. With this in mind, it is highly likely that in the current study, papain reduced the viscosity of ejaculates by targeting P250. This breakdown of P250 allowed spermatozoa to swim more freely after dilution and increased the uniformity of cryoprotectant exposure, ultimately contributing to better protection during the freezing process. This, in turn, saw a benefit for the survival of spermatozoa from the viscous fraction post-thaw, particularly in relation to the noted improvement in cell viability.

Of course, the identity of this protein is the missing part of the puzzle and would help enable future research to confirm our hypothesis. At the time of publication, although good mass spectrometry spectra was generated, Behr and colleagues, and colleagues in 2009 [15] failed to match a protein identity to P250. However, since this time significant development of the rhinoceros genome has occurred, so repeating the search against the genome would be prudent to determine any further leads. It would also be interesting to see if this protein is conserved across other species of rhinoceros, such as the critically endangered black [48], Sumatran [49] and Javan [50] rhinoceroses and vulnerable greater one-horned rhinoceros [51], particularly the Sumatran rhinoceros, given less than 30 individuals remain, and all published reports of semen collection report severely low sperm counts and volume [39,52]. Similarly, comparison against the seminal plasma proteome of the hippopotamus [19], elephants [20,35] and primates [22] would show whether this protein is responsible for causing the production of viscous fractions across a wide variety of species. More so, it would confirm whether papain could be used as a tool to improve the quantity and quality of spermatozoa collected and cryopreserved from wildlife species following electro-ejaculation.

Importantly, when compared to spermatozoa from the sperm-rich fraction, the viability, acrosome integrity, production of reactive oxygen species and lipid membrane disorder was not impacted following treatment with papain. In other words, papain was not detrimental to the functioning of the sperm cell. Similar results were seen in the alpaca where viscosity was reduced with papain [23], and rhino with α-amylase and collagenase [15]; however, no improvement in the ability of rhinoceros spermatozoa to survive freezing was seen either. Sperm membrane characteristics were likely preserved by the addition of the inhibitor, E-64, which prevented any prolonged exposure and eventual target or degradation of the sperm cell during the chilling and freezing process [30]. E-64 is a cysteine protease inhibitor which binds to the active thiol group and reduces the functionality of papain [53,54]. The concentration of papain and E-64 used in the current study mirrored that reported in the alpaca studies [30]. This concentration (0.1 μg/mL) was successful at reducing the viscosity of alpaca semen, with no detriment to freezing or pregnancy rates following AI of females [31]. Pleasingly, 0.1 μg/mL was also successful at reducing the viscosity of rhinoceros samples completely. However, it could be useful to further optimise the protocol in case further improvements could be made, such as the length of enzyme incubation, to reduce overall processing time. Confirmation of pregnancy following AI in the rhinoceros would also be beneficial.

Furthermore, to the best of our knowledge, the current study reports, for the first time, the advanced membrane and molecular characteristics of white rhinoceros spermatozoa, assessed post-thaw using flow cytometry. The viability reported post-thaw is similar to the 48% reported in 2009 [6] which assessed the percentage of live cells using a hypo-osmotic swelling (HOS) test prior to artificial insemination. Lipid membrane disorder or membrane fluidity, as measured by median M540 fluorescence, is considered an early indication of capacitation and apoptotic-like changes [55], a sperm maturation step which occurs within the female and is crucial for fertilisation (for a review please see [56]). Premature capacitation of spermatozoa is a common artifact of freezing and can lead to early apoptosis of the cell [57]. Production of intracellular reactive oxygen species (ROS), as measured by median H2DCFDA fluorescence, is a measure of the harmful free radicals produced during sperm metabolism. If left uncontrolled, excessive ROS levels can lead to cell degradation and DNA damage [58]. Further examination of these rhinoceros sperm cell characteristics is vital to enhance understanding of the molecular nature of rhinoceros spermatozoa during the freezing process, and of how these may influence the outcome of artificial insemination programs, for which the rhinoceros currently records a low success rate [3,4]. Nevertheless, the above results provide an important benchmark for the species.

5. Conclusions

Assisted reproductive technologies are becoming increasingly diverse, complex and revolutionary in their development and application to wildlife species. Cryopreservation and biobanking of gametes, tissues and cell lines from endangered species has the potential to preserve gametes for the long term, maintain genetic diversity and help fight against the threat of extinction. For example, the successful collection of male gametes from the endangered Sumatran rhinoceros in terms of quantity and quality can play a key role in the de-extinction of the species in the future [4]. The current study has contributed towards this goal by demonstrating the ability of papain to ensure high quantities of viable rhinoceros spermatozoa are collected from ejaculates, even if fractions are deemed low sperm and viscous. It has also proven its ability to increase sperm motility and velocity post-thaw, with no detriment to membrane fluidity, acrosome integrity or intracellular ROS levels. In combination, this data identifies papain as a viable tool to improve the success, not only of rhinoceros cryopreservation protocols, but potentially of other wildlife species, where ARTs are limited by viscous ejaculates, the outcomes of which could contribute towards more successful ARTs and the survival of threatened or endangered species for years to come.

Author Contributions: Conceptualization, J.P.R., S.P.d.G. and R.H.; methodology, J.P.R., K.P. and R.H.; formal Analysis, J.P.R.; investigation, J.P.R., K.P., T.P. and R.H.; resources, N.R., M.W., S.P.d.G., T.B.H., F.G. and R.H.; data curation, J.P.R. and R.H.; writing—original draft preparation, J.P.R.; writing—review and editing, J.P.R., S.P.d.G., K.P., N.R., T.P., M.W., T.B.H., F.G. and R.H.; supervision, J.P.R. and R.H.; project administration, N.R. and J.P.R.; funding acquisition, N.R. All authors have read and agreed to the published version of the manuscript.

Funding: This research received no external funding. J.P Rickard was supported by funding from the McCaughey Memorial Institute.

Institutional Review Board Statement: This study was conducted on captive southern white rhinoceros (*Ceratotherium simum*), a species listed on the IUCN Red List of Threatened species™ as Near Threatened [32]. Semen was collected from bulls as part of a clinical examination of their fertility or for use in artificial inseminations. The study was approved by the IACUC animal ethics committee of the Leibniz Institute for Zoo and Wildlife Research (permit number: 2017-08-02).

Informed Consent Statement: Not applicable.

Data Availability Statement: The data presented in this study are openly available in the Sydney eScholarship Repository at http://dx.doi.org/10.25910/2EVA-0569, accessed on 10 January 2022.

Acknowledgments: The authors would like to acknowledge and thank the veterinary, administration and keeper staff at Werribee Open Range Zoo, VIC; Mogo Wildlife Park, NSW; Monarto Safari Park, SA and Halls Gap Zoo, VIC within Australia, as well as Salsburg Zoo, Austria; Kolmarden Zoo, Sweden and Valencia Bio Park, Spain, for their assistance during anaesthesia, semen collection and travel logistics. The authors are also grateful to Minitube Australia, Smythesdale, Victoria for their donation of straws and extenders towards the project. Jette Ziep is acknowledged for her technical assistance, whilst Naomi Bernecic, Michelle Meyers and Evelyn Hall are also thanked for their guidance on the flow cytometry assessment of sperm and statistical analysis of results.

Conflicts of Interest: The authors declare no conflict of interest. The funders had no role in the choice of research project; design of the study; in the collection, analyses or interpretation of data; in the writing of the manuscript or in the decision to publish the results.

References

1. IUCN 2021. The IUCN Red List of Threatened Species Version 2021-2. Available online: https://www.iucnredlist.org (accessed on 4 November 2021).
2. Hermes, R.; Hildebrandt, T. Rhinoceros Theriogenology. In *Fowler's Zoo and Wildlife Medicine Current Therapy*; Miller, R., Fowler, M., Eds.; Elsevier Saunders: Amsterdam, The Netherlands, 2012; Volume 7, pp. 546–561.
3. Stoops, M.A.; Campbell, M.K.; DeChant, C.J.; Hauser, J.; Kottwitz, J.; Pairan, R.D.; Shaffstall, W.; Volle, K.; Roth, T.L. Enhancing captive Indian rhinoceros genetics via artificial insemination of cryopreserved sperm. *Anim. Reprod. Sci.* **2016**, *172*, 60–75. [CrossRef]
4. Hildebrandt, T.B.; Hermes, R.; Goeritz, F.; Appeltant, R.; Colleoni, S.; de Mori, B.; Diecke, S.; Drukker, M.; Galli, C.; Hayashi, K.; et al. The ART of bringing extinction to a freeze-History and future of species conservation, exemplified by rhinos. *Theriogenology* **2021**, *169*, 76–88. [CrossRef] [PubMed]
5. Platz, C.C., Jr.; Seager, S.W.; Bush, M. Collection and analysis of semen from a black rhinoceros. *J. Am. Vet. Med. Assoc.* **1979**, *175*, 1002–1004. [PubMed]
6. Hermes, R.; Goritz, F.; Saragusty, J.; SÛs, E.; Molnar, V.; Reid, C.E.; Schwarzenberger, F.; Hildebrandt, T.B. First successful artificial insemination with frozen-thawed semen in rhinoceros. *Theriogenology* **2009**, *71*, 393–399. [CrossRef]
7. Portas, T.; Johnston, S.D.; Hermes, R.; Arroyo, F.; López-Fernadez, C.; Bryant, B.; Hildebrandt, T.B.; Göritz, F.; Gosalvez, J. Frozen-thawed rhinoceros sperm exhibit DNA damage shortly after thawing when assessed by the sperm chromatin dispersion assay. *Theriogenology* **2009**, *72*, 711–720. [CrossRef]
8. Stoops, M.A.; Atkinson, M.W.; Blumer, E.S.; Campbell, M.K.; Roth, T.L. Semen cryopreservation in the Indian rhinoceros (*Rhinoceros unicornis*). *Theriogenology* **2010**, *73*, 1104–1115. [CrossRef]
9. Reid, C.E.; Hermes, R.; Blottner, S.; Goeritz, F.; Wibbelt, G.; Walzer, C.; Bryant, B.R.; Portas, T.J.; Streich, W.J.; Hildebrandt, T.B. Split-sample comparison of directional and liquid nitrogen vapour freezing method on post-thaw semen quality in white rhinoceroses (*Ceratotherium simum simum* and *Ceratotherium simum cottoni*). *Theriogenology* **2009**, *71*, 275–291. [CrossRef] [PubMed]
10. Pennington, P.M.; Durrant, B.S. Assisted reproductive technologies in captive rhinoceroses. *Mammal Rev.* **2019**, *49*, 1–15. [CrossRef]
11. Hermes, R.; Hildebrandt, T.B.; Göritz, F. Cryopreservation in rhinoceros—Setting a new benchmark for sperm cryosurvival. *PLoS ONE* **2018**, *13*, e0200154. [CrossRef] [PubMed]

12. Roth, T.L.; Stoops, M.A.; Atkinson, M.W.; Blumer, E.S.; Campbell, M.K.; Cameron, K.N.; Citino, S.B.; Maas, A.K. Semen collection in rhinoceroses (*Rhinoceros unicornis, Diceros bicornis, Ceratotherium simum*) by electroejaculation with a uniquely designed probe. *J. Zoo Wildl. Med.* **2005**, *36*, 617–627. [CrossRef]
13. Schaffer, N.; Bryant, W.; Agnew, D.; Meehan, T.; Beehler, B. Ultrasonographic monitoring of artificially stimulated ejaculation in three rhinoceros species (*Ceratotherium simum, Diceros bicornis, Rhinoceros unicornus*). *J. Zoo Wildl. Med.* **1998**, *29*, 386–393.
14. Portas, T. A review of drugs and techniques used for sedation and anaesthesia in captive rhinoceros species. *Aust. Vet. J.* **2004**, *82*, 542–549. [CrossRef] [PubMed]
15. Behr, B.; Rath, D.; Mueller, P.; Hildebrandt, T.B.; Goeritz, F.; Braun, B.C.; Leahy, T.; de Graaf, S.P.; Maxwell, W.M.C.; Hermes, R. Feasibility of sex-sorting sperm from the white and the black rhinoceros (*Ceratotherium simum, Diceros bicornis*). *Theriogenology* **2009**, *72*, 353–364. [CrossRef]
16. Schaffer, N.E.; Foley, G.L.; Gill, S.; Earl Pope, C. Clinical Implications of rhinoceros reproductive tract anatomy and histology. *J. Zoo Wildl. Med.* **2001**, *32*, 31–46.
17. Hermes, R.; Hildebrandt, T.B.; Blottner, S.; Walzer, C.; Silinski, S.; Patton, M.L.; Wibbelt, G.; Schwarzenberger, F.; Göritz, F. Reproductive soundness of captive southern and northern white rhinoceroses (*Ceratotherium simum simum, C.s. cottoni*): Evaluation of male genital tract morphology and semen quality before and after cryopreservation. *Theriogenology* **2005**, *63*, 219–238. [CrossRef]
18. Hermes, R.; Saragusty, J.; Göritz, F.; Bartels, P.; Potier, R.; Baker, B.; Streich, W.J.; Hildebrandt, T.B. Freezing African Elephant Semen as a New Population Management Tool. *PLoS ONE* **2013**, *8*, e57616. [CrossRef] [PubMed]
19. Saragusty, J.; Hildebrandt, T.B.; Bouts, T.; Göritz, F.; Hermes, R. Collection and preservation of pygmy hippopotamus (*Choeropsis liberiensis*) semen. *Theriogenology* **2010**, *74*, 652–657. [CrossRef]
20. Tarmizi, R.; Keng Chee, Y.; Sipangkui, S.; Zainuddin, Z.Z.; Fitri, W.-N. The Comparison of Semen Collection in Electroejaculation, Rectal Massage and Combination of Both Methods in the Critically Endangered Malayan Pangolin, Manis javanica. *Animals* **2020**, *10*, 1948. [CrossRef] [PubMed]
21. Dixson, A.F.; Anderson, M.J. Sexual selection, seminal coagulation and copulatory plug formation in primates. *Folia Primatol.* **2002**, *73*, 63–69. [CrossRef]
22. Kingan, S.B.; Tatar, M.; Rand, D.M. Reduced polymorphism in the chimpanzee semen coagulating protein, semenogelin I. *J. Mol. Evol.* **2003**, *57*, 159–169. [CrossRef]
23. Kershaw-Young, C.M.; Stuart, C.; Evans, G.; Maxwell, W.M.C. The effect of glycosaminoglycan enzymes and proteases on the viscosity of alpaca seminal plasma and sperm function. *Anim. Reprod. Sci.* **2013**, *138*, 261–267. [CrossRef] [PubMed]
24. Kershaw-Young, C.; Maxwell, W. Seminal Plasma Components in Camelids and Comparisons with Other Species. *Reprod. Domest. Anim.* **2012**, *47*, 369–375. [CrossRef]
25. Rodger, J.C.; White, I.G. The collection, handling and some properties of marsupial semen. *Symp. Zool. Soc. Lond.* **1978**, *43*, 289–301.
26. Paris, D.B.B.P.; Taggart, D.A.; Shaw, G.; Temple-Smith, P.D.; Renfree, M.B. Changes in semen quality and morphology of the reproductive tract of the male tammar wallaby parallel seasonal breeding activity in the female. *Reproduction* **2005**, *130*, 367–378. [CrossRef] [PubMed]
27. Hildebrandt, T.B.; Hermes, R.; Jewgenow, K.; Göritz, F. Ultrasonography as an important tool for the development and application of reproductive technologies in non-domestic species. *Theriogenology* **2000**, *53*, 73–84. [CrossRef]
28. Bravo, P.W.; Ccallo, M.; Garnica, J. The effect of enzymes on semen viscosity in Llamas and Alpacas. *Small Rumin. Res.* **2000**, *38*, 91–95. [CrossRef]
29. Giuliano, S.; Carretero, M.; Gambarotta, M.; Neild, D.; Trasorras, V.; Pinto, M.; Miragaya, M. Improvement of llama (*Lama glama*) seminal characteristics using collagenase. *Anim. Reprod. Sci.* **2010**, *118*, 98–102. [CrossRef]
30. Kershaw, C.M.; Evans, G.; Rodney, R.; Maxwell, W.M.C. Papain and its inhibitor E-64 reduce camelid semen viscosity without impairing sperm function and improve post-thaw motility rates. *Reprod. Fertil. Dev.* **2017**, *29*, 1107–1114. [CrossRef]
31. Stuart, C.C.; Vaughan, J.L.; Kershaw, C.M.; de Graaf, S.P.; Bathgate, R. Effect of diluent type, cryoprotectant concentration, storage method and freeze/thaw rates on the post-thaw quality and fertility of cryopreserved alpaca spermatozoa. *Sci. Rep.* **2019**, *9*, 12826. [CrossRef] [PubMed]
32. Emslie, R. *Ceratotherium simum*. The IUCN Red List of Threatened Species 2020. e.T4185A45813880. Available online: http://dx.doi.org/10.2305/IUCN.UK.2012.RLTS.T4185A16980466.en (accessed on 18 September 2021).
33. Evans, G.; Maxwell, W.M.C. *Salamons' Artificial Insemination of Sheep and Goats*; Butterworths: Sydney, Australia, 1987.
34. Saragusty, J.; Hildebrandt, T.B.; Behr, B.; Knieriem, A.; Kruse, J.; Hermes, R. Successful cryopreservation of Asian elephant (*Elephas maximus*) spermatozoa. *Anim. Reprod. Sci.* **2009**, *115*, 255–266. [CrossRef]
35. Susko-Parrish, J.; Parrish, J.J.; Winer, M.A.; First, N.L. Capacitation of Bovine Sperm by Heparin1. *Biol. Reprod.* **1988**, *38*, 1171–1180. [CrossRef]
36. Rijnders, S.; Bolscher, J.G.; McDonnell, J.; Vermeiden, J.P. Filling time of a lamellar capillary-filling semen analysis chamber is a rapid, precise, and accurate method to assess viscosity of seminal plasma. *J. Androl.* **2007**, *28*, 461–465. [CrossRef] [PubMed]
37. Lee, J.A.; Spidlen, J.; Boyce, K.; Cai, J.; Crosbie, N.; Dalphin, M.; Furlong, J.; Gasparetto, M.; Goldberg, M.; Goralczyk, E.M.; et al. MIFlowCyt: The minimum information about a flow cytometry experiment. *Cytom. Part A* **2008**, *73A*, 926–930. [CrossRef] [PubMed]

38. Pool, K.; Rickard, J.; de Graaf, S. Melatonin improves the motility and DNA integrity of frozen-thawed ram spermatozoa likely via suppression of mitochondrial superoxide production. *Domest. Anim. Endocrinol.* **2020**, *74*, 106516. [CrossRef]
39. O'Brien, J.K.; Roth, T.L. Post-coital sperm recovery and cryopreservation in the Sumatran rhinoceros (*Dicerorhinus sumatrensis*) and application to gamete rescue in the African black rhinoceros (*Diceros bicornis*). *J. Reprod. Fertil.* **2000**, *118*, 263–271. [CrossRef]
40. O'Brien, J.K.; Roth, T.L.; Stoops, M.A.; Ball, R.L.; Steinman, K.J.; Montano, G.A.; Love, C.C.; Robeck, T.R. Sperm sex-sorting and preservation for managing the sex ratio and genetic diversity of the southern white rhinoceros (*Ceratotherium simum simum*). *Anim. Reprod. Sci.* **2015**, *152*, 137–153. [CrossRef]
41. Rickard, J.; Schmidt, R.; Maddison, J.; Bathgate, R.; Lynch, G.; Druart, X.; De Graaf, S. Variation in seminal plasma alters the ability of ram spermatozoa to survive cryopreservation. *Reprod. Fertil. Dev.* **2016**, *28*, 516–523. [CrossRef]
42. Moore, A.I.; Squires, E.L.; Graham, J.K. Effect of seminal plasma on the cryopreservation of equine spermatozoa. *Theriogenology* **2005**, *63*, 2372–2381. [CrossRef]
43. Aurich, J.E.; Kühne, A.; Hoppe, H.; Aurich, C. Seminal plasma affects membrane integrity and motility of equine spermatozoa after cryopreservation. *Theriogenology* **1996**, *46*, 791–797. [CrossRef]
44. Bhaskar, R.; Kanaparthi, P.; Sakthivel, R. DNA barcode approaches to reveal interspecies genetic variation of Indian ungulates. *Mitochondrial DNA Part B-Resour.* **2020**, *5*, 938–944. [CrossRef]
45. Druart, X.; Rickard, J.P.; Tsikis, G.; de Graaf, S.P. Seminal plasma proteins as markers of sperm fertility. *Theriogenology* **2019**, *137*, 30–35. [CrossRef]
46. Rodriguez-Martinez, H.; Kvist, U.; Saravia, F.; Wallgren, M.; Johannisson, A.; Sanz, L.; Pena, F.J.; Martinez, E.A.; Roca, J.; Vazquez, J.M.; et al. The physiological roles of the boar ejaculate. *Soc. Reprod. Fertil. Suppl.* **2009**, *66*, 1–21. [CrossRef] [PubMed]
47. Apichela, S.A.; Argañaraz, M.E.; Giuliano, S.; Zampini, R.; Carretero, I.; Miragaya, M.; Miceli, D.C. Llama oviductal sperm reservoirs: Involvement of bulbourethral glands. *Andrologia* **2014**, *46*, 290–295. [CrossRef] [PubMed]
48. Emslie, R. *Diceros bicornis*. The IUCN Red List of Threatened Species 2020. e.T6557A152728945. Available online: https://dx.doi.org/10.2305/IUCN.UK.2020-1.RLTS.T6557A152728945.en (accessed on 21 September 2021).
49. Ellis, S.; Talukdar, B. *Dicerorhinus sumatrensis*. The IUCN Red List of Threatened Species 2020. e.T6553A18493355. Available online: https://dx.doi.org/10.2305/IUCN.UK.2020-2.RLTS.T6553A18493355.en (accessed on 21 September 2021.).
50. Ellis, S.; Talukdar, B. *Rhinoceros sondaicus*. The IUCN Red List of Threatened Species 2020. e.T19495A18493900. Available online: https://dx.doi.org/10.2305/IUCN.UK.2020-2.RLTS.T19495A18493900.en (accessed on 21 September 2021).
51. Ellis, S.; Talukdar, B. *Rhinoceros unicornis*. The IUCN Red List of Threatened Species 2019. e.T19496A18494149. Available online: https://dx.doi.org/10.2305/IUCN.UK.2019-3.RLTS.T19496A18494149.en (accessed on 21 September 2021).
52. Fiúza, A. *The Sabah Rhino Breeding Programme: Reproductive Management of the Critically Endangered Sumatran Rhinoceros of Borneo (Dicerorhinus sumatrensis harrissoni) as Conducted by the IZW-Berlin between 2005 and 2015*; Universidade de Lisboa: Lisboa, Portugal, 2017.
53. Tamai, M.; Hanada, K.; Adachi, T.; Oguma, K.; Kashiwagi, K.; Omura, S.; Ohzeki, M. Papain Inhibitions by Optically Active E-64 Analogs. *J. Biochem.* **1981**, *90*, 255–257. [CrossRef] [PubMed]
54. Barrett, A.J.; Kembhavi, A.A.; Hanada, K. E-64 [L-trans-epoxysuccinyl-leucyl-amido(4-guanidino)butane] and related epoxides as inhibitors of cysteine proteinases. *Acta Biol. Med. Ger.* **1981**, *40*, 1513–1517.
55. Bernecic, N.C.; de Graaf, S.P.; Leahy, T.; Gadella, B.M. HDL mediates reverse cholesterol transport from ram spermatozoa and induces hyperactivated motility. *Biol. Reprod.* **2021**, *104*, 1271–1281. [CrossRef]
56. Gadella, B.M.; Luna, C. Cell biology and functional dynamics of the mammalian sperm surface. *Theriogenology* **2014**, *81*, 74–84. [CrossRef] [PubMed]
57. Bailey, J.L.; Blodeau, J.; Cormier, N. Semen Cryopreservation in Domestic Animals: A Damaging and Capacitating Phenomenon Minireview. *J. Androl.* **2000**, *21*, 1–7. [CrossRef]
58. Sanocka, D.; Kurpisz, M. Reactive oxygen species and sperm cells. *Reprod. Biol. Endocrinol.* **2004**, *2*, 12. [CrossRef]

biology MDPI

Article

Testicular Ultrasound Analysis as a Predictive Tool of Ram Sperm Quality

Melissa Carvajal-Serna, Sara Miguel-Jiménez, Rosaura Pérez-Pe and Adriana Casao *

Grupo BIOFITER-Departamento de Bioquímica y Biología Molecular y Celular, Instituto Universitario de Investigación en Ciencias Ambientales de Aragón (IUCA), Facultad de Veterinaria, Universidad de Zaragoza, 50013 Zaragoza, Spain; melissac@unizar.es (M.C.-S.); 652325@unizar.es (S.M.-J.); rosperez@unizar.es (R.P.-P.)
* Correspondence: adriana@unizar.es

Simple Summary: In animal production, the prediction of male fertility is vital for the success of specific techniques such as artificial insemination. Thus, testicular ultrasound, a non-invasive diagnostic procedure, could be a useful tool. Moreover, recent ultrasound-video analysis and software developments allow the visualization of tissue at the microscopic level. The objective of this work was to establish a possible correlation between testicular ultrasonography and semen quality in rams. For this purpose, the testicles of nine rams were evaluated and the semen was analyzed for one year. The results revealed that the number of white and grey pixels correlated with sperm parameters indicating poor seminal quality. On the other hand, the increase in the seminiferous-tubule density or the lumen area of these tubules was related to a rise in seminal quality. Therefore, ultrasound-video analysis could be a good tool for evaluating the fertility of rams, either for artificial insemination or on the farm.

Abstract: Testicular ultrasound is a non-invasive technique that could be very useful for predicting ram seminal quality. Recent software developments allow macroscopic and microscopic evaluation of testicular parenchyma. Thus, the aim of this study was to evaluate the testicular echotexture using ultrasound-video analysis and investigate its possible correlation with semen quality. Nine rams were evaluated for one year using a portable ultrasound scanner and the echotexture was analyzed with ECOTEXT® software. The number of black (Ec1), white (Ec2), and grey pixels (Ec3), tubular density (TD), lumen area (LA), and lumen diameter (LD) were analyzed. Semen was collected by an artificial vagina the same day and the sperm concentration, morphology, motility, viability, phosphatidylserine (PS) translocation, reactive-oxygen-species (ROS) levels, DNA damage and capacitation state were evaluated. Ec2 and Ec3 correlated positively with "bad quality" sperm parameters (the percentage of spermatozoa with high ROS levels, with PS translocation and proximal cytoplasmic droplets), and negatively with motility. In contrast, TD and LA showed a positive correlation with "good quality" parameters (motility or normal morphology) and a negative correlation with spermatozoa with high ROS levels, with DNA fragmentation, and proximal or distal cytoplasmic droplets. Thus, echotexture analysis by ultrasound-video analysis could be a valuable tool for assessing ram fertility.

Keywords: echotexture; ultrasound; Doppler; seasonality; sperm; testes

Citation: Carvajal-Serna, M.; Miguel-Jiménez, S.; Pérez-Pe, R.; Casao, A. Testicular Ultrasound Analysis as a Predictive Tool of Ram Sperm Quality. *Biology* **2022**, *11*, 261. https://doi.org/10.3390/biology11020261

Academic Editor: Paul S. Cooke

Received: 13 January 2022
Accepted: 5 February 2022
Published: 8 February 2022

Publisher's Note: MDPI stays neutral with regard to jurisdictional claims in published maps and institutional affiliations.

1. Introduction

Seasonality, which is regulated by photoperiod and melatonin secretion [1], could be a limiting factor in the sheep reproduction [2]. In the ram, the effect of seasonality is less marked than in the ewe, and some breeds experience variable degrees of seasonality [3–5], mainly depending on the latitude in which they are located [6]. The study of seasonality changes in the ram has focused on seminal quality [7,8] or testicular measurement [9,10] rather than changes in testicular parenchyma.

Ultrasonographic evaluation is a valuable, non-invasive technique for diagnosing the male genital tract. It can be used as a complementary test for breeding-soundness evaluation [11]. Additionally, changes in the testis echotexture (pixel intensity and uniformity) due to changes in the tissue composition can be evaluated by ultrasonography and assessed with image-analysis software [12,13] that describes the ultrasound pixel intensity in terms of numerical pixel values (NPV). These range from 0 (absolute black) to 255 (absolute white) and provide an indication of tissue echogenicity [14]. Thereby, ultrasound analysis has been used to predict the reproductive soundness of rams [15–17] or the detection of pubertal changes in lambs [18,19]. However, in these previous studies, the ultrasound images were evaluated by non-specific image-analysis software (such as Image ProPlus® or Image J), which could decrease the sensitivity and reproducibility of the results. Recently, a dedicated software package has been specifically developed for studying testicular tissue through macroscopic and microscopic ultrasound-video analysis (Ecotext® software). This software is able to analyze testicle ultrasonograms based on their echotexture properties, measuring pixel intensity as Ec1 (black pixels, the number of pixels with NPV = 0), Ec2 (white pixels, the number of pixels with NPV = 255), and Ec3 (grey level of pixels, the mean value of the pixels with NPV > 0 and <255). It can also evaluate microscopic structures of the testicular parenchyma (up to 200 μm in size) that are usually evaluated by biopsy, such as the seminiferous-tubule (ST) density or the ST lumen area and diameter. This software has already been validated in bulls, rams, stallions, and boars [20–23].

Testicular parenchyma constitutes a tissue of very high metabolic activity, and it is therefore very sensitive to changes in the blood supply by the testicular artery. Furthermore, testicular blood flow has been correlated with spermatogenesis and spermatozoa fertilization ability in several species [24,25]. Thus, a color-Doppler analysis could provide information about the testicular-artery blood flow and vascular integrity of the male reproductive tract [14]. Color-Doppler has previously been used to evaluate changes in blood-flow velocity and the resistance of the blood in the testicular artery of the ram [19,26] and their relationship with semen quality [27,28].

The main goal of testicular ultrasound, other than to discard males with testicular pathology [29], is to predict the seminal quality and, thus, the fertility of sires. In this context, previous works on the ram have found a moderate correlation between testes blood flow, echotexture and sperm quality [16,27,28], whereas others have not found any correlation [17]. In most of these works, only classical sperm parameters, such as volume, motility, concentration or viability, were evaluated. However, recent studies suggest that these classical sperm parameters have a low correlation with field fertility, and parameters related to sperm physiology, such as oxidative stress, apoptotic levels or capacitation state, should be evaluated for ram-fertility prediction [30,31]. Moreover, spermatogenesis in the ram lasts around sixty days, including epididymal storage. For this reason, the results of a semen evaluation performed at any given time do not reflect the current testicular function and, in the same way, the testicular ultrasound analysis likely relates to future semen quality, but not to the sperm parameters obtained on the day of ultrasound scanning [15].

Hence, the aim of this study was (1) to evaluate seasonal changes in testicular echotexture parameters and testicular blood flow using ultrasound-video analysis and color-Doppler analysis, (2) to investigate the possible correlation between changes in ultrasonographic parameters and semen quality (evaluated with classical and non-conventional sperm-quality parameters) at the moment of scanning (present) and 30 and 60 days later.

2. Materials and Methods

Unless otherwise stated, all reagents were purchased from Merck KGaA (Darmstadt, Germany).

2.1. Animals and Semen Collection

Semen was collected from nine *Rasa Aragonesa* rams (2–8 years old) belonging to the *Rasa Aragonesa* National Breeding Association (Asociación Nacional de Criadores de

Ganado Ovino Selecto de Raza Rasa Aragonesa, ANGRA). The rams were kept under uniform feeding conditions at the University of Zaragoza Veterinary School, Spain (latitude 41°41′ N). All experimental procedures were carried out under the project license PI39/17 approved by the Ethics Committee for Animal Experimentation of the University of Zaragoza (approval date: 24 May 2017), according to the Spanish Policy for Animal Protection RD53/2013, which meets the European Union Directive 2010/63 on the protection of animals used for experimental and other scientific purposes.

The rams were subjected to a regimen of continuous semen extraction throughout the year, with two days of abstinence between collections. For this study, a monthly semen evaluation was performed for one year. Second ejaculates from each male were collected individually using an artificial vagina, and semen was maintained at 37 °C until laboratory analysis. Semen samples were diluted 1/100 in a medium containing 0.25 mol/L sucrose, 10 mmol/L Hepes, 2 mmol/L KOH, 5 mmol/L glucose, 0.5 mol/L NaH_2PO_4 and 100 mmol/L EGTA for assessing sperm quality. Sperm concentration was calculated in duplicate using a Neubauer chamber (Marienfeld, Lauda-Konigsofen, Germany).

2.2. *Sperm Motility Evaluation*

Motility parameters were measured using a computer-assisted CASA system (ISAS 1.0.4; Proiser SL, Valencia, Spain). Sperm motility was recorded utilizing a video camera (Basler A312f, Basler AG, Ahrensburg, Germany) mounted on a microscope (Nikon Eclipse 50i, Nikon Instruments Inc, Tokyo, Japan) equipped with a 10x negative-phase contrast lens and a 10x ocular lens. Samples (6 µL of the 1/100 semen dilution) were placed between pre-warmed slides and coverslips and kept at 37 °C in a heated slide holder during analysis. For each sample, five videos at 25 frames/second for 1 s were recorded. The percentages of motile (TM) and progressive motile (PM) spermatozoa were evaluated.

2.3. *Flow Cytometry Analysis*

All the sperm cytometry analyses were performed on a Beckman Coulter FC 500 flow cytometer with CXP software (Beckman Coulter Inc., Brea, CA, USA) equipped with two excitation lasers (Argon-ion laser 488 nm and Red solid-state laser 633 nm) and five absorbance filters (FL1–525, FL2–575, FL3–610, FL4–675 and FL5–755 ± 5 nm each bandpass filter). A minimum of 20,000 events were evaluated in all the experiments. The sperm population was identified for further analysis based on their specific-forward (FS) and side-scatter (SS) properties; thus, other non-sperm events were excluded. A flow rate stabilized at 200–300 cells/sec was used.

2.3.1. Sperm Viability

Three microliters of carboxyfluorescein diacetate (CFDA, 1 mM), 3 µL of propidium iodide (PI, 0.75 mM) and 5 µL of formaldehyde (0.5% v/v in water) were added to 500 µL of sperm samples (6×10^6 cells/mL) based on a modification of the method described by Harrison and Vickers [32]. After 15 min at 37 °C in the dark, the samples were analyzed by flow cytometry. The Argon-ion laser and the 525 (FL1, CFDA) and 675 nm (FL4, PI) filters were used in order to avoid overlapping. The percentage of CFDA+/PI- spermatozoa (viable cells) was evaluated.

2.3.2. Intracellular Reactive Oxygen Species (ROS)

Aliquots of samples (500 µL), prepared at a final concentration of 6×10^6 cells/mL, were stained with 5 µL of H_2DCFDA (20 mM) and PI (1.5 mM). After 15 min at 37 °C in the dark, the samples were fixed with 5 µL formaldehyde (0.5% v/v in water) and analyzed by flow cytometry [33]. The Argon-ion laser and 525 and 675 nm filters were used in order to avoid overlapping. The parameters monitored were FS log, SS log, FL1 (H_2DCFDA), and FL4 (PI). The percentages of viable spermatozoa with low ROS levels and spermatozoa with high ROS levels were evaluated.

2.3.3. Detection of Membrane Phosphatidylserine (PS) Translocation

Annexin V is a calcium-dependent phospholipid-binding protein with a high affinity for PS. FITC-Annexin V (Thermo Fisher Scientific, Waltham, MA, USA) was used simultaneously with PI to detect PS translocation and to differentiate between membrane-intact and damaged cells, with or without PS translocation. Aliquots of 300 μL (4×10^6 cells diluted in binding buffer) were stained with FITC-Annexin V (2 μL) in combination with 7.5 μM PI (3 μL), incubated at 37 °C in the dark for 15 min, and evaluated by flow cytometry. The monitored parameters were FS log, SS log, FL1 (FITC-Annexin V), and FL4 (PI). The percentages of viable spermatozoa without PS translocation (non-apoptotic, Annx-/PI-), and spermatozoa with PS translocation (Annx+) were considered.

2.3.4. DNA Fragmentation—TUNEL Assay

The presence of DNA strand breaks in ram spermatozoa was assessed using the TUNEL assay with fluorescein-isothiocyanate (FITC)-labeled dUTP (In Situ Cell-Death-Detection Kit)). Sperm samples (4×10^7 cells/mL) were fixed with 4% (*w*/*v*) paraformaldehyde in PBS at room temperature (RT) for 1 h. After two washes at 600× *g* with 100 μL PBS, the samples were permeabilized with 0.1% Triton X-100 (*v*/*v*) in 0.1% sodium citrate (*w*/*v*). After centrifugation at 600× *g*, the pellet obtained was incubated with 50 μL of labelling solution containing the TdT enzyme and dUTP for 1 h at 37 °C in the dark. A negative control was prepared for each experimental set by omitting TdT from the reaction mixture. After two consecutive washes with PBS at 600× *g* for 10 min at RT to stop the reaction, flow-cytometry analysis was performed. The monitored parameters were FS log, SS log, and FL1 (TUNEL). The percentage of sperm with DNA fragmentation (TUNEL+ cells) was evaluated.

2.4. Assessment of Capacitation Status by CTC Staining

The capacitation status was determined using chlortetracycline (CTC) staining [34]. A CTC solution (750 μM) was prepared daily in a buffer containing 20 mM Tris, 130 mM NaCl and 5 μM cysteine (pH 7.8) and passed through a 0.22 μm filter. After that, 20 μL of CTC solution and 5 μL of 12.2% (*w*/*v*) paraformaldehyde in 0.5 M Tris–HCl (pH 7.8) were added to a 20 μL sperm sample (4×10^7 cells/mL) and incubated at 4 °C in the dark for at least 30 min. At room temperature and semi-darkness, a 4 μL aliquot of the stained sample was placed on a glass slide and mixed with 2 μL of 0.22 M triethylenediamine (DABCO) in glycerol:PBS (9:1, *v*/*v*). Samples were covered with 24 × 48 mm coverslips, sealed with colorless enamel, and stored in the dark at −20 °C. To evaluate CTC patterns, samples were examined using a Nikon Eclipse E-400 microscope (Kanagawa, Japan) under epifluorescence illumination using a V-2A filter. All samples were processed in duplicate and 200 spermatozoa were scored per slide. Sperm classification followed Gillan et al. [35]: non-capacitated spermatozoa (NC with even yellow fluorescence over the head, with or without a bright equatorial band); capacitated cells (C, with fluorescence on the acrosome) and acrosome-reacted cells (R, without fluorescence on the head and with or without a bright equatorial band).

2.5. Morphological Study by Eosin-Nigrosine Staining

Semen samples (20 μL of 4×10^7 cells/mL) were mixed with 10 μL eosin and 10 μL nigrosine. One droplet of 20 μL of the stained sample was smeared onto a clean slide with the help of another slide. The smears were air-dried and examined by bright-field microscopy at 1000X magnification using a Nikon Eclipse E-400 microscope (Kanagawa, Japan). At least 200 spermatozoa were analyzed. The percentage of cells with normal morphology and abnormal spermatozoa, including primary (detached head) and secondary (proximal or distal droplet, bent tail and coiled tail) abnormalities, were evaluated [36].

2.6. Testes Measurement and Ultrasonography Examination

Scrotal circumference (SC) was measured using a measuring tape positioned around the scrotum's largest circumference. Testis length (TL) and width (TW) were measured using a caliper. TL was measured from the head of the epididymis to the top of the tail of each testis, whereas TW was evaluated in the widest part of the testis. Testicular volume (TV) was determined using the equation proposed by Godfrey et al. [37]: TV = 0.0396 × average testis length of both testes (TL) × SC^2.

The ultrasonography evaluation of the testis was carried out using a portable ultrasound scanner (ExaGo, IMV imaging Angoulême, France) connected to a 7.5 MHz linear probe. The same researcher performed all evaluations, and the ultrasound scanner parameters were adjusted to 60 mm depth, 100% power, 0 dB gain and 60 dB dynamic range. The rams were restrained and no sedatives were used. The probe was positioned transversely to major axis of the testicle. Three videos of 124 frames each were recorded in the upper, medium and lower parts of each testicle for echotexture analyses.

The echotexture analysis was performed with ECOTEXT® software (Humeco, Huesca, Spain). Three testicular parenchymatic characteristics were evaluated at standard resolution: the number of black (ECOTEXT 1, Ec1, referring to the number of pixels with a numerical pixel value of 0), white (ECOTEXT 2, Ec2, referring to the number of pixels with a numerical pixel value of 255), and grey pixels (ECOTEXT 3, Ec3, mean value of the pixels with a numerical pixel value > 0 and <255); another three were evaluated at high resolution: the density of tubules/cm^2 (tubular density, TD), the percentage (%) of the total area occupied by the lumen of the tubules in the parenchyma (lumen area, LA), and the mean diameter (μm) of the lumen of the seminiferous tubules (lumen diameter, LD).

Color-Doppler flow imaging was used to analyze the arterial blood flow of the pampiniform plexus. All color-Doppler scans were performed with a constant gain (20 dB), 50 Hz high pass filter, and 1 mm gate setting (Hedia et al., 2019). The angle between the Doppler beam and each vessel's long axis was ≤60, and the pulse-repetition frequency (2,000 Hz) was adjusted to reduce aliasing. After the spectral pattern of the testicular artery was generated, the frequency (bpm), the resistive index (RI = (maximum velocity−minimum velocity)/maximum velocity) and the pulsatility index (PI = (maximum velocity−minimum velocity)/mean velocity) were calculated [38]. Five measurements were taken along the path of the artery for each testis, and at least five waveforms were recorded per measurement.

2.7. Statistics Analyzes

Monthly and seasonal results are shown as mean ± SEM of the number of samples assessed in each case. Data distribution was analyzed by the Kolmogorov–Smirnov test, and outliers were identified by the Grubbs test. The difference between breeding (B: August to February) and non-breeding seasons (NB: March to July) in terms of concentration and progressive motility, TW, Ec3, TD, LD, frequency and RI were analyzed by the unpaired *t*-test. The rest of the parameters were assessed by the Mann–Whitney test. Differences between the right and left testicles were evaluated by the paired *t*-test for width, Ec3, TD, LD, frequency and RI, and the Wilcoxon matched-pairs signed-rank test for their length, Ec1, Ec2, LA and PI. The correlations between the ultrasound results and the sperm parameters, which were obtained on the same day, 30 and 60 days after the testicular scanning, were evaluated by Spearman's test. All statistical analyses were performed using SPSS (v.15.0) software.

3. Results

In this study, nine rams were subjected to semen collection and sperm-quality evaluation, followed by an ultrasonography examination of the testes every month for one year. All the rams showed suitable sperm quality after evaluation, but one showed calcification spots during the ultrasonography examination; hence, that male was eliminated from the study.

3.1. Changes in Size and Ultrasound Evaluation of the Testes and Ram Sperm Quality between Seasons

When the left and right testes were compared, only the testicular length, tubular density (evaluated by ultrasonography), and frequency (evaluated by Doppler analysis) showed significant differences ($p < 0.05$, Supplemental Table S1). Thus, the mean values of the right and left testicles were used for further evaluation.

No differences were found between the breeding (B) and non-breeding (NB) seasons in terms of testis size (Table 1), although an increase in testicular length and volume during the hottest months of the year can be observed (Supplemental Figure S1A). However, regarding the scanning results, we found seasonal differences ($p < 0.05$) in Ec1 and the seminiferous-tubule lumen diameter (LD), which increased during the B season (Table 1). The rest of the echogenicity parameters did not show any changes. Of the Doppler parameters, only the frequency significantly increased ($p < 0.05$) during the non-breeding season (Table 1), although a non-significant decrease in PI was observed in the coldest months of the year (Supplemental Figure S1B).

Table 1. Seasonal differences in testicular size, echotexture parameters (evaluated by ultrasound-video analysis and ECOTEXT® software) and testicular-artery blood flow (assessed by color-Doppler flow imaging) in *Rasa Aragonesa* rams. Results are shown as mean ± SEM. Different letters indicate $p < 0.05$.

	Breeding Season (n = 56)	Non-Breeding Season (n = 40)
Testicular size		
Scrotal circumference (cm)	35.1 ± 0.4	36.21 ± 0.39
Testicular width (cm)	7.09 ± 0.08	7.21 ± 0.08
Testicular length (cm)	10.54 ± 0.19	10.69 ± 0.2
Testicular volume (cm3)	524.43 ± 18.6	563.3 ± 21.42
ECOTEXT® parameters		
Ec1 (black pixels)	23.24 ± 2.46 ^{a}	14.03 ± 1.88 ^{b}
Ec2 (white pixels)	166.21 ± 19.7	187.29 ± 27.96
Ec3 (grey pixels)	100.96 ± 0.97	103.1 ± 1.3
TD (density of tubules/cm3)	147.97 ± 2.11	151.25 ± 1.91
LA (lumen area, %)	8.54 ± 0.3	7.71 ± 0.32
LD (lumen diameter, µm)	103.68 ± 1.49 ^{a}	97.67 ± 1.58 ^{b}
Doppler parameters		
Frequency (bpm)	90.02 ± 2.51 ^{a}	96.69 ± 2.06 ^{b}
RI (Resistive index)	0.56 ± 0.02	0.55 ± 0.02
PI (Pulsatility index)	0.89 ± 0.04	0.84 ± 0.05

When sperm quality was evaluated, significant differences ($p < 0.05$) between the breeding and non-breeding seasons were observed in the percentages of spermatozoa with high ROS levels (Table 2), with normal morphology, with a detached head (Figure 1), and in the rate of acrosome-reacted cells (Figure 2). Nonetheless, when differences between months were evaluated, significant differences were observed in nearly all the studied parameters, being more marked between March–April and September (Supplemental Figure S2).

Table 2. Seasonal differences in seminal quality parameters in *Rasa Aragonesa* rams. Results are shown as mean ± SEM. Different letters indicate $p < 0.05$.

Sperm Parameter	Breeding Season ($n = 56$)	Non-Breeding Season ($n = 40$)
Sperm concentration ($\times 10^6$ cells/mL)	4203.36 ± 176.74	4561.00 ± 237.38
Total motility (%)	90.36 ± 0.69	91.28 ± 0.66
Progressive motility (%)	41.66 ± 1.24	44.95 ± 1.52
Membrane integrity (Viability %)	75.99 ± 1.44	80.37 ± 1.15
Viable spermatozoa without PS [1] translocation (%)	62.17 ± 2.17	59.98 ± 2.27
Spermatozoa with PS [1] translocation (%)	15.23 ± 1.12	16.85 ± 1.13
Viable spermatozoa with low ROS [2] levels (%)	73.3 ± 1.3	74.55 ± 0.97
Spermatozoa with high ROS [2] levels (%)	7.73 ± 0.35 [a]	9.79 ± 0.4 [b]
Sperm with DNA fragmentation (%)	8.55 ± 0.46	8.78 ± 0.52

[1] PS: phosphatidylserine. [2] ROS: reactive oxygen species.

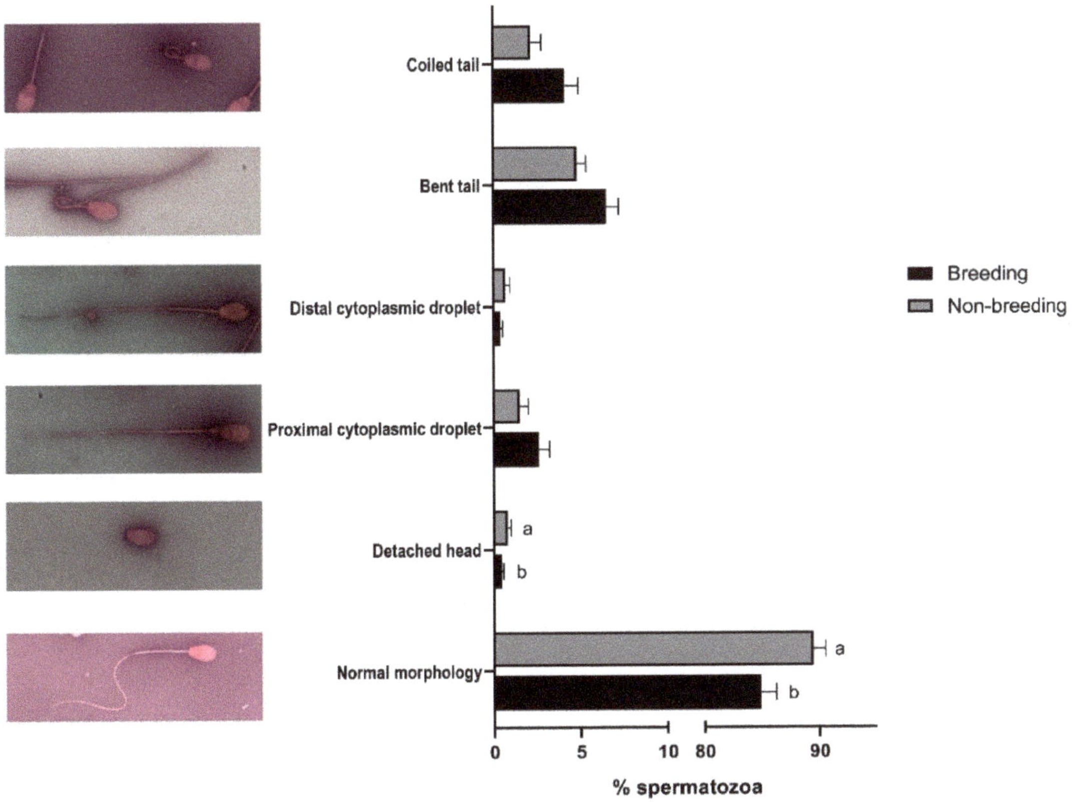

Figure 1. Seasonal differences in sperm morphology in *Rasa Aragonesa* rams. Results are shown as mean ± SEM of $n = 56$ (breeding season) and $n = 40$ (non-breeding season). Different letters indicate $p < 0.05$. Representative images of a normal spermatozoa and sperm abnormalities are also shown.

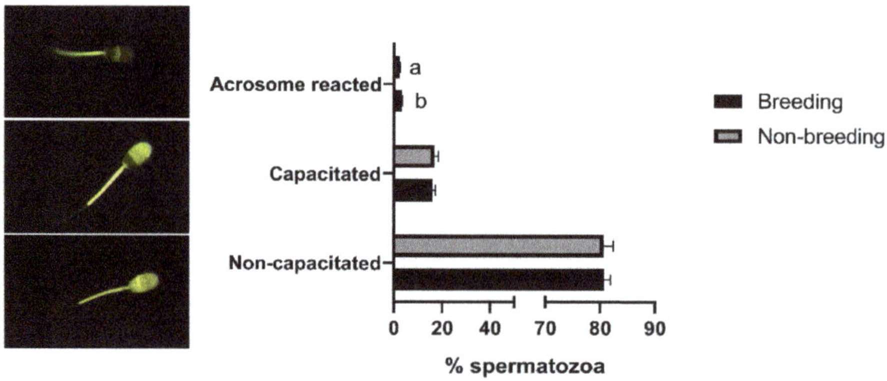

Figure 2. Seasonal differences in sperm capacitation state in *Rasa Aragonesa* rams. Results are shown as mean ± SEM of $n = 56$ (breeding season) and $n = 40$ (non-breeding season). Different letters indicate $p < 0.05$. Representative images of a non-capacitated, capacitated and acrosome-reacted spermatozoa are also shown.

3.2. Correlation between Testicular Parameters and Sperm-Quality Values

We evaluated the correlation between the testicular parameters (testes measurement and ultrasound analysis) and the sperm-quality parameters at the moment of the testis evaluation, and at 30 days or 60 days afterwards. Most correlations were found when the sperm parameters were evaluated simultaneously (Table 3) or 30 days after (Table 4) the testicular analysis. However, most of these relationships were not apparent when sperm quality was evaluated 60 days afterwards (Table 5).

When the correlations between testicular size and sperm parameters were evaluated, the most significant result was a negative one with the viable spermatozoa without PS translocation, which was found during the three periods of time analyzed (0, 30 or 60 days later; Tables 3–5). We also found a negative correlation between testicular length and viability after 30 and 60 days of the measurement and viable spermatozoa with low ROS levels after 60 days (Tables 4 and 5). A positive correlation between scrotal circumference and normal sperm morphology was also found after 30 days (Table 3). Testicular size (length or width) correlated negatively with sperm abnormalities (proximal and distal cytoplasmic droplet or bent tail) the same day and 30 days later (Tables 3 and 4), although testicular length also correlated positively with the percentage of spermatozoa with a detached head (Table 4).

Regarding the echotexture parameters, the macroscopic (standard resolution Ec1-black pixels, Ec2-white pixels and Ec3-grey pixels) correlated mainly with sperm parameters that were evaluated 30 days after the scanning (Table 4). When the analysis was performed on the same-day or 60-day sperm parameters, only a correlation between Ec2 and spermatozoa with high ROS levels (Table 3) or progressive motility (Table 5) was found. In general, Ec2 and Ec3 correlated positively with "bad quality" sperm parameters (the percentage of spermatozoa with high ROS levels (Table 3) or with PS translocation and proximal cytoplasmic droplet (Table 4)), and negatively with "good quality" sperm parameters (total and progressive motility, Tables 4 and 5, respectively). Ec1 only correlated negatively with the percentage of spermatozoa with a detached head 30 days after scanning.

Table 3. Spearman's rank correlation coefficient (Spearman's ρ) between testicular size, echotexture parameters (evaluated by ultrasound-video analysis and ECOTEXT® software) or testicular-artery blood flow (assessed by color-Doppler flow imaging) and some sperm parameters, analyzed the same day of the testes evaluation, in *Rasa Aragonesa* rams ($n = 96$). Bold type indicates statistically significant correlations. * $p < 0.05$ and ** $p < 0.01$.

	Sperm Concentration	Total Motility	Progressive Motility	Viable Spermatozoa without PS Translocation	Total Spermatozoa with High ROS Levels	Normal Morphology	Proximal Cytoplasmic Droplet	Distal Cytoplasmic Droplet	Bent Tail
Testicular size									
Scrotal circumference	−0.141	0.186	0.042	−0.011	0.190	0.177	0.027	−0.138	−0.166
Testicular width	**−0.300 ****	0.124	0.094	0.012	0.148	0.138	−0.139	−0.163	**−0.239 ***
Testicular length	−0.162	0.025	−0.017	**−0.265 ****	−0.024	0.172	**−0.234 ***	0.007	0.047
Testicular volume	−0.173	0.060	0.138	−0.158	0.107	0.190	−0.082	−0.092	−0.091
ECOTEXT® parameters									
Ec1 (black pixels)	−0.032	0.17	0.101	0.070	−0.175	0.051	−0.049	−0.186	0.139
Ec2 (white pixels)	0.168	−0.171	−0.127	−0.108	**0.201 ***	−0.116	0.175	0.037	−0.036
Ec3 (grey pixels)	0.142	−0.17	−0.081	−0.114	0.184	−0.079	0.12	0.004	−0.064
TD (tubular density)	**−0.218 ***	0.2	**0.371 ****	−0.030	**−0.225 ***	**0.233 ***	**−0.338 ****	**−0.263 ****	0.094
LA (lumen area)	−0.101	0.158	0.182	0.133	**−0.238 ***	0.109	−0.167	−0.144	0.155
LD (lumen diameter)	−0.024	0.122	0.027	0.173	−0.182	0.024	−0.006	−0.049	0.123
Doppler parameters									
Frequency	**0.203 ***	**−0.210 ***	−0.078	0.072	0.107	−0.196	−0.005	0.05	**0.283 ****
PI (Pulsatility index)	**−0.310 ****	**0.203 ***	0.054	−0.004	**−0.304 ****	0.019	−0.176	−0.028	−0.009
RI (Resistive index)	**−0.30 3 ****	0.186	0.076	−0.012	**−0.317 ****	0.045	−0.198	−0.043	−0.002

Table 4. Spearman's rank correlation coefficient (Spearman's ρ) between testicular size, echotexture parameters (evaluated by ultrasound-video analysis and ECOTEXT® software) or testicular-artery blood flow (assessed by color-Doppler flow imaging) and some sperm parameters, analyzed 30 days after the testes evaluation, in *Rasa Aragonesa* rams ($n = 96$). Bold type indicates statistically significant correlations. * $p < 0.05$ and ** $p < 0.01$.

	Sperm Concentration	Total Motility	Progressive Motility	Viability	Viable Spermatozoa without PS Translocation	Total Spermatozoa with PS Translocation	Total Spermatozoa with High ROS Levels	Sperm with DNA Fragmentation	Normal Morphology	DETACHED HEAD	Proximal Cytoplasmic Droplet	Distal Cytoplasmic Droplet
Testicular size												
Scrotal circunference	−0.140	0.080	0.039	−0.056	−0.087	0.074	0.129	0.093	**0.203** *	−0.003	−0.029	−0.129
Testicular width	−0.167	**0.205** *	0.007	0.009	−0.062	0.039	0.080	−0.034	0.196	0.032	−0.024	**−0.212** *
Testicular length	−0.195	−0.161	−0.065	**−0.229** *	**−0.284** *	0.081	−0.070	0.115	0.105	**0.255** *	−0.082	−0.042
Testicular volume	−0.172	−0.026	0.003	−0.162	**−0.213** *	0.097	0.046	0.122	0.167	0.110	−0.041	−0.098
ECOTEXT® parameters												
Ec1 (black pixels)	−0.015	0.14	−0.016	−0.052	0.090	−0.099	−0.099	−0.15	−0.140	**−0.251** *	−0.065	−0.174
Ec2 (white pixels)	−0.013	**−0.255** *	−0.124	0.001	−0.071	**0.236** *	0.169	0.137	0.044	0.039	**0.201** *	0.132
Ec3 (grey pixels)	−0.014	**−0.232** *	−0.086	−0.034	−0.104	**0.218** *	0.150	0.12	0.108	0.091	0.147	0.127
TD (tubular density)	−0.134	**0.234** *	**0.295** **	**−0.226** *	−0.052	−0.145	−0.194	**−0.216** *	0.171	−0.087	**−0.291** **	**−0.312** **
LA (lumen area)	−0.043	**0.229** *	0.125	−0.081	0.098	−0.194	−0.140	−0.17	−0.079	−0.157	−0.134	**−0.207** *
LD (lumen diameter)	0.038	0.157	−0.015	0.05	0.130	−0.138	−0.098	−0.102	−0.153	−0.156	−0.105	−0.09
Doppler parameters												
Frequency	**0.232** *	0.091	0.023	0.029	0.083	−0.039	0.050	0.04	0.062	−0.021	−0.095	−0.054
PI (Pulsatility index)	−0.164	0.071	0.032	−0.159	−0.067	−0.001	−0.194	0.021	0.046	0.057	−0.149	−0.101
RI (Resistive index)	−0.179	0.068	0.027	−0.168	−0.081	−0.001	**−0.206** *	0.008	0.086	0.074	−0.184	−0.12

Table 5. Spearman's rank correlation coefficient (Spearman's ρ) between testicular size, echotexture parameters (evaluated by ultrasound-video analysis and ECOTEXT® software) or testicular-artery blood flow (assessed by color-Doppler flow imaging) and some sperm parameters, analyzed 60 days after the testes evaluation, in *Rasa Aragonesa* rams (n = 96). Bold type indicates statistically significant correlations. * $p < 0.05$ and ** $p < 0.01$.

	Sperm Concentration	Progressive Motility	Viability	Viable Spermatozoa without PS Translocation	Viable Spermatozoa with low ROS Levels	Non-Capacitated Sperm	Capacitated Sperm	Distal Cytoplasmic Droplet
Testicular size								
Scrotal circunference	−0.105	−0.030	−0.072	−0.063	−0.156	−0.019	−0.008	−0.156
Testicular width	−0.155	−0.093	0.011	−0.081	−0.057	−0.005	−0.024	−0.056
Testicular length	−0.139	−0.172	**−0.281 ****	**−0.344 ****	**−0.235 ****	−0.042	0.020	−0.053
Testicular volume	−0.124	−0.090	−0.191	**−0.223 ***	**−0.221 ***	−0.022	−0.009	−0.127
ECOTEXT® parameters								
Ec1 (black pixels)	−0.043	0.09	0.087	0.095	0.053	0.063	−0.09	−0.026
Ec2 (white pixels)	−0.052	**−0.231 ***	0.03	0.033	−0.051	0.02	0.014	0.116
Ec3 (grey pixels)	−0.099	−0.178	−0.019	−0.055	−0.083	0.026	0.016	0.066
TD (tubular density)	**−0.329 ****	0.155	**−0.209 ***	−0.125	−0.156	0.001	−0.032	**−0.266 ****
LA (lumen area)	−0.062	0.158	−0.018	0.057	0.013	−0.01	−0.035	−0.126
LD (lumen diameter)	0.081	0.158	0.084	0.112	0.100	0.017	−0.05	0.043
Doppler parameters								
Frequency	0.025	−0.01	−0.193	−0.077	−0.149	**−0.256 ***	**0.296 ****	−0.01
PI (Pulsatility index)	−0.084	0.011	0.001	−0.118	0.014	0.163	−0.158	−0.159
RI (Resistive index)	−0.116	0.021	−0.033	−0.153	0.002	0.137	−0.128	−0.176

Among the microscopic, high-resolution echotexture parameters, the tubular density (TD) showed more correlations with the sperm parameters. Overall, TD showed a positive correlation with "good quality" sperm parameters, such as total (Table 3) and progressive motility (Tables 3 and 4), and normal morphology (Table 3), and a negative correlation with the "bad quality" sperm parameters: the percentage of spermatozoa with high ROS levels (Table 3), with DNA fragmentation (Table 4) or with proximal (Tables 3 and 4) or distal cytoplasmic droplet (Tables 3–5). However, a negative correlation between TD and sperm concentration (Tables 3 and 5) or viability (Tables 4 and 5) was also found. The lumen area correlated negatively with total spermatozoa with high ROS levels that were obtained the same day of scanning and positively with the total motility of the sample obtained 30 days later. Lumen diameter did not correlate with any sperm parameters.

The main correlations for the Doppler parameters were found with sperm parameters that were evaluated the same day of the ultrasound analysis (Table 3). The frequency correlated positively with sperm concentration and the percentage of spermatozoa with a bent tail and negatively with total motility, whereas PI and RI correlated negatively with sperm concentration and spermatozoa with high ROS levels. PI was also positively related to total motility. These correlations were not apparent when the semen quality was studied 30 days after Doppler evaluation, and only the frequency–sperm-concentration and RI–high-ROS-level correlations were maintained (Table 4). Finally, a negative correlation between the frequency and non-capacitated spermatozoa and a positive correlation with capacitated sperm were found when the sperm parameters were evaluated after 60 days (Table 5).

3.3. Correlation between Ultrasonographic and Morphometric Parameters

All testicular-measurement parameters correlated negatively with Ec1, LA and LD and positively with Ec3 (except for testicular width, which showed a negative correlation with LD). Ec2 correlated positively with testicular volume and scrotal circumference (Table 6). No correlation was found between testicular measures and tubular density.

On the other hand, when the correlation between the Doppler parameters and echotexture was evaluated, the frequency showed a negative correlation with Ec1 and LA (Table 7). PI and RI correlated negatively with Ec2 and Ec3 and positively with Ec1, TD, LA and LD. No correlation was found between the Doppler parameters and testicular size (Table 7).

Table 6. Spearman's rank correlation coefficient (Spearman's ρ) between testicular size and echotexture parameters (evaluated by ultrasound-video analysis and ECOTEXT® software) in *Rasa Aragonesa* rams ($n = 96$). Bold type indicates statistically significant correlations. * $p < 0.05$ and ** $p < 0.01$.

	Ec1 (Black Pixels)	Ec2 (White Pixels)	Ec3 (Grey Pixels)	TD (Tubular Density)	LA (lumen AREA)	LD (Lumen Diameter)
Scrotal circumference	**−0.249 ***	**0.245 ***	**0.299 ****	0.093	**−0.227 ***	**−0.365 ****
Testicular width	−0.188	0.145	0.171	0.118	−0.134	**−0.245 ***
Testicular length	**−0.271 ****	0.184	**0.236 ***	0.176	**−0.210 ***	**−0.361 ****
Testicular volume	**−0.273 ****	**0.226 ***	**0.287 ****	0.155	**−0.225 ***	**−0.391 ****

Table 7. Spearman's rank correlation coefficient (Spearman's ρ) between testicular size or echotexture parameters (evaluated by ultrasound-video analysis and ECOTEXT® software) and testicular-artery blood flow (assessed by color-Doppler flow imaging) in *Rasa Aragonesa* rams ($n = 96$). Bold type indicates statistically significant correlations. * $p < 0.05$ and ** $p < 0.01$.

	Frequency	**Pulsatility Index (PI)**	**Resistive Index (RI)**
Testicular size			
Scrotal circumference	−0.067	−0.102	−0.111
Testicular width	0.160	−0.026	−0.024
Testicular length	0.041	0.083	0.109
Testicular volume	−0.037	−0.005	−0.002
ECOTEXT® parameters			
Ec1 (black pixels)	**−0.226** *	**0.314** **	**0.308** **
Ec2 (white pixels)	0.171	**−0.422** **	**−0.430** **
Ec3 (grey pixels)	0.179	**−0.369** **	**−0.371** **
TD (tubular density)	−0.120	**0.490** **	**0.520** **
LA (lumen area)	**−0.211** *	**0.386** **	**0.387** **
LD (lumen diameter)	−0.192	**0.225** *	**0.218** *

4. Discussion

Testicular ultrasonography has been postulated, along with testes measurement, as a non-invasive method to assess the reproductive soundness of sires. Moreover, recent software and ultrasound-probe developments have allowed the evaluation of microscopic structures within the testis parenchyma. Thus, in this work, we have investigated seasonal variations in testicular echotexture parameters, evaluated at the macroscopic and microscopic level by ultrasound-video analysis, and their relationship with testicular size and sperm-quality parameters.

The mean values obtained after echotexture evaluation with ECOTEXT® software were similar to those previously reported for the ram [39]. When the effect of the season was evaluated, a decrease in Ec1 (black pixels) and the seminiferous-tubule lumen diameter (LD) during the non-breeding season were found. Previous studies had shown an increase in testicular numerical-pixel values (NPVs) and pixel heterogeneity during the NB season [15,16] that could have been related to changes in the seminiferous tubules [13,14]. Our work revealed that these seasonal differences were due to a decrease in the number of black pixels and the diameter of the ST lumen, without changes in ST density or the percentage of the lumen area. Regarding the Doppler analysis, we only found seasonal differences in pulse frequency, in contrast with Hedia et al. [27], who recorded a marked increase in RI and PI values during the spring and summer. However, despite the lack of seasonal differences, we similarly detected significant differences in PI between the coldest and the hottest months of the year. Thus, the changes in Doppler parameters in the ram could be due to environmental temperature changes and not to physiological variations due to seasonal reproduction, as has been suggested for bucks [26].

We also studied testicular size and volume throughout the year because they are commonly considered a good index of sperm production [40] and an important feature in male selection [41]. Although previous studies suggested a seasonal variation in ram testicular size and volume [4,8,42], our study showed no significant differences in any measurements analyzed between B and NB seasons. However, we found some differences between the hottest and coldest months of the year, as in other sheep breeds [27,43]. Moreover, in other studies, a significant positive correlation between testicle volume and air temperature has also been observed [40,43].

When the correlation between testicular measurement and ultrasound values was evaluated, we found a negative correlation between testis size and volume and Ec1 and a positive correlation with Ec2 and Ec3. Thus, as testicular size increases, black pixels decrease, and white and grey pixels increase. This result suggests that the testicular size increase is caused by the rise in the number of cells within the seminiferous tubules and not by an increase in tubule density or lumen size. This hypothesis is corroborated by the negative correlation between testicular size and lumen-related parameters (lumen area and diameter) and the lack of correlation with tubular density. Previous studies showed that testicular lipid-content variations could be evaluated by testicular ultrasonography [44]. Lipids are highly echogenic, so increasing cellular density would increase the grey and white pixels due to the lipids present in the cell membranes. Other studies have also demonstrated that the increase in the testes echogenicity during puberty in bull calves was due to a rise in the numbers of cells within the seminiferous tubules [45].

We did not find any correlation between testicular measurements and the Doppler parameters, in contrast with previous studies [27] in which a negative correlation between testicular volume and RI and PI was found. However, the method used in those studies for the testicular-volume calculation was different. Nonetheless, when the correlation between echotexture and the Doppler parameters was evaluated, PI and RI correlated negatively with Ec2 and Ec3 and positively with Ec1 and all of the microscopic ultrasound parameters. RI and PI are inversely linked to the blood-flow perfusion, so our results suggest that an increase in blood flow in the testis would increase the number of cells within the seminal tubule, thus reducing the lumen area as other authors suggested [45].

Nonetheless, the main goal of our study was to determine the predictive capacity of testicular echography (both echotexture evaluation and testicular-artery Doppler analysis) and testicular measurement on sperm quality, evaluated with classical and non-conventional parameters.

We only found seasonal differences in the percentage of spermatozoa with high ROS levels, acrosome-reacted spermatozoa and sperm morphology. These results contrast with other studies, which found differences in motility, concentration, sperm morphology or membrane integrity [7,16,46]. It is probable that our continuous regime of semen extraction can overcome the loss of seminal quality provoked by the non-reproductive season in the ram, although differences in breed or geographical location could also influence the response to seasonality [47].

The correlation analysis was performed between ultrasound parameters and seminal quality at the moment of scanning and 30 and 60 days later. Most correlations were found with semen that was obtained the same day or 30 days after ultrasonography. Ram spermatogenesis in the seminiferous tubule lasts 47 days [48]. ECOTEXT® software evaluates ST density and lumen characteristics, which could be more closely related to the final phase of the spermatogenesis, i.e., the spermiogenesis. Other similar studies using the pixel intensity and pixel heterogeneity of testicular parenchyma had observed correlations between testicular echotexture and the quality of semen obtained two or four weeks after ultrasound examination in bulls [49], and the same day [16] or 60 days afterwards in rams [15]. This suggests that testicular pixel values might be a good predictor of future sperm quality. Our study, in which we could differentiate between black, white and grey pixels, revealed that Ec2 and Ec3 (white and grey pixels, respectively) correlated positively with "bad quality" sperm parameters and negatively with "good quality" ones, especially between those parameters related to apoptosis and the sperm collected 30 days after ultrasound evaluation. Thus, an increase in Ec2 and Ec3 could indicate a present or near-future decrease in seminal quality. In this study, we also found that the increase in testicular size increased Ec2 and Ec3, and a negative correlation between testicular dimensions and viability, viable sperm without PS translocation, and low ROS levels. Thus, if an increase in white and grey pixels indicates an increase in the number of cells within the seminiferous tubules, as studies in prepubertal males suggest [13,45,50], then this excess of cells could initiate the apoptosis process or induce oxidative stress that could compromise future seminal quality and fertility [31,51]. Regarding microscopic echotexture parameters, TD and LA correlated positively with "good quality" sperm parameters and negatively with "bad" ones, especially with those related to sperm morphology and ROS or apoptotic damage, which suggests that both of them could be predictors of good seminal quality. Although other studies have previously observed a correlation between testicular echotexture and sperm morphology [15,16] or DNA fragmentation [28], this is the first time, to the best of our knowledge, that testicular echotexture has been correlated to sperm physiological parameters such as oxidative stress or apoptotic state.

Finally, when the Doppler results were analyzed, the majority of correlations were found with seminal-quality parameters that were evaluated the same day. We found a negative correlation between both PI and RI with sperm concentration, consistent with the results obtained by Hedia et al. [27] and Ntemka et al. [28], but also with spermatozoa with high ROS levels. Thus, although the semi-hypoxic environment in the testis could prevent oxygen-radical damage to sperm [52], changes in blood-flow perfusion and oxygen tension could affect sperm quality at the moment of ejaculation or even a month later [53].

5. Conclusions

In conclusion, echotexture analysis by ultrasound-video analysis could be a valuable tool for assessing the breeding soundness of rams. An increase in Ec2 and Ec3 could indicate a decrease in seminal quality, and tubular density and lumen area could be predictors of good seminal quality.

Supplementary Materials: The following supporting information can be downloaded at: https://www.mdpi.com/article/10.3390/biology11020261/s1, Supplemental Table S1: Differences in testicular size, echotexture parameters (evaluated by ultrasound-video analysis and ECOTEXT® software) and testicular-artery blood flow (assessed by color-Doppler flow imaging) between left and right testicles in *Rasa Aragonesa* rams. Results are shown as mean ± SEM of $n = 96$. Different letters indicate $p < 0.05$. Supplemental Figure S1: Monthly differences in testicular size (A), echotexture (C and D) and testicular-artery blood supply (B) in *Rasa Aragonesa* rams. Results are shown as mean ± SEM of $n = 8$. Different letters indicate $p < 0.05$. Supplemental Figure S2: Monthly differences in seminal quality parameters in Rasa Aragonesa rams. Results are shown as mean ± SEM of $n = 8$. Different letters indicate $p < 0.05$.

Author Contributions: Conceptualization, R.P.-P. and A.C.; methodology, R.P.-P. and A.C.; validation, M.C.-S., R.P.-P. and A.C.; formal analysis, M.C.-S.; investigation, M.C.-S. and S.M.-J.; resources, R.P.-P.; data curation, M.C.-S. and S.M.-J.; writing—original draft preparation, M.C.-S.; writing—review and editing, R.P.-P. and A.C.; visualization, A.C.; supervision, R.P.-P. and A.C.; project administration, R.P.-P.; funding acquisition, R.P.-P. All authors have read and agreed to the published version of the manuscript.

Funding: This research was funded by CICYT AGL-2017-83799-R and DGA A07 17R. M.C.-S. has a predoctoral contract from MICINN PRE2018-085198. S.M.-J. has a predoctoral contract from the Diputación General de Aragón (DGA).

Institutional Review Board Statement: The animal study protocol was approved by the University of Zaragoza Ethics Committee for Animal Experiments (project license PI39/17 and date of approval: 24 May 2017), following the Spanish Policy for Animal Protection RD53/2013, which meets the European Union Directive 2010/63 on the protection of animals used for experimental and other scientific purposes.

Informed Consent Statement: Not applicable.

Data Availability Statement: The data presented in this study are openly available in FigShare at https://doi.org/10.6084/m9.figshare.18102599 (accessed on 4 February 2022).

Acknowledgments: The authors would like to acknowledge the use of the Servicio General de Apoyo a la Investigación-SAI (Universidad de Zaragoza) and ANGRA for supplying the sires.

Conflicts of Interest: The authors declare no conflict of interest.

References

1. Malpaux, B.; Viguie, C.; Skinner, D.C.; Thiery, J.C.; Pelletier, J.; Chemineau, P. Seasonal breeding in sheep: Mechanism of action of melatonin. *Anim. Reprod. Sci.* **1996**, *42*, 109–117. [CrossRef]
2. Chemineau, P.; Guillaume, D.; Migaud, M.; Thiery, J.C.; Pellicer-Rubio, M.T.; Malpaux, B. Seasonality of reproduction in mammals: Intimate regulatory mechanisms and practical implications. *Reprod. Domest. Anim.* **2008**, *43* (Suppl. 2), 40–47. [CrossRef]
3. D'Occhio, M.J.; Schanbacher, B.D.; Kinder, J.E. Profiles of luteinizing hormone, follicle-stimulating hormone, testosterone and prolactin in rams of diverse breeds: Effects of contrasting short (8L:16D) and long (16L:8D) photoperiods. *Biol. Reprod.* **1984**, *30*, 1039–1054. [CrossRef] [PubMed]
4. Lincoln, G.A.; Lincoln, C.E.; McNeilly, A.S. Seasonal cycles in the blood plasma concentration of FSH, inhibin and testosterone, and testicular size in rams of wild, feral and domesticated breeds of sheep. *J. Reprod. Fertil.* **1990**, *88*, 623–633. [CrossRef] [PubMed]
5. Poulton, A.L.; Robinson, T.J. The response of rams and ewes of three breeds to artificial photoperiod. *J. Reprod. Fertil.* **1987**, *79*, 609–626. [CrossRef] [PubMed]
6. Jackson, G.L.; Jansen, H.; Kao, C. Continuous exposure of Suffolk ewes to an equatorial photoperiod disrupts expression of the annual breeding-season. *Biol. Reprod.* **1990**, *42*, 63–73. [CrossRef] [PubMed]
7. Karagiannidis, A.; Varsakeli, S.; Alexopoulos, C.; Amarantidis, I. Seasonal variation in semen characteristics of Chios and Friesian rams in Greece. *Small Rumin. Res.* **2000**, *37*, 125–130. [CrossRef]
8. Mandiki, S.N.M.; Derycke, G.; Bister, J.L.; Paquay, R. Influence of season and age on sexual maturation parameters of Texel, Suffolk and Ile-de-France rams: 1. Testicular size, semen quality and reproductive capacity. *Small Rumin. Res.* **1998**, *28*, 67–79. [CrossRef]
9. Mickelsen, W.D.; Paisley, L.G.; Dahmen, J.J. The effect of season on the scrotal circumference and sperm motility and morphology in rams. *Theriogenology* **1981**, *16*, 45–51. [CrossRef]
10. Coyan, K.; Kaya, A.; Karaca, F.; Ataman, M.B.; Yildiz, C. The effect of melatonin on sperm quality and testicular size of normospermic and pathospermic rams in the anoestrous season. *Wien. Tierarztl. Monatsschr.* **1998**, *85*, 383–388.
11. Gouletsou, P.G.; Amiridis, G.S.; Cripps, P.J.; Lainas, T.; Deligiannis, K.; Saratsis, P.; Fthenakis, G.C. Ultrasonographic appearance of clinically healthy testicles and epididymides of rams. *Theriogenology* **2003**, *59*, 1959–1972. [CrossRef]
12. Gouletsou, P.G. Ultrasonographic examination of the scrotal contents in rams. *Small Rumin. Res.* **2017**, *152*, 100–106. [CrossRef]
13. Giffin, J.L.; Bartlewski, P.M.; Hahnel, A.C. Correlations among ultrasonographic and microscopic characteristics of prepubescent ram lamb testes. *Exp. Biol. Med.* **2014**, *239*, 1606–1618. [CrossRef] [PubMed]
14. Giffin, J.L.; Franks, S.E.; Rodriguez-Sosa, J.R.; Hahnel, A.; Bartlewski, P.M. A Study of Morphological and Haemodynamic Determinants of Testicular Echotexture Characteristics in the Ram. *Exp. Biol. Med.* **2009**, *234*, 794–801. [CrossRef] [PubMed]
15. Ahmadi, B.; Lau, C.P.-S.; Giffin, J.; Santos, N.; Hahnel, A.; Raeside, J.; Christie, H.; Bartlewski, P. Suitability of epididymal and testicular ultrasonography and computerized image analysis for assessment of current and future semen quality in the ram. *Exp. Biol. Med.* **2012**, *237*, 186–193. [CrossRef] [PubMed]
16. Hedia, M.; El-Belely, M.; Ismail, S.; Abo El-Maaty, A. Seasonal changes in testicular ultrasonogram pixel-intensity and their association with semen characteristics in rams. *Asian Pac. J. Reprod.* **2020**, *9*, 49–54. [CrossRef]
17. Urt, M.A.G.; Bakarji, E.W.B.; Faria, F.J.C.; Santos, E.F.; Costa, D.S. Lack of relationship between testicular echotexture and seminal characteristics in adult Texel rams. *Agric. Sci.* **2018**, *9*, 936–946. [CrossRef]
18. Andrade, A.K.G.; Soares, A.T.; Freitas, F.F.; Silva, S.V.; Pena-Alfaro, C.E.; Batista, A.M.; Guerra, M.M.P. Testicular and epididymal ultrasonography in Santa Ines lambs raised in Brazil. *Anim. Reprod.* **2014**, *11*, 110–118.
19. Camela, E.S.C.; Nociti, R.P.; Santos, V.J.C.; Macente, B.I.; Murawski, M.; Vicente, W.R.R.; Bartlewski, P.M.; Oliveira, M.E.F. Changes in testicular size, echotexture, and arterial blood flow associated with the attainment of puberty in Dorper rams raised in a subtropical climate. *Reprod. Domest. Anim.* **2019**, *54*, 131–137. [CrossRef]
20. Echegaray, A.; Marcantonio, S.; Maraboli, C.; Munoz, I.; Escartin, N.; Gnemmi, G. New echotexture parameters to evaluate the testicular parenchyma in bulls. In *Reproduction in Domestic Animals, Proceedings of the 22nd Annual Conference of the European Society for Domestic Animal Reproduction (ESDAR), Cordoba, Spain, 27–29 September 2018*; Wiley Blackwell: Hoboken, NJ, USA, 2018; p. 129.
21. Escartin, N.; Munoz, I.; Gil, A.; Echegaray, A. New echotexture parameters to evaluate the testicular parenchyma in rams. In *Reproduction in Domestic Animals, Proceedings of the 22nd Annual Conference of the European Society for Domestic Animal Reproduction (ESDAR), Cordoba, Spain, 27–29 September 2018*; Wiley Blackwell: Hoboken, NJ, USA, 2018; pp. 130–131.
22. Lafuente, A.; Lafuente, M.; Escartin, N.; Munoz, I.; Echegaray, A. New echotexture parameters to evaluate the testicular parenchyma in stallions. In *Reproduction in Domestic Animals, Proceedings of the 22nd Annual Conference of the European Society for Domestic Animal Reproduction (ESDAR), Cordoba, Spain, 27–29 September 2018*; Wiley Blackwell: Hoboken, NJ, USA, 2018; pp. 155–156.
23. Munoz, I.; Escartin, N.; Zamora, A.; Echegaray, A. New echotexture parameters to evaluate the testicular parenchyma in boars. In *Reproduction in Domestic Animals, Proceedings of the 22nd Annual Conference of the European Society for Domestic Animal Reproduction (ESDAR), Cordoba, Spain, 27–29 September 2018*; Wiley Blackwell: Hoboken, NJ, USA, 2018; pp. 169–170.
24. Kastelic, J.P.; Rizzoto, G.; Thundathil, J. Review: Testicular vascular cone development and its association with scrotal thermoregulation, semen quality and sperm production in bulls. *Animal* **2018**, *12*, S133–S141. [CrossRef]

25. Zelli, R.; Troisi, A.; Ngonput, A.E.; Cardinali, L.; Polisca, A. Evaluation of testicular artery blood flow by Doppler ultrasonography as a predictor of spermatogenesis in the dog. *Res. Vet. Sci.* **2013**, *95*, 632–637. [CrossRef] [PubMed]
26. Samir, H.; Nyametease, P.; Nagaoka, K.; Watanabe, G. Effect of seasonality on testicular blood flow as determined by color Doppler ultrasonography and hormonal profiles in Shiba goats. *Anim. Reprod. Sci.* **2018**, *197*, 185–192. [CrossRef] [PubMed]
27. Hedia, M.G.; El-Belely, M.S.; Ismail, S.T.; Abo El-Maaty, A.M. Monthly changes in testicular blood flow dynamics and their association with testicular volume, plasma steroid hormones profile and semen characteristics in rams. *Theriogenology* **2019**, *123*, 68–73. [CrossRef] [PubMed]
28. Ntemka, A.; Kiossis, E.; Boscos, C.; Theodoridis, A.; Kourousekos, G.; Tsakmakidis, I. Effects of testicular hemodynamic and echogenicity changes on ram semen characteristics. *Reprod. Domest. Anim.* **2018**, *53*, 50–55. [CrossRef] [PubMed]
29. Ahmad, N.; England, G.C.W.; Noakes, D.E. Ultrasonography of spontaneous lesions of the genital system of three rams, and their influence on semen quality. *Vet. Rec.* **2000**, *146*, 10–15. [CrossRef]
30. Del Valle, I.; Mendoza, N.; Casao, A.; Cebrián-Pérez, J.A.; Pérez-Pé, R.; Muiño-Blanco, T. Significance of Non-conventional Parameters in the Evaluation of Cooling-induced Damage to Ram Spermatozoa Diluted in Three Different Media. *Reprod. Domest. Anim.* **2010**, *45*, 260–268. [CrossRef]
31. Mendoza, N.; Casao, A.; Domingo, J.; Quintín, F.; Laviña, A.; Fantova, E.; Cebrián-Pérez, J.Á.; Muiño-Blanco, T.; Pérez-Pe, R. Influence of Non-conventional Sperm Quality Parameters on Field Fertility in Ovine. *Front. Vet. Sci.* **2021**, *8*, 413. [CrossRef]
32. Harrison, R.A.; Vickers, S.E. Use of fluorescent probes to assess membrane integrity in mammalian spermatozoa. *J. Reprod. Fertil.* **1990**, *88*, 343–352. [CrossRef]
33. Guthrie, H.D.; Welch, G.R. Determination of intracellular reactive oxygen species and high mitochondrial membrane potential in Percoll-treated viable boar sperm using fluorescence-activated flow cytometry1. *J. Anim. Sci.* **2006**, *84*, 2089–2100. [CrossRef]
34. Ward, C.R.; Storey, B.T. Determination of the time course of capacitation in mouse spermatozoa using a chlortetracycline fluorescence assay. *Dev. Biol.* **1984**, *104*, 287–296. [CrossRef]
35. Gillan, L.; Evans, G.; Maxwell, W.M. Capacitation status and fertility of fresh and frozen-thawed ram spermatozoa. *Reprod. Fertil. Dev.* **1997**, *9*, 481–487. [CrossRef] [PubMed]
36. Hafez, E.S.E.; Hafez, B. *Reproduction in Farm Animals*, 7th ed.; Wiley Blackwell: Hoboken, NJ, USA, 2000; p. 509.
37. Godfrey, R.W.; Collins, J.R.; Gray, M.L. Evaluation of sexual behavior of hair sheep rams in a tropical environment. *J. Anim. Sci.* **1998**, *76*, 714–717. [CrossRef] [PubMed]
38. Batissaco, L.; Celeghini, E.C.C.; Pinaffi, F.L.V.; Oliveira, B.M.M.d.; Andrade, A.F.C.d.; Recalde, E.C.S.; Fernandes, C.B. Correlations between testicular hemodynamic and sperm characteristics in rams. *Braz. J. Vet. Res. Anim. Sci.* **2013**, *50*, 384–395. [CrossRef]
39. Abecia, J.A.; Carvajal-Serna, M.; Casao, A.; Palacios, C.; Pulinas, L.; Keller, M.; Chemineau, P.; Delgadillo, J.A. The continuous presence of ewes in estrus in spring influences testicular volume, testicular echogenicity and testosterone concentration, but not LH pulsatility in rams. *Animal* **2020**, *14*, 2554–2561. [CrossRef] [PubMed]
40. Sarlós, P.; Egerszegi, I.; Balogh, O.; Molnár, A.; Cseh, S.; Rátky, J. Seasonal changes of scrotal circumference, blood plasma testosterone concentration and semen characteristics in Racka rams. *Small Rumin. Res.* **2013**, *111*, 90–95. [CrossRef]
41. Matos, C.A.P.; Thomas, D.L. Physiology and genetics of testicular size in sheep: A review. *Livest. Prod. Sci.* **1992**, *32*, 1–30. [CrossRef]
42. Lindsay, D.R.; Pelletier, J.; Pisselet, C.; Courot, M. Changes in photoperiod and nutrition and their effect on testicular growth of rams. *Reproduction* **1984**, *71*, 351–356. [CrossRef]
43. Avdi, M.; Banos, G.; Stefos, K.; Chemineau, P. Seasonal variation in testicular volume and sexual behavior of Chios and Serres rams. *Theriogenology* **2004**, *62*, 275–282. [CrossRef]
44. Ahmadi, B.; Mirshahi, A.; Giffin, J.; Oliveira, M.E.F.; Gao, L.; Hahnel, A.; Bartlewski, P.M. Preliminary assessment of the quantitative relationships between testicular tissue composition and ultrasonographic image attributes in the ram. *Vet. J.* **2013**, *198*, 282–285. [CrossRef]
45. Evans, A.C.O.; Pierson, R.A.; Garcia, A.; McDougall, L.M.; Hrudka, F.; Rawlings, N.C. Changes in circulating hormone concentrations, testes histology and testes ultrasonography during sexual maturation in beef bulls. *Theriogenology* **1996**, *46*, 345–357. [CrossRef]
46. Kafi, M.; Safdarian, M.; Hashemi, M. Seasonal variation in semen characteristics, scrotal circumference and libido of Persian Karakul rams. *Small Rumin. Res.* **2004**, *53*, 133–139. [CrossRef]
47. Malpaux, B.; Migaud, M.; Tricoire, H.; Chemineau, P. Biology of mammalian photoperiodism and the critical role of the pineal gland and melatonin. *J. Biol. Rhythm.* **2001**, *16*, 336–347. [CrossRef] [PubMed]
48. Senger, P.L. *Pathways to Pregnancy and Parturition*, 2nd ed.; Current Conceptions, Inc.: Pullman, WA, USA, 2003.
49. Arteaga, A.A.; Barth, A.D.; Brito, L.F.C. Relationship between semen quality and pixel–intensity of testicular ultrasonograms after scrotal insulation in beef bulls. *Theriogenology* **2005**, *64*, 408–415. [CrossRef]
50. Bartlewski, P.M.; Giffin, J.L.; Oluwole, O.A.; Hahnel, A.C. Prospective ultrasonographic and endocrine predictors of spermatogenic onset in ram lambs. *Anim. Reprod. Sci.* **2017**, *179*, 44–48. [CrossRef] [PubMed]
51. Asadi, A.; Ghahremani, R.; Abdolmaleki, A.; Rajaei, F. Role of sperm apoptosis and oxidative stress in male infertility: A narrative review. *Int. J. Reprod. Biomed.* **2021**, *19*, 493–504. [CrossRef]

52. Aitken, R.J.; Roman, S.D. Antioxidant Systems and Oxidative Stress in the Testes. *Oxid. Med. Cell Longev.* **2008**, *1*, 616821. [CrossRef]
53. Balci, A.; Karazincir, S.; Gorur, S.; Sumbas, H.; Egilmez, E.; Inandi, T. Long-term effect of varicocele repair on intratesticular arterial resistance index. *J. Clin. Ultrasound* **2008**, *36*, 148–152. [CrossRef]

biology

MDPI

Article

Effect of Motility Factors D-Penicillamine, Hypotaurine and Epinephrine on the Performance of Spermatozoa from Five Hamster Species

Maximiliano Tourmente [1,2,3,*], Ana Sanchez-Rodriguez [1] and Eduardo R. S. Roldan [1,*]

1 Department of Biodiversity and Evolutionary Biology, Museo Nacional de Ciencias Naturales, Spanish Research Council (CSIC), 28006 Madrid, Spain; anasanro.vet@gmail.com

2 Centro de Biología Celular y Molecular, Facultad de Ciencias Exactas, Físicas y Naturales, Universidad Nacional de Córdoba, Cordoba X5016GCA, Argentina

3 Instituto de Investigaciones Biológicas y Tecnológicas (IIByT), Consejo Nacional de Investigaciones Científica y Técnicas (CONICET), Cordoba X5016GCA, Argentina

* Correspondence: maximiliano.tourmente@unc.edu.ar (M.T.); roldane@mncn.csic.es (E.R.S.R.)

Simple Summary: Analysis of sperm performance under in vitro conditions provides a good indication of fertilizing potential. Parameters such as motility, swimming kinetics, acrosome integrity, or ATP content are thus examined in efforts to characterize such potential. Hamster species are a good model to study sperm parameters that are key determinants of fertilizing capacity because these species are at the higher end of the diversity of mammalian sperm morphology and performance. In vitro functional studies demand that sperm remain viable during a long period of time under conditions that resemble those in the female tract. Sperm from certain species require supplementation of the incubation medium with factors that stimulate viability and swimming, or that promote acquisition of fertilizing capacity. Molecules important for sperm performance in hamsters have been identified, namely D-penicillamine, hypotaurine and epinephrine (PHE). In the present study, we investigated the effect of PHE on spermatozoa from five hamster species incubated for up to 4 h. Our results revealed that PHE maintains sperm performance in the golden hamster, whereas it improves sperm quality in the Chinese hamster. In contrast, it does not seem to have any effect on sperm from the Siberian (Djungarian), Roborovski and Campbell's dwarf hamsters. These results are valuable to understand the different regulatory mechanisms of sperm motility and survival in different species.

Abstract: Assessments of sperm performance are valuable tools for the analysis of sperm fertilizing potential and to understand determinants of male fertility. Hamster species constitute important animal models because they produce sperm cells in high quantities and of high quality. Sexual selection over evolutionary time in these species seems to have resulted in the largest mammalian spermatozoa, and high swimming and bioenergetic performances. Earlier studies showed that golden hamster sperm requires motility factors such as D-penicillamine, hypotaurine and epinephrine (PHE) to sustain survival over time, but it is unknown how they affect swimming kinetics or ATP levels and if other hamster species also require them. The objective of the present study was to examine the effect of PHE on spermatozoa of five hamster species (*Mesocricetus auratus*, *Cricetulus griseus*, *Phodopus campbelli*, *P. sungorus*, *P. roborovskii*). In sperm incubated for up to 4 h without or with PHE, we assessed motility, viability, acrosome integrity, sperm velocity and trajectory, and ATP content. The results showed differences in the effect of PHE among species. They had a significant positive effect on the maintenance of sperm quality in *M. auratus* and *C. griseus*, whereas there was no consistent effect on spermatozoa of the *Phodopus* species. Differences between species may be the result of varying underlying regulatory mechanisms of sperm performance and may be important to understand how they relate to successful fertilization.

Keywords: sperm performance; hamster; sperm motility; acrosome integrity; ATP content; *Mesocricetus auratus*; *Cricetulus griseus*; *Phodopus sungorus*; *Phodopus roborovskii*; *Phodopus campbelli*

Citation: Tourmente, M.; Sanchez-Rodriguez, A.; Roldan, E.R.S. Effect of Motility Factors D-Penicillamine, Hypotaurine and Epinephrine on the Performance of Spermatozoa from Five Hamster Species. *Biology* **2022**, *11*, 526. https://doi.org/10.3390/biology11040526

Academic Editors: Pilar Santolaria Blasco, Hiromitsu Tanaka, Rosaura Pérez-Pé and Jessica Rickard

Received: 3 February 2022
Accepted: 28 March 2022
Published: 30 March 2022

1. Introduction

Sperm performance, a major determinant of male fertility, can be dissected into a series of traits that are intricately connected to sperm fertilizing potential [1,2]. Therefore, a combination of various performance parameters is usually assessed to evaluate the quality of sperm samples and predict its fertilizing potential in man and other animals [2–10]. Sperm viability, motility and kinetics, as well as acrosome integrity, are all linked to sperm survival. The proportion of motile sperm and the velocity of spermatozoa are essential for sperm to swim along the cervix, uterus and the uterotubal junction, and cells that show higher values in these parameters also have higher chances to reach and fertilize the ovum [11]. The assessment of sperm motility may be carried out subjectively, estimating the percentage of motile sperm by microscopic visualization, and objectively through the quantification of sperm swimming parameters by computer aided sperm analysis (CASA) [1,2,10,12,13].

The sperm acrosome contains enzymes that are released during exocytosis, an essential step required for penetration of the oocyte's vestments [14]. The timing of release is important, so the integrity of the acrosome during sperm transport in the female tract and, in particular, along the oviduct, has to be preserved in order to ensure fertilizing potential [15]. Controversy still exists as to which is the site where acrosomal exocytosis takes place or which is the physiological ligand (or ligands) responsible for initiating exocytosis [16–19]. Nevertheless, since this is a key event in the processes leading to completion of fertilization, acrosomal status has become a valuable assessment of fertilizing potential [20].

Other relevant sperm parameters relate to sperm bioenergetics, with sperm ATP content serving as an indicator of the balance between sperm ATP production and consumption [21]. High ATP levels are positively correlated to sperm swimming velocity in rodents [22] and mammals in general [23]. Moreover, assessments of the variation in sperm ATP content and sperm traits over time among rodent species revealed that maintenance of high performance in species with high competitive ability is associated with high concentrations of intracellular ATP over time [22,24,25].

Exogenous factors could influence sperm performance in vivo and under in vitro conditions [26]. Besides the provision of energetic substrates that are fundamental for the production of ATP for sperm motility and survival, conditions in the female tract promote the acquisition of sperm's fertilizing ability, a process known as capacitation [27]. Among physical factors, extracellular pH, temperature, and viscosity are known to affect the survival and performance of sperm cells [28–35]. Several biological factors are also known to have important roles in sperm survival in vivo and in vitro. Early efforts to achieve in vitro fertilization had to rely on homologous or heterologous serum to ensure sperm survival and the acquisition of fertilizing ability [36,37]. Better definition of media was possible with replacement of serum by bovine serum albumin. However, these media were still not completely defined [38]. Components of the extracellular milieu may be required to sustain sperm motility, and variations between species may exist regarding the nature of such components. For golden hamster (*Mesocricetus auratus*) spermatozoa, one set of important motility factors are catecholamines, which maintain and stimulate sperm motility in vitro [39]. Within the group of catecholamines, epinephrine is an essential co-factor for golden hamster sperm as it activates motility as well as Na^+/K^+ ATPase and Ca^{2+}-ATPase [40], and it is also involved in the acquisition of fertilizing ability [41]. However, epinephrine is not able to maintain golden hamster sperm motility on its own, and a second factor, hypotaurine, is also essential. Hypotaurine is a superoxide scavenger that functions inhibiting lipid peroxidation and superoxide dismutase [42], which prevents motility loss. Hypotaurine and epinephrine together also cause a mild increase of the acrosome reaction in hamster sperm [43]. Other factors have been shown to influence hamster sperm function [44–46].

Another molecule of interest, with regards to hamster sperm survival, is D-penicillamine. This is an α-amino acid which acts as a cation chelator, protecting sperm from oxidation in several species [40,47–50]. As a zinc chelator, D-penicillamine facilitates capacitation, acting in the early step of this process because it removes most of the zinc within the first ten minutes of its addition, but it is not enough to support full capacitation [51]. D-penicillamine also prolongs hyperactivated motility [52]. This amino acid together with epinephrine and hypotaurine (PHE: penicillamine + hypotaurine + epinephrine) act as motility factors, necessary for the maintenance of golden hamster sperm motility in vitro [51]. The addition of PHE to incubation media maintains golden hamster sperm motility within the first hour [53,54] and also promotes sperm capacitation in this [54] and other species [55]. The synergistic effect of the three components of PHE can reactivate immotile spermatozoa of golden hamsters [56].

Many early studies of sperm function have used the golden hamster as a model, particularly because of the ease of examining acrosomal status in motile spermatozoa [57]. Studies of sperm behavior in a related species, the Chinese hamster (*Cricetulus griseus*), have been performed [58–60] and these studies have also identified the need to support sperm viability and motility over time to achieve fertilization. Similar requirements seem to exist for Siberian hamster (*Phodopus sungorus*) spermatozoa because low success was achieved in in vitro fertilization with a variety of media and supplements [61], but characterization of these requirements has not been undertaken. Hamsters are a valuable model for studies of sperm-oocyte recognition and interaction. Hybridization has been observed among hamsters (*Mesocricetus* species: [62–66]; *Phodopus* species: [67]; *Allocricetulus* species: [68]), and cross-fertilization in vitro between golden and Chinese hamsters has been reported [37]. Despite the potential of this group of species as a model for sperm biology, little is still known about the spermatozoa and fertilization of most species in this group. Comparative and evolutionary studies are needed to understand diversity in function and underlying mechanisms of sperm survival, capacitation and acrosomal exocytosis. Hamster species are attractive because they are at the higher end of the range of mammalian sperm dimensions [69,70], have high sperm swimming velocities [22] and exhibit high levels of sperm ATP [22,24]. Such traits, characterizing these species as high performers, may be the result of intense sperm competition [71]. Detailed comparative studies of sperm performance may therefore be rewarding in hamsters of the subfamily *Cricetinae* for which good background information of phylogenetic relations currently exists based on molecular studies [72] and chromosome evolution [73,74].

In the present study, we evaluated whether several biological molecules had a role in the performance of spermatozoa from five different species of hamster: *Mesocricetus auratus*, *Cricetulus griseus*, *Phodopus campbelli*, *P. sungorus* and *P. roborovskii*, focusing on sperm motility, viability, sperm kinematics, acrosome integrity, and bioenergetics. Spermatozoa from these species are at the higher end of the range of sperm quality parameters among rodent species [22,24]; thus, a detailed analysis of modulators of sperm function in these species could help understand determinants of male fertility.

2. Materials and Methods

2.1. Reagents

Unless stated otherwise, reagents were purchased from Sigma or Merck (both of Madrid, Spain).

2.2. Animals and Sperm Collection

Adult males (4–6 months old) of *Cricetulus griseus* (n = 5), *Mesocricetus auratus* (n = 6), *Phodopus campbelli* (n = 5), *P. sungorus* (n = 7), and *P. roborovskii* (n = 5) were kept in captivity in our animal facilities. Animals were maintained under standard conditions (14 h light–10 h darkness, 22–24 °C), with food and water provided ad libitum. Each male to be used in this study was housed alone (i.e., in individual cages) for at least one month before sperm collection. Males were sacrificed by cervical dislocation and weighed immediately. Testes were then removed and weighed. Relative testes size was calculated using the potential equation defined for rodents by Kenagy and Trombulak [75].

The caudae epididymides were excised after removing all blood vessels, fat, and surrounding connective tissues. Each cauda epididymis was placed in a Petri dish containing one of two variants of culture medium pre-warmed to 37 °C and spermatozoa were collected by performing three to five incisions in the distal region of the cauda, and allowing them to swim out for 5 min. As standard medium ("control" treatment), we used a Hepes-buffered modified Tyrode medium with albumin (see mT-H in [76]), with the addition of lactate and pyruvate (mTALP: 120.89 mM NaCl, 2.68 mM KCl, 0.49 mM $MgCl_2 \cdot 6H_2O$, 0.36 mM $NaH_2PO_4 \cdot 2H_2O$, 20 mM Hepes, 1.80 mM $CaCl_2$, 5.56 mM glucose, 1 mM sodium pyruvate, 10 mM sodium lactate, 4 mg mL^{-1} bovine serum albumin) under air (pH 7.4). A modified medium (PHE treatment) was mTALP supplemented with 20 μM D-penicillamine, 100 μM hypotaurine, and 1 μM epinephrine. Sperm suspensions were placed in plastic tubes, where sperm concentration was estimated by using a modified Neubauer chamber and adjusted to 20×10^6 sperm mL^{-1} by diluting with medium. Sperm parameters were assessed in the sperm suspensions corresponding to each treatment immediately after adjusting the concentration (hereafter referred as 0 h), and after 2, 3, and 4 h of incubation at 37 °C under air. Large-bore pipette tips were used to minimize damage to spermatozoa in all procedures.

2.3. *Sperm Motility, Viability and Acrosomal Integrity*

Sperm motility was evaluated by examining 10 μL of a previously diluted sperm suspension, placed between a pre-warmed slide and a coverslip, at 100× magnification under phase-contrast optics. The percentage of motile sperm was estimated by at least two independent, experienced observers, whose estimations were averaged and rounded to the nearest 5% value.

The assessment of sperm viability and acrosome integrity was performed by staining sperm first with eosin-nigrosin and subsequently with Giemsa [77]. Briefly, 5 μL sperm suspension and 10 μL eosin-nigrosin solution were mixed on a glass slide placed on a stage at 37 °C and 30 s later the mix was smeared and allowed to air-dry. Smears were fixed by immersion during 10 min in a solution of 4% formaldehyde in TPB buffer. After fixation, the smears were stained with Giemsa solution and mounted with DPX. Slides were examined at 1000× under bright field and 200 spermatozoa per male were examined to evaluate sperm viability and integrity of the acrosome. Viable spermatozoa were those excluding eosin (from the eosin-nigrosin stain). Acrosome integrity was reported as the percentage of sperm with intact acrosomes, excluding cells that showed damaged or missing acrosomes.

2.4. *Sperm Velocity and Trajectory*

To assess sperm swimming velocity and trajectory, an aliquot of sperm suspension was placed in a pre-warmed microscopy chamber with a depth of 20 μm (Leja, Nieuw-Vennep, The Netherlands) and filmed using a phase contrast microscope with pseudo-negative phase connected to a digital video camera (Basler A312fc, Vision Technologies, Glen Burnie, MD, USA). A 4× objective was used instead of the 10× objective traditionally used for CASA analysis on sperm from humans and domestic animals. This resulted in larger field of observation, which allowed for tracking of the unusually large and fast hamster sperm for longer, and a deeper focal plane to account for the depth of the observation chamber. Sperm trajectories were assessed using a computer aided sperm analyzer (Sperm Class Analyzer—SCA v.4.0, Microptic, Barcelona, Spain), and the following swimming parameters were estimated for each track: curvilinear velocity (VCL, μm s^{-1}), straight-line velocity (VSL, μm s^{-1}), average path velocity (VAP, μm s^{-1}), linearity (LIN = VSL/VCL), straightness (STR = VSL/VAP), wobble (WOB = VAP/VCL), amplitude of lateral head displacement (ALH, μm), and beat-cross frequency (BCF, Hz). The software was set with frame rate 50 s^{-1}, maximum particle size 500 μm, minimum particle size 50 μm, connectivity 30, contrast 400, and brightness 160. All video captures were compared to their overlaying analyzed tracks, and trajectories that did not belong to sperm were removed. In addition, trajectories with VAP values lower than 20 μm s^{-1}, and LIN, STR and WOB values of 100 in the post-capture analysis were discarded as these are typical of drifting but immotile sperm or other non-sperm particles. Since spermatozoa of *Cricetulus griseus* are significantly larger than those of the other species studied (about 250 μm [69]), and their head has a slender falciform shape that makes it almost undistinguishable from the midpiece in the moving sperm, the SCA software was not able to obtain accurate sperm trajectories. Thus, data on sperm velocity and trajectory for *C. griseus* are not presented.

2.5. *Sperm ATP Content*

Sperm ATP content was measured using a luciferase-based ATP bioluminescent assay kit (Roche, ATP Bioluminescence Assay Kit HS II), as previously described [22,24]. A 100 μL-aliquot of diluted sperm suspension was mixed with 100 μL of Cell Lysis Reagent, vortexed and incubated at room temperature for 5 min. The resulting cell lysate was centrifuged at 12,000× *g* for 2 min, and the supernatant was recovered and frozen in liquid N_2. Bioluminescence was measured in triplicate in 96-well plates using a luminometer (Varioskan Flash, Thermo Fisher Scientific Inc., Waltham, MA, USA). In each well, 50 μL of Luciferase reagent were added to 50 μL of sample (via auto-injection), and, following a 1 s delay, light emission was measured over a 10 s integration period. Standard curves were constructed using solutions containing known concentrations of ATP diluted in mTALP and Cell Lysis Reagent in a proportion equivalent to that of the samples. ATP content was expressed as amol sperm^{-1}.

2.6. *Data Analysis*

2.6.1. Principal Component Analyses for Sperm Velocity Parameters

Since swimming parameters tend to be highly correlated [78], principal component analyses (PCA) were performed to construct variables that summarize the information obtained through CASA. The variables were divided in two groups defining sperm velocity (VCL, VSL, VAP) and sperm trajectory (LIN, STR, WOB, ALH, BCF) and one independent PCA was carried out for each group. The first principal component for the velocity group (VPC1) accounted for 91.8% of the variability in the three summarized variables (VCL, VSL, VAP), while the second principal component (VPC2) only accounted for 7.6% of the variability (Figure 1A). The three sperm velocity descriptors showed extremely high correlation coefficients with VPC1 and were weakly or non-significantly correlated with VPC2 (Table 1). For the trajectory, the first principal component (TPC1) accounted for 58.4% of the variability, while the second principal component (TPC2) represented 26.6% of the variability (Figure 1B). Although the five variables included in this group were correlated with TPC1, the strength of the correlation was higher for LIN, STR, and WOB (Table 1). On the other hand, ALH and BCF showed a stronger correlation with TPC2 (Table 1). Thus, VPC1, TPC1 and TPC2 values for each treatment and species were used as our integrated sperm velocity and trach shape measures.

2.6.2. Statistical Analyses

The effect of incubation with PHE on sperm parameters over time was analyzed with a two-factor repeated-measures ANOVA for each species, using treatment (2 levels: control and PHE) and time (4 levels: 0, 2, 3, 4 h) as factors. Differences between conditions were analyzed through a post-hoc DGC multiple comparisons test [79]. All variables were $\log_{10}$-transformed for statistical purposes, with the exception of percentages (motility, viability, acrosome integrity, LIN, STR and WOB), which were arcsine-transformed. The statistical analyses were performed using SPSS Statistics (SPSS v.23.0.0.0; SPSS, IBM Corporation, Somers, NY, USA), and InfoStat v.2015p (Grupo Infostat, Universidad Nacional de Córdoba, Córdoba, Argentina) with $\alpha = 0.05$.

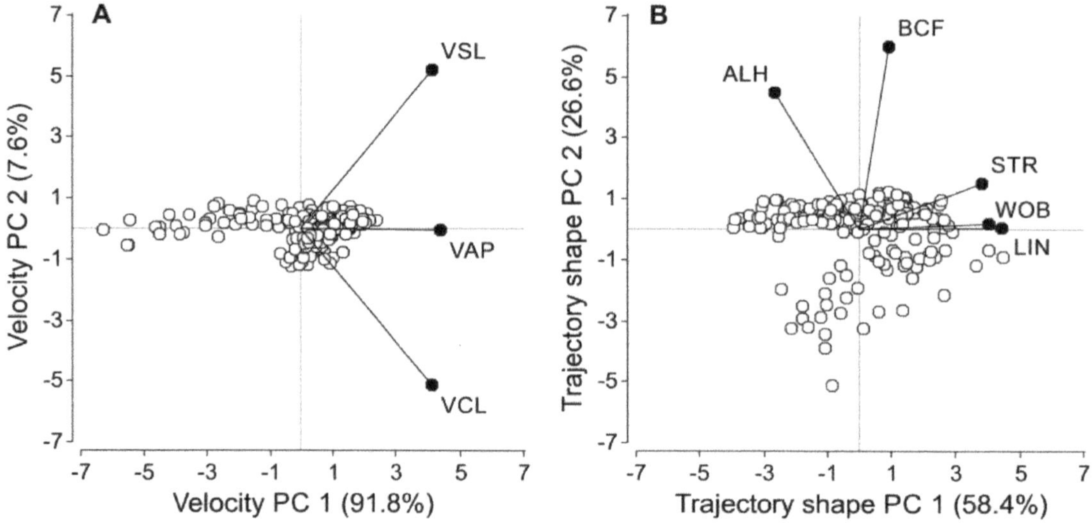

Figure 1. Biplots detailing the relationships between principal components of sperm velocity (**A**) and trajectory shape (**B**) and their constituent variables. Variable values were $\log_{10}$ or arcsine transformed prior to analysis. VCL: curvilinear velocity. VSL: straight-line velocity. VAP: average path velocity. LIN: linearity. STR: straightness. WOB: wobble coefficient. ALH: amplitude of the lateral head displacement. BCF: beat-cross frequency.

Table 1. Loadings and correlation of sperm traits with principal components of sperm velocity and trajectory shape in five hamster species. Values presented are Pearson's correlation coefficients. Significant correlation coefficients ($p < 0.05$) are shown in bold. PC1: principal component 1. PC2: principal component 2. Variable values were $\log_{10}$ or arcsine transformed prior to analysis.

Variables	Factor Loadings		Factor Correlation	
	PC1	PC2	PC1	PC2
Sperm velocity principal components:				
Curvilinear velocity	0.5667	−0.7023	**0.9405**	**−0.3345**
Straight-line velocity	0.5658	0.7118	**0.9390**	0.3390
Average path velocity	0.5989	−0.0079	**0.9938**	−0.0038
Sperm trajectory shape principal components				
Linearity	0.5800	0.0041	**0.9914**	0.0047
Straightness	0.5008	0.1996	**0.8560**	0.2302
Wobble coefficient	0.5261	0.0217	**0.8992**	0.0251
Amplitude of lateral head displacement	−0.3498	0.5869	**−0.5980**	**0.6767**
Beat-cross frequency	0.1167	0.7844	**0.1995**	**0.9045**

3. Results

3.1. Relative Testes Size and Sperm Numbers

Body mass, testes mass, relative testes size and sperm numbers per individual are presented in Table 2. The mass of testes in relation to body mass was high in all species when compared with other species of muroid rodents (see [22,77]). Among the species examined here, *Cricetulus griseus* showed the highest relative testes size (*sensu* [75]), while *Phodopus sungorus* presented the smallest testes in relation to body mass.

3.2. Sperm Motility, Viability, Acrosome Integrity and ATP Content

Initial sperm motility (~80–90%), viability (~95–99%), and acrosome integrity (~75–90%) were high for all species in both media (Table 3). In addition, sperm of the five species showed relatively high initial ATP content in media without and with PHE (Table 3) when compared with other muroid rodent species (see [24,80]). In general, sperm showed a significant gradual decline in their motility and acrosome integrity throughout the 4 h of incubation (Table 4, Figure 2) with the exception of *Mesocricetus auratus* spermatozoa which exhibited a more pronounced decrease in motility between 0 and 2 h of incubation, with a slight decrease afterwards (Figure 2D). Sperm ATP content also decreased significantly with time in the five species (Table 4), with a pronounced phase of decline between 0 and 2 h of incubation in four of the five species (Figure 2), and a more constant decrease over the 4 h of incubation in the case of *C. griseus* (Figure 2C). Sperm viability did not change significantly during incubation time in any of the five species, maintaining high values (>90%) in all species with both treatments (Table 4, Figure S1).

Table 2. Corporal measurements and sperm numbers in five hamster species. Values indicate mean ± standard error. RTS: relative testes size calculated according to Kenagy and Trombulak [75].

Species	Body Mass (g)	Testes Mass (g)	RTS	Sperm Numbers ($\times 10^6$)
Cricetulus griseus	33.72 ± 0.38	1.78 ± 0.04	3.83 ± 0.11	88.00 ± 07.65
Mesocricetus auratus	125.00 ± 1.63	3.50 ± 0.12	2.75 ± 0.11	585.88 ± 36.81
Phodopus campbelli	48.55 ± 3.90	2.01 ± 0.08	3.30 ± 0.21	317.82 ± 25.12
Phodopus sungorus	46.82 ± 1.25	0.94 ± 0.06	1.58 ± 0.13	160.76 ± 27.07
Phodopus roborovskii	25.72 ± 1.15	1.06 ± 0.04	2.82 ± 0.12	175.64 ± 26.59

Table 3. Sperm motility, acrosome integrity, viability and ATP content in five hamster species. Values correspond to mean and standard error (SE) for spermatozoa in the absence (control) or presence of penicillamine, hypotaurine and epinephrine (PHE) at the start of the incubation period (time zero).

Species	Variable	Control		PHE	
		Mean	SE	Mean	SE
Cricetulus griseus	Motility (%)	80.00	1.58	83.00	2.00
	Acrosome Integrity (%)	88.60	2.62	89.20	2.92
	Viability (%)	94.00	1.41	95.60	1.29
	ATP content (amol cell^{-1})	449.34	51.76	489.47	53.91
Mesocricetus auratus	Motility (%)	79.17	2.01	84.17	2.01
	Acrosome Integrity (%)	86.83	2.21	85.50	1.61
	Viability (%)	98.67	0.42	97.83	0.75
	ATP content (amol cell^{-1})	437.68	63.39	368.09	16.23
Phodopus campbelli	Motility (%)	88.00	1.22	84.00	2.45
	Acrosome Integrity (%)	88.60	1.69	86.86	1.09
	Viability (%)	99.40	0.40	99.40	0.40
	ATP content (amol cell^{-1})	339.83	20.64	379.83	24.23
Phodopus sungorus	Motility (%)	84.29	1.70	81.43	1.80
	Acrosome Integrity (%)	79.71	3.89	74.86	3.74
	Viability (%)	98.43	0.48	98.29	0.36
	ATP content (amol cell^{-1})	369.20	59.19	435.80	83.28
Phodopus roborovskii	Motility (%)	89.00	1.00	87.00	1.22
	Acrosome Integrity (%)	86.80	3.14	87.40	2.62
	Viability (%)	96.40	0.93	96.40	1.69
	ATP content (amol cell^{-1})	459.57	33.29	467.05	32.20

The incubation in a medium with PHE showed a significantly positive effect on sperm motility in *C. griseus* and *M. auratus* (Table 4). The effect was stronger and appeared earlier in *M. auratus* (2 h of incubation, Figure 2D) than in *C. griseus* (4 h of incubation, Figure 2A).

In *C. griseus*, sperm viability was significantly higher in the PHE treatment than in the control after 2 h of incubation (Table 4, Figure S1A). Conversely, *M. auratus* sperm showed differences favoring the control over the PHE treatment at 3 and 4 h (Table 4, Figure S1B). In the case of *Phodopus* species, the addition of PHE to the incubation medium showed no significant effects on sperm motility, viability, and acrosome integrity (Table 4, Figure 2 and Figure S1). Sperm ATP content was not affected by the addition of PHE to the incubation medium (Table 4, Figure 2) with the exception of *C. griseus* in which PHE promoted a less pronounced decline in ATP content (Table 4, Figure 2C).

Table 4. Effect of penicillamine, hypotaurine, and epinephrine (PHE) on sperm motility, acrosome integrity, viability, and ATP content, in five hamster species. *F* and *p* values correspond to repeated measures ANOVA using time of incubation and treatment (control vs. PHE) as independent variables and sperm parameters as dependent variables. Significant differences between treatments and times of incubation ($p < 0.05$) are shown in boldface.

Species	Variable	Time		Treatment		Interaction	
		F	*p*	*F*	*p*	*F*	*p*
Cricetulus griseus	Motility (%)	34.27	**<0.001**	9.14	**0.005**	1.34	0.280
	Acrosome Integrity (%)	44.62	**<0.001**	0.33	0.570	1.54	0.225
	Viability (%)	0.61	0.617	34.45	**<0.001**	2.53	0.078
	ATP content (amol cell^{-1})	35.39	**<0.001**	15.47	**0.001**	4.01	**0.007**
Mesocricetus auratus	Motility (%)	254.11	**<0.001**	110.40	**<0.001**	6.98	**0.001**
	Acrosome Integrity (%)	86.67	**<0.001**	6.66	**0.014**	0.24	0.870
	Viability (%)	2.10	0.118	6.04	**0.019**	2.00	0.132
	ATP content (amol cell^{-1})	120.43	**<0.001**	0.83	0.368	1.94	0.141
Phodopus campbelli	Motility (%)	16.81	**<0.001**	2.44	0.129	1.06	0.381
	Acrosome Integrity (%)	29.84	**<0.001**	5.17	0.085	0.22	0.880
	Viability (%)	1.69	0.192	1.13	0.298	1.97	0.142
	ATP content (amol cell^{-1})	43.92	**<0.001**	1.89	0.180	0.09	0.966
Phodopus sungorus	Motility (%)	13.17	**<0.001**	0.08	0.775	1.49	0.231
	Acrosome Integrity (%)	24.93	**<0.001**	1.54	0.261	0.92	0.442
	Viability (%)	0.42	0.743	0.98	0.327	0.55	0.648
	ATP content (amol cell^{-1})	37.52	**<0.001**	1.34	0.253	0.46	0.709
Phodopus roborovskii	Motility (%)	38.87	**<0.001**	1.06	0.312	0.44	0.727
	Acrosome Integrity (%)	18.66	**<0.001**	0.03	0.868	0.08	0.972
	Viability (%)	0.44	0.726	0.07	0.790	0.21	0.888
	ATP content (amol cell^{-1})	14.43	**<0.001**	1.46	0.237	0.12	0.950

3.3. Sperm Velocity and Trajectory

Regarding sperm velocity and trajectory parameters, all species presented similar values regardless of treatment at the beginning of the incubation period (0 h) (Table 5). As a general trend, velocity and trajectory parameters, as well as their estimated principal components (VPC1, TPC1, and TPC2) exhibited a time-related decline in both control and PHE treatment (Table 6, Figure 3). However, several variables did not show this time-related effect in some species. The sperm of *M. auratus* that were incubated in the presence of PHE maintained their linearity, straightness, and amplitude of lateral head displacement throughout the incubation period (Table 6, Figure 3B,C). In *P. sungorus*, straightness, amplitude of lateral head displacement, beat-cross frequency, and TPC1 were similar during the 4 h incubation regardless of treatment (Table 6, Figure 3H). This pattern was also observed in the values of linearity, and TPC2 but only for the control treatment (Table 6, Figure 3I). In *P. roborovskii*, sperm straight-line velocity, average path velocity, and VPC1 did not decrease over time (Table 6, Figure 3J). Notably, in this species, sperm linearity, straightness, wobble coefficient, and TPC1 registered a gradual increase during incubation time (Table 6, Figure 3K).

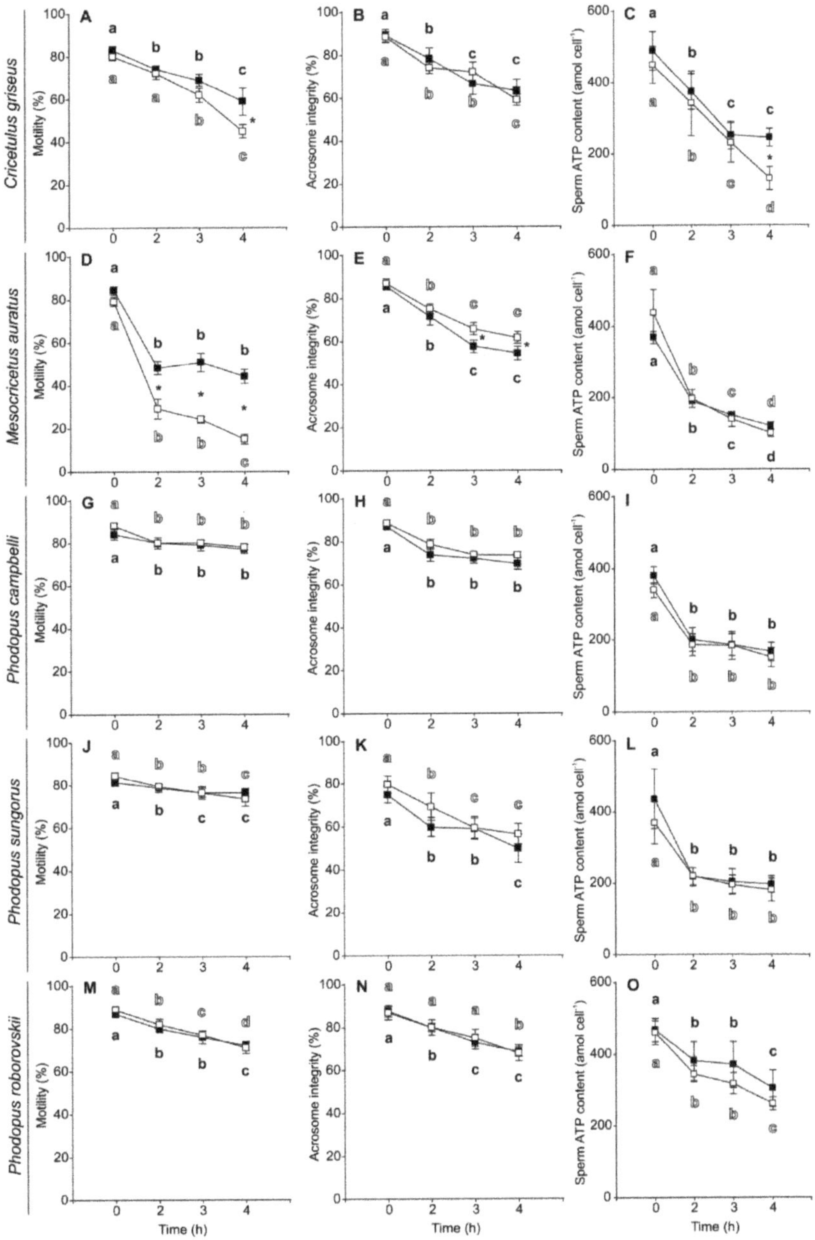

Figure 2. Changes in motility, acrosome integrity, and ATP content in spermatozoa from five hamster species during incubation without or with penicillamine, hypotaurine and epinephrine (PHE). Spermatozoa were collected in medium with (black squares) or without PHE (white squares) and incubated at 37 °C under air for up to 4 h. (**A**,**D**,**G**,**J**,**M**) Percentage of motile sperm. (**B**,**E**,**H**,**K**,**N**) Percentage of spermatozoa with an intact acrosome. (**C**,**F**,**I**,**L**,**O**) Sperm ATP content (amol $\times$ cell^{-1}). Values are means $\pm$ standard errors. Different letters between times of incubation for the same treatment indicate statistically significant differences ($p < 0.05$) in a DGC post-hoc test. Asterisks indicate statistical differences ($p < 0.05$) between treatments for the same time in a DGC post-hoc test.

Table 5. Sperm velocity and trajectory shape parameters in four hamster species. Values correspond to means and standard errors for spermatozoa in the absence (control) or with penicillamine, hypotaurine and epinephrine (PHE) at the start of the incubation period (time zero). VCL: curvilinear velocity. VSL: straight-line velocity. VAP: average path velocity. LIN: linearity. STR: straightness. WOB: wobble coefficient. ALH: amplitude of lateral head movement. BCF: beat-cross frequency.

Species	Variable	Control		PHE	
		Mean	SE	Mean	SE
Mesocricetus auratus	VCL ($\mu m\ s^{-1}$)	114.71	2.40	123.45	3.99
	VSL ($\mu m\ s^{-1}$)	54.66	2.33	56.76	2.27
	VAP ($\mu m\ s^{-1}$)	76.64	2.02	82.22	2.26
	LIN (VSL/VCL)	0.47	0.02	0.46	0.02
	STR (VSL/VAP)	0.67	0.02	0.66	0.01
	WOB (VAP/VCL)	0.67	0.01	0.68	0.01
	ALH (μm)	5.17	0.11	5.67	0.22
	BCF (Hz)	6.32	0.18	6.27	0.18
Phodopus campbelli	VCL ($\mu m\ s^{-1}$)	162.79	2.25	162.80	2.01
	VSL ($\mu m\ s^{-1}$)	75.25	1.27	71.62	1.93
	VAP ($\mu m\ s^{-1}$)	102.05	1.48	98.86	1.33
	LIN (VSL/VCL)	0.45	0.01	0.43	0.01
	STR (VSL/VAP)	0.70	0.01	0.69	0.01
	WOB (VAP/VCL)	0.62	0.01	0.61	0.01
	ALH (μm)	6.48	0.12	6.40	0.12
	BCF (Hz)	9.83	0.15	9.70	0.12
Phodopus sungorus	VCL ($\mu m\ s^{-1}$)	152.02	2.24	157.70	2.29
	VSL ($\mu m\ s^{-1}$)	68.81	2.36	69.10	2.19
	VAP ($\mu m\ s^{-1}$)	92.19	1.91	92.81	2.13
	LIN (VSL/VCL)	0.44	0.01	0.42	0.01
	STR (VSL/VAP)	0.69	0.01	0.69	0.01
	WOB (VAP/VCL)	0.60	0.01	0.58	0.01
	ALH (μm)	6.34	0.11	6.56	0.11
	BCF (Hz)	8.70	0.34	9.21	0.28
Phodopus roborovskii	VCL ($\mu m\ s^{-1}$)	152.85	2.91	146.54	1.01
	VSL ($\mu m\ s^{-1}$)	62.13	1.85	56.36	2.42
	VAP ($\mu m\ s^{-1}$)	87.26	1.68	80.80	1.50
	LIN (VSL/VCL)	0.40	0.01	0.38	0.02
	STR (VSL/VAP)	0.68	0.01	0.67	0.02
	WOB (VAP/VCL)	0.57	0.01	0.55	0.01
	ALH (μm)	6.96	0.12	7.14	0.11
	BCF (Hz)	8.40	0.35	7.53	0.48

The sperm of *M. auratus* exhibited a strong significant response to the exposure to PHE during the incubation, in regard to their velocity and trajectory parameters. All velocity and trajectory variables, including their estimated principal components, showed significant higher values in the PHE treatment (Table 6, Figure 3A–C). The addition of PHE to the medium appears to prevent the time-related decline of sperm kinetics in this species. Conversely, the addition of PHE did not appear to have any positive effect on the kinetics of the sperm of any of the three species of the genus *Phodopus*. As a general trend, there were no significant differences in the values of most of the sperm velocity and trajectory variables in *Phodopus* species (Table 6, Figure 3D,F–I,K,L). In the few cases in which significant differences were observed (*P. campbelli*: VSL, LIN, WOB, ALH, and TPC1; *P. sungorus*: LIN, WOB; *P. roborovskii*: VCL, VSL, VAP, WOB, VPC1), these differences were of relatively low magnitude, were not maintained over time, and values were always higher in the control treatment (Table 6, Figure 3E,J).

Table 6. Effect of penicillamine, hypotaurine and epinephrine (PHE) on sperm velocity and trajectory shape parameters in four hamster species. F and p values correspond to repeated measures ANOVA using time of incubation and treatment (control vs. PHE) as independent variables, and sperm parameters as dependent variables. VCL: curvilinear velocity. VSL: straight-line velocity. VAP: average path velocity. LIN: linearity. STR: straightness. WOB: wobble coefficient. ALH: amplitude of lateral head movement. BCF: beat-cross frequency. VPC1: velocity principal component 1. TPC1: trajectory shape principal component 1. TPC2: trajectory shape principal component 2. Significant differences between treatments and times of incubation ($p < 0.05$) are shown in boldface.

Species	Variable	Time		Treatment		Interaction	
		F	p	F	p	F	p
Mesocricetus auratus	VCL (μm s^{-1})	28.78	**<0.001**	67.30	**<0.001**	3.54	**0.024**
	VSL (μm s^{-1})	14.62	**<0.001**	46.25	**<0.001**	4.43	**0.011**
	VAP (μm s^{-1})	29.51	**<0.001**	50.80	**<0.001**	3.26	**0.031**
	LIN	2.75	0.056	17.82	**<0.001**	2.83	0.074
	STR	1.04	0.384	19.68	**<0.001**	3.80	**0.017**
	WOB	15.50	**<0.001**	10.48	**0.002**	1.28	0.295
	ALH (μm)	0.68	0.573	26.39	**<0.001**	1.10	0.361
	BCF (Hz)	101.82	**<0.001**	93.48	**<0.001**	15.10	**<0.001**
	VPC1	25.98	**<0.001**	59.27	**<0.001**	4.09	**0.014**
	TPC1	5.12	**0.004**	9.61	**0.004**	2.45	0.077
	TPC2	7.08	**0.001**	73.25	**<0.001**	5.73	**0.003**
Phodopus campbelli	VCL (μm s^{-1})	12.90	**<0.001**	0.22	0.644	0.52	0.669
	VSL (μm s^{-1})	99.74	**<0.001**	4.53	**0.042**	0.09	0.964
	VAP (μm s^{-1})	83.46	**<0.001**	1.57	0.221	0.24	0.867
	LIN	61.32	**<0.001**	4.98	**0.034**	0.52	0.673
	STR	17.99	**<0.001**	1.51	0.230	0.29	0.832
	WOB	196.44	**<0.001**	14.20	**0.001**	1.11	0.363
	ALH (μm)	60.78	**<0.001**	7.96	**0.009**	2.29	0.100
	BCF (Hz)	43.82	**<0.001**	2.29	0.141	0.70	0.559
	VPC1	98.05	**<0.001**	2.15	0.154	0.15	0.931
	TPC1	70.65	**<0.001**	6.23	**0.019**	0.62	0.608
	TPC2	7.02	**0.001**	0.10	0.751	0.25	0.862
Phodopus sungorus	VCL (μm s^{-1})	8.19	**<0.001**	3.47	0.069	0.92	0.441
	VSL (μm s^{-1})	7.41	**<0.001**	2.46	0.124	0.63	0.603
	VAP (μm s^{-1})	9.70	**<0.001**	0.76	0.387	0.63	0.600
	LIN	2.75	0.055	6.27	**0.016**	0.32	0.814
	STR	1.88	0.148	2.75	0.105	0.59	0.626
	WOB	2.95	**0.043**	6.38	**0.015**	0.27	0.846
	ALH (μm)	1.02	0.394	5.36	0.060	0.09	0.966
	BCF (Hz)	1.85	0.153	0.02	0.899	1.11	0.355
	VPC1	9.79	**<0.001**	0.49	0.488	0.81	0.497
	TPC1	2.24	0.098	3.55	0.108	0.26	0.854
	TPC2	5.19	**0.004**	3.15	0.083	1.34	0.274
Phodopus roborovskii	VCL (μm s^{-1})	17.20	**<0.001**	4.75	**0.038**	0.22	0.879
	VSL (μm s^{-1})	2.61	0.071	8	**0.009**	0.60	0.620
	VAP (μm s^{-1})	0.89	**0.460**	9.34	**0.005**	0.42	0.739
	LIN	21.86	**<0.001**	2.99	0.095	0.29	0.833
	STR	9.20	**<0.001**	0.15	0.702	0.28	0.839
	WOB	20.12	**<0.001**	4.99	**0.035**	0.22	0.884
	ALH (μm)	30.78	**<0.001**	0.48	0.493	0.17	0.914
	BCF (Hz)	4.76	**0.008**	2.81	0.105	1.82	0.166
	VPC1	0.86	0.426	8.65	**0.006**	0.49	0.694
	TPC1	22.57	**<0.001**	2.33	0.138	0.27	0.843
	TPC2	11.21	**<0.001**	0.56	0.461	0.65	0.588

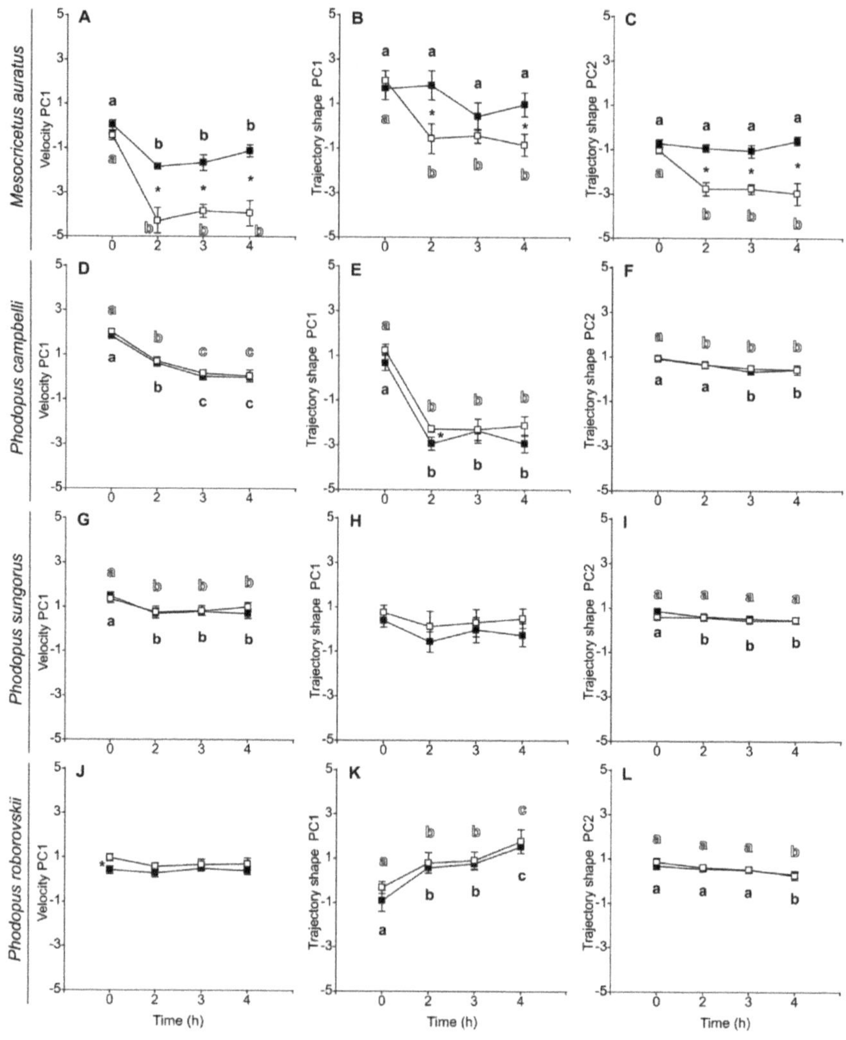

Figure 3. Changes in velocity or trajectory of spermatozoa from five hamster species during incubation in the absence (control) or the presence of penicillamine, hypotaurine and epinephrine (PHE). Spermatozoa were collected in medium with (dark squares) or without PHE (white squares) and incubated at 37 °C under air for up to 4 h. Principal component analysis returned one main velocity component (velocity PC1) and two trajectory components (trajectory shape PC1, trajectory shape PC2). (**A**,**D**,**G**,**J**) Velocity PC1. (**B**,**E**,**H**,**K**) Trajectory shape PC1. (**C**,**F**,**I**,**L**) Trajectory shape PC2. Values are means ± standard errors. Different letters between times of incubation for the same treatment indicate statistically significant differences ($p < 0.05$) in a DGC post-hoc test. Asterisks indicate statistical differences ($p < 0.05$) between treatments for the same time in a DGC post-hoc test.

4. Discussion

The results of this study show that the addition of D-penicillamine, hypotaurine and epinephrine (PHE) to the incubation medium seems to be necessary to sustain the performance (motility, sperm swimming velocity and trajectory) and, to a lesser extent, acrosome integrity and viability, of *M. auratus* sperm throughout incubation. Moreover, PHE appears to have a positive effect on the maintenance of sperm quality (motility, viability, and ATP content) in *C. griseus*. In contrast, the presence of PHE had no consistent effect on spermatozoa of the *Phodopus* species.

Golden hamster sperm is known to be very susceptible to in vitro dilution for more than 15–30 min [81,82] unless the medium is supplemented with motility factors. The addition of epinephrine, taurine, hypotaurine, penicillamine, or their combination, to hamster sperm has been studied mainly in *M. auratus* [39,40,56,81–85]. In the present work, a combination of D-penicillamine (20 μM), hypotaurine (100 μM) and epinephrine (1 μM) was selected to analyze their effects on sperm performance. Our results showed an improvement in motility, integrity of the acrosome and velocity and trajectory parameters from 2 h onwards after addition of PHE in *M. auratus*. Several studies showed that epinephrine and hypotaurine have a synergistic effect in maintaining sperm motility and promoting the acrosome reaction [40,43]. These two molecules caused a slight enhancement in motility, acrosome reaction and fertility when added separately, and these effects were increased when used together [83]. Moreover, the effect of hypotaurine occurs within the first 2 h [83], in agreement with the present study, where sperm motility, velocity and trajectory showed better results 2 h after the addition of PHE in *M. auratus*.

Epinephrine is a catecholamine that stimulates Na^+/K^+ ATPase and Ca^{2+}-ATPase, and inhibits certain phosphodiesterases [40,86,87]. The inhibition of Na^+/K^+ ATPase decreased the acrosome reaction, whereas the inhibition of phosphodiesterase increased it [85]. Hypotaurine is a β-amino acid present in ejaculated sperm and oviductal fluid of mammals [43]. It is an intracellular scavenger that protects from lipid peroxidation and inactivation of superoxide dismutase by superoxidation, preventing sperm motility loss [84]. D-penicillamine acts as a divalent cation chelator, increasing the effects of epinephrine by protecting it from oxidation [40]. It was found that this α-amino acid could maintain sperm motility in hamsters at lower concentrations (10 μM) and stimulate sperm capacitation at higher concentrations (50 μM) [51]. In the present study, the concentration of penicillamine was 20 μM, which seems to be adequate in *M. auratus* to maintain or improve sperm traits for 4 h, in combination with epinephrine and hypotaurine. Taken together, results lead to the conclusion that the addition of PHE to *M. auratus* spermatozoa improves sperm quality over time, probably through the interaction with ATPases and phosphodiesterases and by protecting sperm from oxidation.

Changes in sperm traits in *C. griseus* after the addition of PHE are different from those observed in *M. auratus*. In *C. griseus* sperm motility and ATP content are enhanced at a later time, i.e., after 4 h of incubation with PHE. Moreover, sperm viability improves from 2 h onwards. In an earlier study [58], it was found that only 20–30% of cauda epididymal *C. griseus* sperm exhibit motility, which is significantly lower than the results in our study. This discrepancy may be due to differences in media used to collect and incubate spermatozoa because in the earlier study [58] the medium contained epinephrine and taurine (instead of hypotaurine, as in our study), and lacked penicillamine. The improvement in sperm survival and performance observed in our work may be useful to improve the reduced in vitro fertilization success observed in this species [58,88]. Indeed, better fertilization rates were obtained when sperm were pre-exposed to PHE and to the Ca^{2+} ionophore A23187, while doubling the usual Ca^{2+} concentrations [60].

Different patterns of sperm motility have been found when comparing *M. auratus* with *C. griseus* [58]. Whereas *M. auratus* sperm flex the entire length of the tail when moving, *C. griseus* sperm move by vibrating the tails firmly. This difference in swimming patterns may explain the diverse results in both species. Despite these differences, the addition of PHE seems to promote an improvement of sperm performance in both species. There does not seem to be a relation between the presence or absence of PHE, ATP levels, and motility when comparing these two species. In *M. auratus* PHE sustains motility better and for longer times than in controls, but there does not seem to be an effect on ATP levels. In *C. griseus*, there seems to be a parallel improvement of PHE on both motility and ATP levels, but only at the end of incubation.

The results in *Phodopus* species are somewhat surprising because we did not find differences over time in variables assessed in sperm incubated in the absence or presence of PHE. To the best of our knowledge, this is the first report examining the effects of PHE in these three species of hamsters. Since previous studies found that epinephrine acts through α and β-adrenergic receptors [41] and hypotaurine activates ATPases and inhibits phosphodiesterases, differences between sperm from *Phodopus*, *C. griseus* and *M. auratus* may be explained by differences in receptors or signaling mechanisms. Thus, one possibility could be that targets of PHE action vary in concentration between species, and thus effects are not visible. Another possibility is that responses to each factor vary, since different concentrations of penicillamine, hypotaurine and epinephrine are known to have different effects in somatic and sperm cells [51,86,87]. More studies are necessary to elucidate if changes in concentrations of PHE may affect sperm traits in *Phodopus* species. In any case, sperm quality in these hamsters is extremely high when compared to other hamster species after 4 h of incubation without PHE, indicating that the addition of these molecules or other motility factors may not be required to sustain sperm performance.

5. Conclusions

The results of this study revealed that supplementation with a combination of D-penicillamine, hypotaurine and epinephrine maintains or improves the performance of spermatozoa from five hamster species in different manners, depending on the species. Further studies are needed to understand the different mechanisms underlying the stimulatory effects in some species and lack of stimulation in others, which suggest divergence in regulatory mechanisms of motility and survival.

Supplementary Materials: The following are available online at https://www.mdpi.com/article/10.3390/biology11040526/s1, Figure S1. Changes in sperm viability in five hamster species during incubation without or with penicillamine, hypotaurine and epinephrine (PHE). (A) *Cricetulus griseus*. (B) *Mesocricetus auratus*. (C) *Phodopus campbelli*. (D) *P. sungorus*. (E) *P. roborovskii*. Spermatozoa were collected in medium with (black squares) or without PHE (white squares) and incubated at 37 °C under air for up to 4 h. Values are means ± standard errors. Different letters between times of incubation for the same treatment indicate statistically significant differences ($p < 0.05$) in a DGC post hoc test. Asterisks indicate statistical differences ($p < 0.05$) between treatments for the same time in a DGC post-hoc test.

Author Contributions: M.T. and E.R.S.R. conceived the study. M.T. performed experiments. M.T., A.S.-R. and E.R.S.R. analysed data. M.T., A.S.-R. and E.R.S.R. wrote the paper. All authors have read and agreed to the published version of the manuscript.

Funding: This work was supported by the Spanish Ministry of Science and Innovation (projects CGL2016-80577-P and PID2019-108649GB-I00). This research also received support from the SYNTHESYS+ Project which is financed by the European Commission via the H2020 Research Infrastructure programme ID (823827) at the Museo Nacional de Ciencias Naturales (CSIC). M.T and A.S.-R. held "Juan de la Cierva" postdoctoral fellowships from the Spanish Ministry of Science and Innovation.

Institutional Review Board Statement: This study was carried out in strict accordance with the recommendations of the Code of Good Scientific Practices of the Spanish Research Council (CSIC) and was approved by the Animal Experimentation Ethics Committee of the National Museum of Natural Sciences (CSIC) and Comunidad de Madrid, Spain (28079-47-A). Animal handling complied with Spanish Animal Protection Regulation RD53/2013, which conforms to European Union Regulation 2010/63.

Informed Consent Statement: Not applicable.

Data Availability Statement: Data are included in the article.

Acknowledgments: We thank Juan Antonio Rielo for supervising animal facilities and Esperanza Navarro for animal care at the Museo Nacional de Ciencias Naturales in Madrid.

Conflicts of Interest: The authors declare no conflict of interest. The funders had no role in the design of the study; in the collection, analyses, or interpretation of data; in the writing of the manuscript, or in the decision to publish the results.

References

1. Roldan, E.R.S. Male fertility overview. In *Encyclopedia of Reproduction*, 2nd ed.; Skinner, M.K., Ed.; Academic Press: Cambridge, MA, USA, 2018; Volume 1, pp. 408–415.
2. Roldan, E.R.S. Assessments of sperm quality integrating morphology, swimming patterns, bioenergetics and cell signalling. *Theriogenology* **2020**, *150*, 388–395. [CrossRef] [PubMed]
3. Rodriguez-Martinez, H. Laboratory semen assessment and prediction of fertility: Still utopia? *Reprod. Domest. Anim.* **2003**, *38*, 312–318. [CrossRef]
4. Björndahl, L.; Mortimer, D.; Barratt, C.L.R.; Castilla, J.A.; Roelof, M.; Kvist, U.; Álvarez, J.G.; Haugen, T.B. *A Practical Guide to Basic Laboratory Andrology*; Cambridge University Press: Cambridge, UK, 2010; p. 336.
5. Chenoweth, P.J.; Lorton, P.S. *Animal Andrology: Theories and Applications*; CABI: Wallingford, UK, 2014; p. 584.
6. Srivastava, N.; Pande, M. *Protocols in Semen Biology (Comparing Assays)*; Springer: Singapore, 2017; p. 288.
7. Oehninger, S.; Ombelet, W. Limits of current male fertility testing. *Fertil. Steril.* **2019**, *111*, 835–841. [CrossRef] [PubMed]
8. Aitken, R.J.; Mortimer, D.; Kovacs, D. *Male and Sperm Factors that Maximize IVF Success*; Cambridge University Press: Cambridge, UK, 2020; p. 244.
9. Turner, K.A.; Rambhatla, A.; Schon, S.; Agarwal, A.; Krawetz, S.A.; Dupree, J.M.; Avidor-Reiss, T. Male infertility is a women's health issue-research and clinical evaluation of male infertility is needed. *Cells* **2020**, *9*, 990. [CrossRef]
10. van der Horst, G. Status of sperm functionality assessment in wildlife species: From fish to primates. *Animals* **2021**, *11*, 1491. [CrossRef]
11. Malo, A.F.; Garde, J.J.; Soler, A.J.; Garcia, A.J.; Gomendio, M.; Roldan, E.R.S. Male fertility in natural populations of red deer is determined by sperm velocity and the proportion of normal spermatozoa. *Biol. Reprod.* **2005**, *72*, 822–829. [CrossRef] [PubMed]
12. Yaniz, J.L.; Silvestre, M.A.; Santolaria, P.; Soler, C. CASA-Mot in mammals: An update. *Reprod. Fertil. Dev.* **2018**, *30*, 799–809. [CrossRef]
13. van der Horst, G. Computer aided sperm analysis (CASA) in domestic animals: Current status, three D tracking and flagellar analysis. *Anim. Reprod. Sci.* **2020**, *220*, 106350. [CrossRef]
14. Buffone, M.G. *Sperm Acrosome Biogenesis and Function during Fertilization*; Springer: New York, NY, USA, 2016; p. 172.
15. Suarez, S.S.; Pacey, A.A. Sperm transport in the female reproductive tract. *Hum. Reprod. Update* **2006**, *12*, 23–37. [CrossRef]
16. Hirohashi, N. Site of mammalian sperm acrosome reaction. *Adv. Anat. Embryol. Cell Biol.* **2016**, *220*, 145–158. [CrossRef]
17. La Spina, F.A.; Puga Molina, L.C.; Romarowski, A.; Vitale, A.M.; Falzone, T.L.; Krapf, D.; Hirohashi, N.; Buffone, M.G. Mouse sperm begin to undergo acrosomal exocytosis in the upper isthmus of the oviduct. *Dev. Biol.* **2016**, *411*, 172–182. [CrossRef]
18. Guidobaldi, H.A.; Hirohashi, N.; Cubilla, M.; Buffone, M.G.; Giojalas, L.C. An intact acrosome is required for the chemotactic response to progesterone in mouse spermatozoa. *Mol. Reprod. Dev.* **2017**, *84*, 310–315. [CrossRef]
19. Hirohashi, N.; Yanagimachi, R. Sperm acrosome reaction: Its site and role in fertilization. *Biol. Reprod.* **2018**, *99*, 127–133. [CrossRef]
20. Balestrini, P.A.; Jablonski, M.; Schiavi-Ehrenhaus, L.J.; Marin-Briggiler, C.I.; Sanchez-Cardenas, C.; Darszon, A.; Krapf, D.; Buffone, M.G. Seeing is believing: Current methods to observe sperm acrosomal exocytosis in real time. *Mol. Reprod. Dev.* **2020**, *87*, 1188–1198. [CrossRef] [PubMed]
21. Tourmente, M.; Varea-Sanchez, M.; Roldan, E.R.S. Faster and more efficient swimming: Energy consumption of murine spermatozoa under sperm competitiondagger. *Biol. Reprod.* **2019**, *100*, 420–428. [CrossRef] [PubMed]
22. Tourmente, M.; Villar-Moya, P.; Varea-Sanchez, M.; Luque-Larena, J.J.; Rial, E.; Roldan, E.R.S. Performance of rodent spermatozoa over time is enhanced by increased ATP concentrations: The role of sperm competition. *Biol. Reprod.* **2015**, *93*, 64. [CrossRef] [PubMed]
23. Tourmente, M.; Roldan, E.R.S. Mass-specific metabolic rate influences sperm performance through energy production in mammals. *PLoS ONE* **2015**, *10*, e0138185. [CrossRef]
24. Tourmente, M.; Rowe, M.; Gonzalez-Barroso, M.M.; Rial, E.; Gomendio, M.; Roldan, E.R.S. Postcopulatory sexual selection increases ATP content in rodent spermatozoa. *Evolution* **2013**, *67*, 1838–1846. [CrossRef]
25. Sansegundo, E.; Tourmente, M.; Roldan, E.R.S. Energy metabolism and hyperactivation of spermatozoa from three mouse species under capacitating conditions. *Cells* **2022**, *11*, 220. [CrossRef] [PubMed]
26. Tvrda, E.; Benko, F.; Slanina, T.; du Plessis, S.S. The role of selected natural biomolecules in sperm production and functionality. *Molecules* **2021**, *26*, 5196. [CrossRef]
27. Gervasi, M.G.; Visconti, P.E. Chang's meaning of capacitation: A molecular perspective. *Mol. Reprod. Dev.* **2016**, *83*, 860–874. [CrossRef] [PubMed]
28. Sato, M.; Ishikawa, A. Room temperature storage of mouse epididymal spermatozoa: Exploration of factors affecting sperm survival. *Theriogenology* **2004**, *61*, 1455–1469. [CrossRef] [PubMed]
29. Contri, A.; Gloria, A.; Robbe, D.; Valorz, C.; Wegher, L.; Carluccio, A. Kinematic study on the effect of pH on bull sperm function. *Anim. Reprod. Sci.* **2013**, *136*, 252–259. [CrossRef] [PubMed]

30. Zhou, J.; Chen, L.; Li, J.; Li, H.; Hong, Z.; Xie, M.; Chen, S.; Yao, B. The semen pH affects sperm motility and capacitation. *PLoS ONE* **2015**, *10*, e0132974. [CrossRef]
31. Perez-Cerezales, S.; Lopez-Cardona, A.P.; Gutierrez-Adan, A. Progesterone effects on mouse sperm kinetics in conditions of viscosity. *Reproduction* **2016**, *151*, 501–507. [CrossRef]
32. Mata-Martinez, E.; Darszon, A.; Trevino, C.L. pH-dependent Ca(+2) oscillations prevent untimely acrosome reaction in human sperm. *Biochem. Biophys. Res. Commun.* **2018**, *497*, 146–152. [CrossRef]
33. Giojalas, L.C.; Guidobaldi, H.A. Getting to and away from the egg, an interplay between several sperm transport mechanisms and a complex oviduct physiology. *Mol. Cell. Endocrinol.* **2020**, *518*, 110954. [CrossRef]
34. Ruiz-Diaz, S.; Luongo, C.; Fuentes-Albero, M.C.; Abril-Sanchez, S.; Sanchez-Calabuig, M.J.; Barros-Garcia, C.; De la Fe, C.; Garcia-Galan, A.; Ros-Santaella, J.L.; Pintus, E.; et al. Effect of temperature and cell concentration on dolphin (Tursiops truncatus) spermatozoa quality evaluated at different days of refrigeration. *Anim. Reprod. Sci.* **2020**, *212*, 106248. [CrossRef]
35. Yan, Y.; Zhang, B.; Fu, Q.; Wu, J.; Liu, R. A fully integrated biomimetic microfluidic device for evaluation of sperm response to thermotaxis and chemotaxis. *Lab Chip* **2021**, *21*, 310–318. [CrossRef]
36. Yanagimachi, R.; Chang, M.C. In vitro fertilization of golden hamster ova. *J. Exp. Zool.* **1964**, *156*, 361–375. [CrossRef]
37. Roldan, E.R.S.; Yanagimachi, R. Cross-fertilization between Syrian and Chinese hamsters. *J. Exp. Zool.* **1989**, *250*, 321–328. [CrossRef] [PubMed]
38. Bavister, B.D. A consistently successful procedure for in vitro fertilization of golden hamster eggs. *Gamete Res.* **1989**, *23*, 139–158. [CrossRef] [PubMed]
39. Bavister, B.D.; Chen, A.F.; Fu, P.C. Catecholamine requirement for hamster sperm motility in vitro. *J. Reprod. Fertil.* **1979**, *56*, 507–513. [CrossRef]
40. Meizel, S.; Working, P.K. Further evidence suggesting the hormonal stimulation of hamster sperm acrosome reactionsby catecholamines in vitro. *Biol. Reprod.* **1980**, *22*, 211–216. [CrossRef] [PubMed]
41. Cornett, L.E.; Meizel, S. Stimulation of in vitro activation and the acrosome reaction of hamster spermatozoa by catecholamines. *Proc. Natl. Acad. Sci. USA* **1978**, *75*, 4954–4958. [CrossRef]
42. Alvarez, J.G.; Storey, B.T. Spontaneous lipid peroxidation in rabbit and mouse epididymal spermatozoa: Dependence of rate on temperature and oxygen concentration. *Biol. Reprod.* **1985**, *32*, 342–351. [CrossRef]
43. Meizel, S.; Lui, C.W.; Working, P.K.; Mrsny, R.J. Taurine and Hypotaurine: Their effects on motility, capacitation and the acrosome reaction of hamster sperm in vitro and their presence in sperm and reproductive tract fluids of several mammals. *Dev. Growth Differ.* **1980**, *22*, 483–494. [CrossRef]
44. Fujinoki, M. Regulation and disruption of hamster sperm hyperactivation by progesterone, 17beta-estradiol and diethylstilbestrol. *Reprod. Med. Biol.* **2014**, *13*, 143–152. [CrossRef]
45. Fujinoki, M.; Takei, G.L. gamma-Aminobutyric acid suppresses enhancement of hamster sperm hyperactivation by 5-hydroxytryptamine. *J. Reprod. Dev.* **2017**, *63*, 67–74. [CrossRef]
46. Sakamoto, C.; Fujinoki, M.; Kitazawa, M.; Obayashi, S. Serotonergic signals enhanced hamster sperm hyperactivation. *J. Reprod. Dev.* **2021**, *67*, 241–250. [CrossRef]
47. Brogan, P.T.; Beitsma, M.; Henning, H.; Gadella, B.M.; Stout, T.A. Liquid storage of equine semen: Assessing the effect of d-penicillamine on longevity of ejaculated and epididymal stallion sperm. *Anim. Reprod. Sci.* **2015**, *159*, 155–162. [CrossRef] [PubMed]
48. Leahy, T.; Rickard, J.P.; Aitken, R.J.; de Graaf, S.P. Penicillamine prevents ram sperm agglutination in media that support capacitation. *Reproduction* **2016**, *151*, 167–177. [CrossRef] [PubMed]
49. Fernandez, M.C.; Yu, A.; Moawad, A.R.; O'Flaherty, C. Peroxiredoxin 6 regulates the phosphoinositide 3-kinase/AKT pathway to maintain human sperm viability. *Mol. Hum. Reprod.* **2019**, *25*, 787–796. [CrossRef] [PubMed]
50. Uribe, P.; Merino, J.; Bravo, A.; Zambrano, F.; Schulz, M.; Villegas, J.V.; Sanchez, R. Antioxidant effects of penicillamine against *in vitro*-induced oxidative stress in human spermatozoa. *Andrologia* **2020**, *52*, e13553. [CrossRef] [PubMed]
51. Andrews, J.C.; Nolan, J.P.; Hammerstedt, R.H.; Bavister, B.D. Role of zinc during hamster sperm capacitation. *Biol. Reprod.* **1994**, *51*, 1238–1247. [CrossRef]
52. Ruiz-Diaz, S.; Oseguera-Lopez, I.; De La Cuesta-Diaz, D.; Garcia-Lopez, B.; Serres, C.; Sanchez-Calabuig, M.J.; Gutierrez-Adan, A.; Perez-Cerezales, S. The presence of D-penicillamine during the in vitro capacitation of stallion spermatozoa prolongs hyperactive-like motility and allows for sperm selection by thermotaxis. *Animals* **2020**, *10*, 1467. [CrossRef]
53. Cornwall, G.A.; Smyth, T.B.; Vindivich, D.; Harter, C.; Robinson, J.; Chang, T.S. Induction and enhancement of progressive motility in hamster caput epididymal spermatozoa. *Biol. Reprod.* **1986**, *35*, 1065–1074. [CrossRef]
54. Andrews, J.C.; Bavister, B.D. Capacitation of hamster spermatozoa with the divalent cation chelators D-penicilamine, L-histidine, and L-cysteine in a protein-free culture medium. *Gamete Res.* **1989**, *23*, 159–170. [CrossRef]
55. Kang, S.S.; Koyama, K.; Huang, W.; Yang, Y.; Yanagawa, Y.; Takahashi, Y.; Nagano, M. Addition of D-penicillamine, hypotaurine, and epinephrine (PHE) mixture to IVF medium maintains motility and longevity of bovine sperm and enhances stable production of blastocysts in vitro. *J. Reprod. Dev.* **2015**, *61*, 99–105. [CrossRef]
56. Boatman, D.E.; Bavister, B.D.; Cruz, E. Addition of hypotaurine can reactivate immotile golden hamster spermatozoa. *J. Androl.* **1990**, *11*, 67–72.

57. Hirose, M.; Ogura, A. The golden (Syrian) hamster as a model for the study of reproductive biology: Past, present, and future. *Reprod. Med. Biol.* **2019**, *18*, 34–39. [CrossRef] [PubMed]
58. Yanagimachi, R.; Kamiguchi, Y.; Sugawara, S.; Mikamo, K. Gametes and fertilization in the Chinese hamster. *Gamete Res.* **1983**, *8*, 97–117. [CrossRef]
59. Tateno, H.; Kamiguchi, Y. In vitro fertilisation of Chinese hamster oocytes by spermatozoa that have undergone ionophore A23187-induced acrosome reaction, and their subsequent development into blastocysts. *Zygote* **1996**, *4*, 93–99. [CrossRef] [PubMed]
60. Tateno, H.; Tamura-Nakano, M.; Kusakabe, H.; Hirohashi, N.; Kawano, N.; Yanagimachi, R. Sperm acrosome status before and during fertilization in the Chinese hamster (*Cricetulus griseus*), and observation of oviductal vesicles and globules. *Mol. Reprod. Dev.* **2021**, *88*, 793–804. [CrossRef]
61. Parkening, T.A. In vitro fertilization of Siberian hamster oocytes. *J. Exp. Zool.* **1990**, *254*, 305–312. [CrossRef] [PubMed]
62. Raicu, P.; Bratosin, S. Interspecific reciprocal hybrids between *Mesocricetus auratus* and *M. newtoni*. *Genet Res.* **1968**, *11*, 113–114. [CrossRef] [PubMed]
63. Raicu, P.; Ionescu-Varo, M.; Duma, D. Interspecific crosses between the Rumanian and Syrian hamster. Cytogenetic and histological studies. *J. Hered.* **1969**, *60*, 149–152. [CrossRef]
64. Raicu, P.; Ionescu-Varo, M.; Nicolaescu, M.; Kirillova, M. Interspecific hybrids between Romanian and Kurdistan hamsters. *Genetica* **1972**, *43*, 223–230. [CrossRef]
65. Raicu, P.; Nicolaescu, M.; Kirillova, M. The interspecific hybrids between Kurdistan hamster (*Mesocricetus brandti*) and golden hamster (*Mesocricetus auratus*). *Rev. Roum. Biol.* **1973**, *18*, 451–455.
66. Todd, N.B.; Nixon, C.W.; Mulvaney, D.A.; Connelly, M.E. Karyotypes of *Mesocricetus brandti* and hybridization within the genus. *J. Hered.* **1972**, *63*, 73–77. [CrossRef]
67. Ishishita, S.; Matsuda, Y. Interspecific hybrids of dwarf hamsters and Phasianidae birds as animal models for studying the genetic and developmental basis of hybrid incompatibility. *Genes Genet. Syst.* **2016**, *91*, 63–75. [CrossRef] [PubMed]
68. Gureeva, A.V.; Feoktistova, N.Y.; Matveevsky, S.N.; Kolomiets, O.L.; Surov, A.V. Speciation of Eversmann and Mongolian hamsters (*Allocricetulus*, Cricetinae): Experimental hybridization. *Biol. Bull.* **2017**, *43*, 736–742. [CrossRef]
69. Cummins, J.M.; Woodall, P.F. On mammalian sperm dimensions. *J. Reprod. Fertil.* **1985**, *75*, 153–175. [CrossRef] [PubMed]
70. Tourmente, M.; Gomendio, M.; Roldan, E.R.S. Sperm competition and the evolution of sperm design in mammals. *BMC Evol. Biol.* **2011**, *11*, 12. [CrossRef]
71. Teves, M.E.; Roldan, E.R.S. Sperm bauplan and function and underlying processes of sperm formation and selection. *Physiol. Rev.* **2022**, *102*, 7–60. [CrossRef]
72. Lebedev, V.S.; Bannikova, A.A.; Neumann, K.; Ushakova, M.V.; Ivanova, N.V.; Surov, A.V. Molecular phylogenetics and taxonomy of dwarf hamsters *Cricetulus* Milne-Edwards, 1867 (Cricetidae, Rodentia): Description of a new genus and reinstatement of another. *Zootaxa* **2018**, *4387*, 331–349. [CrossRef] [PubMed]
73. Romanenko, S.A.; Volobouev, V.T.; Perelman, P.L.; Lebedev, V.S.; Serdukova, N.A.; Trifonov, V.A.; Biltueva, L.S.; Nie, W.; O'Brien, P.C.; Bulatova, N.; et al. Karyotype evolution and phylogenetic relationships of hamsters (Cricetidae, Muroidea, Rodentia) inferred from chromosomal painting and banding comparison. *Chromosome Res.* **2007**, *15*, 283–297. [CrossRef]
74. Romanenko, S.A.; Lebedev, V.S.; Bannikova, A.A.; Pavlova, S.V.; Serdyukova, N.A.; Feoktistova, N.Y.; Jiapeng, Q.; Yuehua, S.; Surov, A.V.; Graphodatsky, A.S. Karyotypic and molecular evidence supports the endemic Tibetan hamsters as a separate divergent lineage of Cricetinae. *Sci. Rep.* **2021**, *11*, 10557. [CrossRef]
75. Kenagy, G.J.; Trombulak, S.C. Size and function of mammalian testes in relation to body size. *J. Mammal.* **1986**, *67*, 1–22. [CrossRef]
76. Shi, Q.X.; Roldan, E.R.S. Bicarbonate/CO_2 is not required for zona pellucida- or progesterone-induced acrosomal exocytosis of mouse spermatozoa but is essential for capacitation. *Biol. Reprod.* **1995**, *52*, 540–546. [CrossRef]
77. Gomez Montoto, L.; Magana, C.; Tourmente, M.; Martin-Coello, J.; Crespo, C.; Luque-Larena, J.J.; Gomendio, M.; Roldan, E.R.S. Sperm competition, sperm numbers and sperm quality in muroid rodents. *PLoS ONE* **2011**, *6*, e18173. [CrossRef] [PubMed]
78. Gomez Montoto, L.; Varea Sanchez, M.; Tourmente, M.; Martin-Coello, J.; Luque-Larena, J.J.; Gomendio, M.; Roldan, E.R.S. Sperm competition differentially affects swimming velocity and size of spermatozoa from closely related muroid rodents: Head first. *Reproduction* **2011**, *142*, 819–830. [CrossRef] [PubMed]
79. Di Rienzo, J.A.; Guzmán, A.W.; Casanoves, F. A multiple comparisons method based on the distribution of the root node distance of a binary tree. *J. Agric. Biol. Environ. Stat.* **2002**, *7*, 129–142. [CrossRef]
80. Tourmente, M.; Villar-Moya, P.; Rial, E.; Roldan, E.R.S. Differences in ATP generation via glycolysis and oxidative phosphorylation and relationships with sperm motility in mouse species. *J. Biol. Chem.* **2015**, *290*, 20613–20626. [CrossRef]
81. Bavister, B.D. The effect of variations in culture conditions on the motility of hamster spermatozoa. *J. Reprod. Fertil.* **1974**, *38*, 431–440. [CrossRef] [PubMed]
82. Bavister, B.D.; Yanagimachi, R. The effects of sperm extracts and energy sources on the motility and acrosome reaction of hamster spermatozoa in vitro. *Biol. Reprod.* **1977**, *16*, 228–237. [CrossRef]
83. Leibfried, M.L.; Bavister, B.D. Effects of epinephrine and hypotaurine on in-vitro fertilization in the golden hamster. *J. Reprod. Fertil.* **1982**, *66*, 87–93. [CrossRef]
84. Alvarez, J.G.; Storey, B.T. Taurine, hypotaurine, epinephrine and albumin inhibit lipid peroxidation in rabbit spermatozoa and protect against loss of motility. *Biol. Reprod.* **1983**, *29*, 548–555. [CrossRef]

85. Mrsny, R.J.; Meizel, S. Inhibition of hamster sperm Na^+, K^+-ATPase activity by taurine and hypotaurine. *Life Sci.* **1985**, *36*, 271–275. [CrossRef]
86. Hexum, T.D. The effect of catecholamines on transport (Na,K) adenosine triphosphatase. *Biochem. Pharmacol.* **1977**, *26*, 1221–1227. [CrossRef]
87. Fagan, J.B.; Racker, E. Reversible inhibition of (Na^+, K^+) ATPase by Mg^{2+}, adenosine triphosphate, and K^+. *Biochemistry* **1977**, *16*, 152–158. [CrossRef] [PubMed]
88. Pickworth, S.; Chang, M.C. Fertilization of Chinese hamster eggs in vitro. *J. Reprod. Fertil.* **1969**, *19*, 371–374. [CrossRef] [PubMed]

biology

MDPI

Article

Ability of the ISAS3Fun Method to Detect Sperm Acrosome Integrity and Its Potential to Discriminate between High and Low Field Fertility Bulls

Jesús L. Yániz [1,*], Inmaculada Palacín [1], Miguel A. Silvestre [2], Carlos Olegario Hidalgo [3], Carolina Tamargo [3] and Pilar Santolaria [1]

1 BIOFITER Research Group, Institute of Environmental Sciences (IUCA), University of Zaragoza, 22071 Huesca, Spain; ipalacin@unizar.es (I.P.); psantola@unizar.es (P.S.)
2 Departamento de Biología Celular, Biología Funcional y Antropología Física, Universidad de Valencia, 46100 Burjassot, Spain; miguel.silvestre@uv.es
3 Animal Genetics and Reproduction Area, Regional Agrifood Research and Development Service (SERIDA), 33394 Gijón, Spain; cohidalgo@serida.org (C.O.H.); ctamargo@serida.org (C.T.)
* Correspondence: jyaniz@unizar.es

Simple Summary: The objective of the present study was to investigate whether differences in bull fertility are associated with variations of sperm quality. Differences between high- and low-fertility bulls were found mainly in parameters related to sperm acrosome integrity when using a new fluorescence method that allows clear and precise detection of the sperm plasma membrane and acrosome: the ISAS3Fun method. It was concluded that the simultaneous assessment of sperm viability and acrosome integrity with the ISAS3Fun method is precise and seems to have a greater potential to discriminate between high- and low-fertility bulls than the more conventional in vitro sperm quality test. These results may help to predict the breeding soundness of bulls used in artificial insemination, which is important for the dairy industry.

Abstract: The objective of the present study was to investigate whether fertility differences in bulls are reflected in variations of sperm quality when analysing only one ejaculate per male. Two experiments were performed. In the first experiment, frozen semen samples from 20 adult bulls were tested; 10 bulls had high field fertility and 10 bulls had low field fertility. Analyses of sperm motility, membrane integrity, and membrane–acrosome integrity with the ISAS3Fun method were performed. Sperm morphometry of the fluorescence sperm subpopulations obtained with the ISAS3Fun method was also analysed. Significant differences between high- and low-fertility groups were only found with the ISAS3Fun technique, specifically in sperm acrosome integrity, the proportion of spermatozoa with an intact acrosome and damaged membrane, and in sperm head width of spermatozoa with intact structures. Discriminant analyses allowed us to correctly classify 90% of sperm samples in their fertility group using sperm quality parameters. Given that only the results obtained with the ISAS3Fun technique were related to bull fertility, we performed a second experiment aimed to validate the efficacy of this technique to detect the acrosomal integrity of bull spermatozoa, comparing them with the conventional FITC-PNA/propidium iodide (PNA/PI) combination under capacitating conditions. The results indicated that the ISAS3Fun combination provided an accurate assessment of both viability and acrosomal integrity for ejaculated spermatozoa, while the PNA/PI combination underestimated the extension of acrosomal damage due to false negatives. It was concluded that the simultaneous assessment of sperm plasma membranes and acrosome integrity with the ISAS3Fun method is precise and seems to have a greater potential to discriminate between high- and low-fertility bulls than more conventional in vitro sperm quality tests.

Keywords: *Bos taurus*; spermiogram; male fertility; capacitation; acrosome reaction

Citation: Yániz, J.L.; Palacín, I.; Silvestre, M.A.; Hidalgo, C.O.; Tamargo, C.; Santolaria, P. Ability of the ISAS3Fun Method to Detect Sperm Acrosome Integrity and Its Potential to Discriminate between High and Low Field Fertility Bulls. *Biology* **2021**, *10*, 1135. https://doi.org/10.3390/biology10111135

Academic Editors: Paul S. Cooke and Hiroetsu Suzuki

Received: 13 September 2021
Accepted: 1 November 2021
Published: 4 November 2021

1. Introduction

The prediction of breeding soundness of bulls and of the fertilising ability of semen used in artificial insemination (AI) is important for the dairy industry [1]. However, achieving satisfactory predictions based on the in vitro evaluation of sperm quality remains a challenge and still heavily relies on the simple assessment of sperm motility and morphology [2]. Beyond these important parameters, a motile sperm may not be able to fertilise an oocyte if it does not have an intact acrosome or may give rise to a viable embryo if its DNA is damaged. After spermatogenesis, bull sperm appears to be quite resistant to DNA fragmentation [3], but the acrosome is a delicate membranous structure that can be severely damaged during freezing and thawing procedures, which determines the success or failure of AI [4].

The development of acrosome-specific fluorescent probes, among which lectins are the most widely used, has facilitated the determination of acrosomal damage, even using flow cytometry. *Pisum sativum* (edible pea) agglutinin (PSA) and *Arachis hypogaea* (peanut) agglutinin (PNA) are the most commonly used lectins for this purpose [5]. They bind to glycosidic residues in different parts of the acrosomal membrane and, being conjugated with a fluorochrome such as fluorescein isothiocyanate (FITC), sperm with damaged acrosomes may be easily visualised. Nevertheless, extensive acrosomal damage can produce false negatives due to missing binding sites. [5]. The use of acidotropic probes, with an affinity for organelles such as the acrosome, has been shown to be ineffective due to non-specific (LysoSensor™ Green DND-189) or heterogeneous (LysoTracker® DND-99 and dapoxyl 2-aminoethyl sulphonamide) labelling [5]. Although the pH of the acrosome is acidic, it gradually increases during the course of sperm capacitation [6].

In previous studies [7,8], we described a fluorochrome combination (ISAS3Fun) that allowed us to discriminate sperm viability, acrosomal integrity, and sperm functionality in bulls. This study was designed to investigate whether differences in the field fertility of bulls were reflected in variations of sperm quality parameters, including those obtained with the ISAS3Fun test, when analysing only one ejaculate per male. Given that only the results obtained with the ISAS3Fun test were related to bull fertility, we conducted a study in which the efficacy of this technique was compared with the conventional FITC-PNA/PI (PNA/PI) combination to detect the acrosomal integrity of bull spermatozoa before and after incubation under capacitating conditions and the induction of an acrosome reaction.

2. Materials and Methods

Unless otherwise indicated, all chemicals were obtained from Sigma (Merck Life Science S.L.U., Madrid, Spain).

2.1. Semen Collection and Processing

This study analysed cryopreserved semen samples collected from commercial Holstein–Friesian bulls. Ejaculates with a total sperm motility or with morphologically normal spermatozoa lower than 70% were discarded. The ejaculates were extended in BullXcell (IMV Technologies, Humeco, Huesca, Spain) to a final concentration of 1.3×10^7 sperm per 0.25 mL semen straw (IMV Technologies). All straws were cryopreserved following standard procedures [9]. Before use, straws were thawed for 1 min at 37 °C in a water bath and processed.

2.2. Experimental Design

To achieve the proposed objectives, two experiments were performed.

Experiment 1: Relationship between in vitro sperm quality test and field fertility of bulls.

Frozen semen samples from 20 bulls, 10 of high field fertility and 10 with low field fertility, were selected for this experiment with an age range between 2 and 6 years. Field fertility was determined by a retrospective study using data from the first and second AI postpartum results of at least 100 Holstein–Friesian cows in their second lactation over a period of 5 y (2009 to 2014) in different herds. Then, 353 bulls were evaluated and, between them, the 10 bulls with the highest (>53%) and 10 bulls with the lowest (<39%) fertility were included in the study. Cow fertility (pregnancy rate) was determined through pregnancy diagnosis using ultrasonography or rectal palpation of the genital tract at 35–45 days post-AI. Two frozen samples from the same ejaculate of each bull were collected between those used for the field fertility analysis. Semen samples were thawed, pooled, and processed for sperm quality assessment as detailed below.

2.2.1. Sperm Motility (CASA-Mot)

Within a few minutes after thawing, a diluted semen sample was placed in a pre-warmed Makler chamber (Sefi-Medical Instruments Ltd., Haifa, Israel), and a computer-assisted sperm analyser (CASA-Mot; ISAS®, version 1.32; PROISER, Paterna, Spain) was used to assess sperm motility [10]. Spermatozoa were classified as progressive if VCL > 20 μm/s and STR > 80%.

2.2.2. Sperm Plasma Membrane Integrity

Sperm plasma membrane integrity (viability) was determined using an acridine orange and propidium iodide combination [11]. The number of total spermatozoa and the percentages of membrane-intact spermatozoa were determined using the OpenCASA software [12]. At least 300 sperm cells were analysed per sample.

2.2.3. Sperm Plasma/Acrosome Membrane Integrity

Combined sperm plasma/acrosome membrane integrity was determined using the ISAS3Fun technique (PROISER), as described in a previous work [8]. The labelling mix included propidium iodide (PI), Hoechst 33342, and carboxyfluorescein diacetate (CFDA). Two hundred spermatozoa were examined on each slide, and the staining patterns were classified into four categories (Figure 1), aided with v2 of the OpenCASA software [12]: (a) those with intact acrosomes and plasma membranes (IAIM); (b) those with intact acrosomes and damaged plasma membranes (IADM); (c) those with damaged acrosomes and intact plasma membranes (DAIM); and (d) those with damaged acrosomes and plasma membranes (DADM) [7].

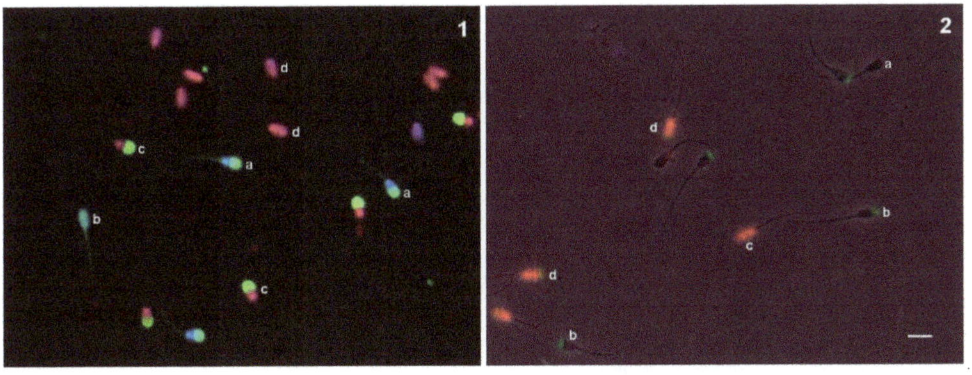

Figure 1. Fluorescence image of bull spermatozoa stained with ISAS3Fun kit (**1**) and fluorescence-phase contrast image of bull spermatozoa stained with PI/PNA. (**2**): a, Intact plasma membrane and acrosome; b, intact plasma membrane and damaged acrosome; c, damaged plasma membrane and intact acrosome; d, damaged plasma membrane and acrosome. Scale bar = 10 μm.

2.2.4. Sperm Morphometry (CASA-Morph)

Sperm head morphometry of the fluorescent sperm subpopulations was determined as described in [7]. Briefly, spermatozoa were stained with the fluorochromes of the ISAS3Fun kit, and the morphometry of the sperm head, nucleus, and acrosome of the different fluorescent subpopulations was determined using a specific plug-in module created on ImageJ.

Experiment 2: Validation of the use of the ISAS3Fun method for the assessment of sperm acrosome integrity in bulls.

Frozen semen samples from 18 bulls were thawed in a water bath, centrifuged (600× *g*, 10 min), and resuspended in S-TALP (sperm Tyrode's albumin, lactate, and pyruvate) medium [13] with 10 IU/mL sodium heparin salt to induce sperm capacitation. The spermatozoa were allowed to capacitate for 4 h at 39 °C under 5% CO_2. Ionomycin (7 µM final concentration) was added to the sperm suspension to induce the acrosome reaction, and the incubation period was extended for 1 h in the same conditions.

Aliquots were evaluated at 0 h, 4 h (after induced capacitation), and 5 h (an hour after the induced acrosome reaction) for sperm motility with the CASA system and for sperm plasma/acrosome membrane integrity as detailed below.

For each incubation time, a sample aliquot was mixed in a vial with PNA (10 µg/mL) for 10 min and PI (15 µM) for 5 min at 37 °C. Then, 3 µL of the labelled sample was placed on a prewarmed slide, covered, and pressed with the Trumorph® system [14]. Stained spermatozoa were examined under a fluorescence/phase contrast microscope (DM4500B, Leica, Wetzlar, Germany) equipped with a warmed stage and a 20× plan apochromatic objective, and a standard green/red filter set (G/R, excitation: 490–520, 575–630 nm) was used to obtain digital images of the fluorescence-labelled sperm. The staining patterns of the spermatozoa were first examined using fluorescence microscopy, and then the localisation was confirmed using phase contrast optics. Two hundred spermatozoa were examined on each slide, and the staining patterns were classified into four categories (Figure 1): (a) those unstained (intact acrosome and plasma membrane, IAIM); (b) those with stained nuclei (intact acrosome and damaged plasma membrane, IADM); (c) those with a stained acrosome (damaged acrosome and intact plasma membrane, DAIM); and (d) those with a stained acrosome and nuclei (damaged acrosome and plasma membrane, DADM).

Another sample aliquot was labelled with the ISAS3Fun kit (PROISER), as detailed above.

2.3. Statistical Analysis

Statistical analyses were performed using the SPSS package, version 23.0 (SPSS Inc., Chicago, IL, USA). Statistical significance was set at $p < 0.05$. Distribution normality and the homogeneity of variance of the median score for each dataset were checked using the Kolmogorov–Smirnov and Levene tests, respectively. As data were normally distributed, parametric analysis was performed throughout. In the first experiment, differences in the sperm quality results between high- and low-fertility bulls were evaluated by Wilk's lambda ANOVA. A discriminant analysis was also performed with the linear stepwise procedure to identify the most useful parameters for the classification of high- and low-fertility bulls. Wilk's lambda ANOVA was used to compare the fraction of the total dispersion of data not accounted for.

In the second experiment, the effects of time (0 h, 4 h, and 5 h) using the ISAS3Fun method vs. the PNA/PI combination on the sperm membrane and acrosome integrity (IM and IA, respectively), as well as on the distribution of sperm subpopulations (IAIM, DAIM, IADM, and DADM), were analysed by General Linear Models (GLM) repeated measures analysis of variance.

3. Results

3.1. Experiment 1

Significant differences between high- and low-fertility bulls were observed in acrosome integrity, the proportion of IADM sperm and the head width of the IAIM sperm subpopulation (Table 1). Highly fertile bulls had a higher acrosome integrity, percentage of IADM spermatozoa, and head width of the IAIM subpopulation ($p < 0.05$) than those of low-fertility bulls. No statistical differences were found for the other sperm quality parameters analysed.

Table 1. Differences in sperm quality parameters between high- and low-fertility bulls. Data are represented as mean ± S.E.M.

Sperm Parameter	High Fertility	Low Fertility
Acrosome integrity (%)	46.20 ± 2.87	35.14 ± 4.43
IADM (%)	24.84 ± 3.14	14.20 ± 2.25
Head width of IAIM (μm)	5.38 ± 0.05	5.25 ± 0.03

IADM: spermatozoa with an intact acrosome and a damaged plasma membrane; IAIM: spermatozoa with an intact acrosome and plasma membrane. The sperm subpopulations were obtained using the ISAS3Fun fluorescence method. The three sperm quality parameters were significantly different between high- and low-fertility groups at $p < 0.05$.

A discriminant analysis was performed to determine if it was possible to distinguish to which group of bull fertility each individual semen sample belonged. The results indicated that the percentage of the IADM sperm subpopulation, the head width, and the acrosome perimeter of the IAIM sperm subpopulation were the variables selected as the best discriminators of high and low field fertility. The matrix of classification obtained gave the Fisher's discriminant linear functions for each class (Table 2, $p < 0.001$), with a globally correct assignment of 90% of the samples (Table 3).

Table 2. Discriminant classification matrix showing Fisher's linear discriminant functions.

	Coefficient of Function of Classification	
	High Fertility	Low Fertility
IADM	−4.907	−5.268
Head width of IAIM (μm)	111.692	96.572
Acrosome perimeter of IAIM (μm)	187.255	195.212
Constant	−1974.853	−2035.336

Values obtained by linear stepwise discriminant analysis. IADM: spermatozoa with an intact acrosome and a damaged plasma membrane; IAIM: spermatozoa with an intact acrosome and plasma membrane. The sperm subpopulations were obtained using the ISAS3Fun fluorescence method.

Table 3. Percentage of sperm samples assigned to each fertility group by discriminant analysis.

	Number Allocated to Group by Discriminant Analysis	
	High Fertility	Low Fertility
High fertility	9	1
Low fertility	1	9

90.0% of the samples were classified correctly.

3.2. *Experiment 2*

There was a reduction in sperm motility between 0 h (35.88% ± 3.50) and 4 h (27.86% ± 3.25) of incubation, and most spermatozoa were immotile at 5 h. The main model of the GLM repeated measures analysis of factors affecting sperm membrane and acrosome integrity is depicted in Table 4. There was a significant effect of time, time x fluorescent method and the fluorescent method used. These effects are shown in Figure 2. The treatment induced a slight decrease in plasma membrane integrity between 0 h and 4 h of incubation under capacitating conditions, but sperm viability was preserved afterwards, as determined using both fluorescence combinations (Figure 2a). The fluorescent methods clearly differed in the results of acrosome integrity; the percentage of spermatozoa with intact acrosomes was higher using the PNA/PI combination than with the ISAS3Fun method for all incubation periods (Figure 2b). Using phase contrast microscopy, it was observed that spermatozoa without a clearly distinguishable acrosome were not stained with the PNA, generating false negatives that led to an underestimation of the extent of acrosomal damage. Furthermore, using this technique, the reduction in acrosome integrity was only evident at 5 h, while a clear reduction was observed at 4 h with the ISAS3Fun method, followed by a more pronounced acrosomal loss at 5 h after ionomycin treatment (Figure 2b).

Table 4. Main model of the GLM repeated measurement analysis for factors affecting evolution of sperm quality variables over time using frozen/thawed bull semen (n = 14).

Subject Effects	Factor	IM	IA	IAIM	DAIM	IADM	DADM
Within	Time	0.0240	<0.0001	<0.0001	<0.0001	0.3340	0.0100
	Time × method	<0.0001	<0.0001	0.3780	<0.0001	<0.0001	0.2050
Between	Method	<0.0001	<0.0001	0.486	<0.0001	<0.0001	<0.0001

IM: spermatozoa with an intact plasma membrane; IA: spermatozoa with an intact acrosome; IAIM: spermatozoa with an intact acrosome and plasma membrane; DAIM: spermatozoa with a damaged acrosome and an intact plasma membrane; IADM: spermatozoa with an intact acrosome and a damaged plasma membrane; DADM: spermatozoa with a damaged acrosome and plasma membrane.

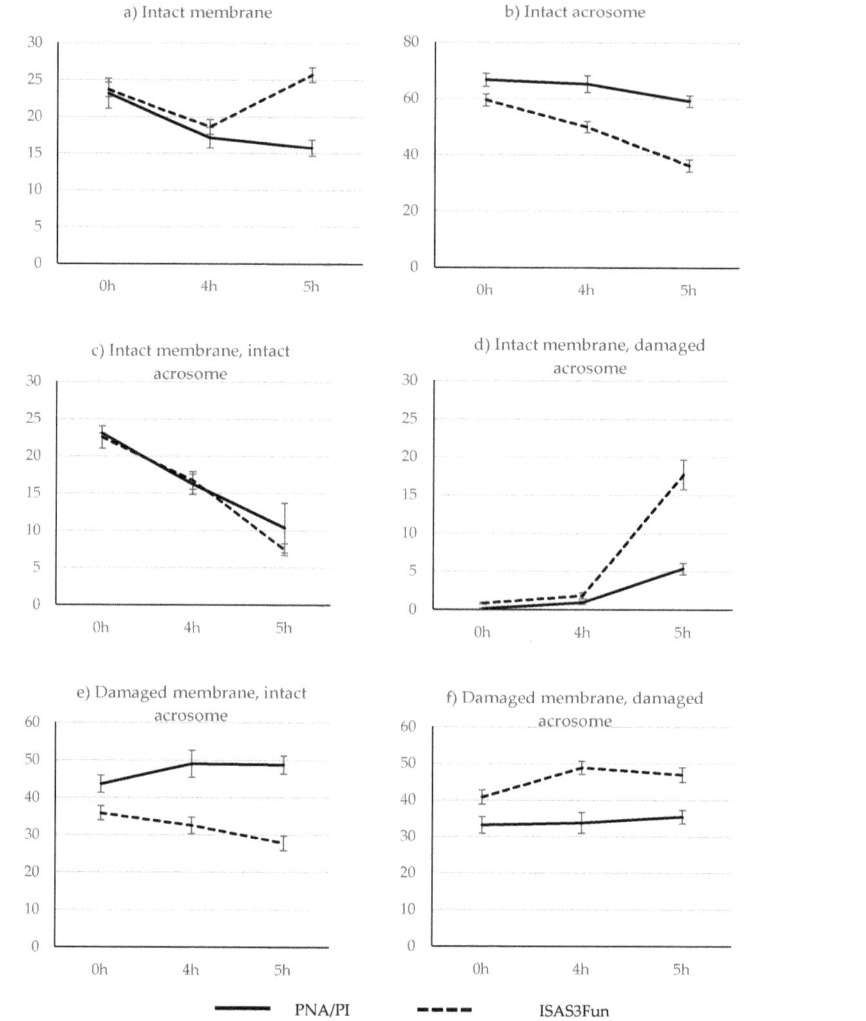

Figure 2. Time evolution of sperm membrane and acrosome integrity using the PNA/PI and the ISAS3Fun fluorescence methods after capacitation (4 h) and ionomycin (5 h) treatment of frozen/thawed bull semen samples (**a**–**f**).

In the fluorescence sperm subpopulations, spermatozoa with damaged acrosomes and intact plasma membranes (DAIM) were barely observed at 0 h, then increased slightly at 4 h, and finally increased noticeably at 5 h (Figure 2d). This increase was significantly greater using the ISAS3Fun method than with the PNA/PI combination. In contrast, the proportion of spermatozoa with intact acrosomes and damaged plasma membranes (IADM) was much higher for PNA/PI than for the ISAS3Fun method for all incubating periods and only decreased with the latter method over time (Figure 2e).

4. Discussion

Male fertility assessment based on test inseminations is expensive and time consuming [10], and it would be useful to have simple in vitro test for predicting potential bull fertility. In this very restrictive study, in which only one random ejaculate from each bull was used, only the results of the ISAS3Fun technique were significantly related to bull fertility. Spermatozoa from bulls with superior field fertility displayed increased acrosomal integrity and head width. In all likelihood, the repeated analysis of more ejaculates would increase the chance of finding associations between in vitro sperm quality and bull fertility. However, although such relationships were certainly less likely to be observed when using a single ejaculate, the fact that they were found could indicate a consistent relationship with fertility that should be confirmed by further studies. In agreement with our study, Bernecic et al. [15] have described that acrosome integrity, together with viability, were the only sperm attributes that were significantly different between high- and low-fertility bulls. Unlike our work, the evaluation of the acrosome integrity in this study was carried out using PNA, which showed a lower sensitivity than that of the ISAS3Fun technique.

In porcine, it was observed that the number of spermatozoa with reacted acrosomes after capacitation was correlated with prolificacy, and the difference in the rate of reacted acrosome sperm before and after capacitation was correlated with fertility [16]. Therefore, an accurate assessment of the acrosome structure is essential in order to acquire new knowledge of acrosomal exocytosis and to predict the fertilising ability of sperm.

In a previous study in rams, we found clear differences in motility and plasma membrane integrity between high- and low-fertility males, which were not observed in this study of bulls. Unlike in rams, AI in cows is highly efficient, with the semen deposited deeply into the uterus. Consequently, spermatozoa might not have difficulties reaching the site of fertilisation; therefore, high motility and viability might not be as restrictive as in the ewe. Another possible explanation for the differences between these two studies is that the experimental design was different. In the study of rams, fresh semen samples of only eight rams with extreme fertility (four high and four low) were checked weekly during six consecutive weeks. The analysis of frozen/thawed doses of a single ejaculate per bull and the selection of more males per fertility group in the current study would probably reduce the chances of finding differences in more sperm quality parameters. In fact, Bernecic et al. [15] observed that both viability and acrosome integrity could serve as bull fertility biomarkers in the field, while Kjaestad et al. [17] described an association between sperm motility and velocity with bull fertility.

For the establishment of a viable pregnancy, spermatozoa must reach the site of fertilisation, then penetrate and fertilise the oocyte, and, finally, produce a viable embryo. Acrosomal exocytosis is a crucial and irreversible process key in ovum fertilisation that occurs when the acrosomal outer membrane fuses with the plasma membrane at multiple points, releasing proteolytic enzymes [18,19]. Until a few years ago, it was thought that a sperm that had suffered from acrosomal exocytosis could not penetrate the zona pellucida. However, recent studies have orchestrated new scenarios, since it has been observed that spermatozoa that had previously undergone the acrosome reaction were able to penetrate the zona pellucida [18].

Given the difficulty in clearly discerning the sperm acrosome in most animal species, its evaluation is usually limited to research or to occasional studies of sperm quality. In a recent work, we described a simple method that allows the simultaneous visualisation of sperm plasma and acrosomal membrane integrity in bulls by combining the fluorochromes Hoechst 33342, CFDA, and PI (ISAS3Fun) [8]. To facilitate the analysis using this test, a new module was added to the OpenCASA software, which allows for the automated analysis of these sperm quality parameters [12].

CFDA and PI have been combined to assess sperm membrane integrity in several species [11,20–24]. This fluorochrome combination allows the acrosomes of spermatozoa with a damaged plasma membrane to be distinguished, whose red nuclei (PI stained) contrast with the green acrosome (carboxyfluorescein stained) [11,20]. However, live spermatozoa accumulated carboxyfluorescein throughout both the head and the flagellum, and the acrosome is not clearly distinguishable [20,25]. In a previous study [8], we added a third fluorochrome to this combination, Hoechst 33342, allowing the discrimination of acrosomal integrity both in live and dead spermatozoa (ISAS3Fun). The staining pattern of the ISAS3Fun method is opposite to that of the lectins. Only sperm with intact acrosomes contain esterases able to hydrolyse CFDA to carboxyfluorescein, emitting an intense green fluorescence. This fluorochrome has a low molecular weight and is therefore rapidly lost after damage is sustained by the acrosomal membrane, avoiding false negatives. As a consequence, the ISAS3Fun method seems particularly sensitive for the evaluation of acrosomal integrity in bulls. Even in frozen/thawed semen samples (0 h in Experiment 2), the PNA/PI combination underestimated the extent of acrosomal damage when compared to the ISAS3Fun kit.

During sperm capacitation, hyperpolarisation of the plasma membrane occurs, and this hyperpolarisation has been attributed to an increase in calcium during acrosomal exocytosis [18]. Calcium ionophores have long been known to increase intracellular calcium and induce acrosomal exocytosis in mammalian sperm [26]. In the present study, the calcium ionophore challenge validated the use of the ISAS3Fun method to quantify the acrosome integrity in bulls, showing higher sensitivity than traditional PNA/PI staining. In agreement with the findings of other authors [5], it seems that lectin-binding sites disappear after extensive acrosomal damage, producing false negatives that cause an underestimation of the degree of acrosomal injury. Consequently, the fluorescent methods have clearly differed in their results of acrosome integrity, such that the percentage of spermatozoa with intact acrosomes was higher using the PNA/PI combination than with the ISAS3Fun method in most cases. Furthermore, the differences increased with time, particularly after the ionomycin treatment, likely because of the presence of a higher number of spermatozoa with extensive acrosomal loss.

The study of the evolution of the fluorescence sperm subpopulations allowed us to clarify the effect of the treatments. Live spermatozoa showed a clear induced acrosome reaction, with extensive acrosomal damage at 5 h after calcium ionophore was added. This evolution was more clearly appreciated when using the ISAS3Fun combination, with a sudden increase in the proportion of live spermatozoa with damaged acrosomes after ionomycin treatment. The evolution of dead spermatozoa was less evident, showing a reduction in spermatozoa with intact acrosomes when using the ISAS3Fun method, while the proportion of this subpopulation even increased with time when using PNA/PI, likely because the number of spermatozoa with complete acrosomal loss was higher in dead spermatozoa.

Surprisingly, the induction of the acrosome reaction with ionomycin caused an apparent increase in spermatozoa with intact plasma membranes when using the ISAS3Fun kit (Figure 2a). We repeated the experiment several times and observed this increase repeatedly. There is also no clear correspondence between the decrease in the IADM subpopulation (Figure 2e) and the expected increase in the DADM subpopulation (Figure 2f) between 4 h and 5 h. Both events may be related to the increase in the DAIM subpopulation (Figure 2d). At 4 h, we observed the presence of spermatozoa with intact acrosomes and partial permeability to the propidium iodide, which stained red only in the head base, and they were classified as IADM. It is possible that these spermatozoa with intact acrosomes and only partially damaged plasma membranes were still capable of experiencing the acrosomal reaction. At 5 h, we observed a clear increase in the spermatozoa with increased fluorescence intensity at the head and flagellum using the ISAS3Fun kit (IFI subpopulation described in [8], data not shown). Given that the acrosome and the nucleus of most of these spermatozoa were not distinguishable, they were classified as DAIM spermatozoa. However, the increase in fluorescence intensity associated with the acrosome reaction could have hidden the partial staining of the nucleus with propidium iodide, and, therefore, these sperm could have damaged plasma membranes. Despite this, given that the proportion of the IFI sperm subpopulation in the freshly thawed semen is very low, this observation does not seem to limit the options of the ISAS3Fun method to predict potential bull fertility.

5. Conclusions

In conclusion, the simultaneous assessment of sperm plasma membranes and acrosome integrity with the ISAS3Fun method seems to have a higher ability to discriminate between high- and low-fertility bulls than a conventional in vitro sperm quality test. Moreover, the use of this technique to quantify acrosome integrity was validated in bulls, with a higher sensitivity than traditional PNA/PI staining.

Author Contributions: Conceptualisation, J.L.Y., I.P. and P.S.; methodology, J.L.Y., I.P., C.O.H., C.T., M.A.S. and P.S.; validation, J.L.Y., I.P., C.O.H., C.T., M.A.S. and P.S.; formal analysis, J.L.Y., I.P. and P.S.; investigation, J.L.Y., I.P. and P.S.; resources, J.L.Y., I.P., C.O.H., C.T., M.A.S. and P.S.; data curation, J.L.Y., I.P. and P.S.; writing—original draft preparation, J.L.Y., I.P. and P.S.; writing—review and editing, J.L.Y., I.P., C.O.H., C.T., M.A.S. and P.S.; supervision, J.L.Y. and P.S.; project administration, J.L.Y. and P.S.; funding acquisition, J.L.Y. and P.S. All authors have read and agreed to the published version of the manuscript.

Funding: This research was funded by the Spanish Ministry of Economy and Finance (MINECO) (grant AGL2017-85030-R), the European Territorial Cooperation Operational Program—Spain, France, and the Andorra Area 2014–20 (Program DietaPYR2 EFA144/16), the University of Valencia (UV-INV-AE-1563487), and by the DGA-FSE (grant A07_17R) and the European FEDER Funds.

Institutional Review Board Statement: Ethical review and approval were waived for this study because did not involve a prospective evaluation, did not involve laboratory animals and only involved the use of frozen commercial semen samples obtained using standard procedures.

Informed Consent Statement: Not applicable.

Data Availability Statement: The data are not publicly available due to them containing information that could compromise research.

Acknowledgments: We would like to acknowledge Asturiana de Control Lechero (ASCOL Soc. Coop.) for providing bull semen doses and the use of the Research Support Service-SAI, University of Zaragoza.

Conflicts of Interest: The authors declare no conflict of interest.

References

1. Perez-Pe, R.; Marti, J.I.; Sevilla, E.; Fernandez-Sanchez, M.; Fantova, E.; Altarriba, J.; Cebrian-Perez, J.A.; Muino-Blanco, T. Prediction of fertility by centrifugal countercurrent distribution (CCCD) analysis: Correlation between viability and heterogeneity of ram semen and field fertility. *Reproduction* **2002**, *123*, 869–875. [CrossRef]
2. Sellem, E.; Broekhuijse, M.L.W.J.; Chevrier, L.; Camugli, S.; Schmitt, E.; Schibler, L.; Koenen, E.P.C. Use of combinations of in vitro quality assessments to predict fertility of bovine semen. *Theriogenology* **2015**, *84*, 1447–1454. [CrossRef] [PubMed]
3. Carreira, J.T.; Trevizan, J.T.; Carvalho, I.R.; Kipper, B.; Rodrigues, L.H.; Silva, C.; Perri, S.H.V.; Drevet, J.R.; Koivisto, M.B. Does sperm quality and DNA integrity differ in cryopreserved semen samples from young, adult, and aged Nellore bulls? *Basic Clin. Androl.* **2017**, *27*, 12. [CrossRef] [PubMed]
4. Kishida, K.; Sakase, M.; Minami, K.; Arai, M.M.; Syoji, R.; Kohama, N.; Akiyama, T.; Oka, A.; Harayama, H.; Fukushima, M. Effects of acrosomal conditions of frozen-thawed spermatozoa on the results of artificial insemination in Japanese Black cattle. *J. Reprod. Dev.* **2015**, *61*, 519–524. [CrossRef] [PubMed]
5. Martinez-Pastor, F.; Mata-Campuzano, M.; Alvarez-Rodriguez, M.; Alvarez, M.; Anel, L.; de Paz, P. Probes and Techniques for Sperm Evaluation by Flow Cytometry. *Reprod. Domest. Anim.* **2010**, *45*, 67–78. [CrossRef] [PubMed]

6. Nakanishi, T.; Ikawa, M.; Yamada, S.; Toshimori, K.; Okabe, M. Alkalinization of acrosome measured by GFP as a pH indicator and its relation to sperm capacitation. *Dev. Biol.* **2001**, *237*, 222–231. [CrossRef]
7. Palacin, I.; Santolaria, P.; Alquezar-Baeta, C.; Soler, C.; Silvestre, M.A.; Yaniz, J. Relationship of sperm plasma membrane and acrosomal integrities with sperm morphometry in Bos taurus. *Asian J. Androl.* **2020**, *22*, 578–582. [CrossRef]
8. Yaniz, J.L.; Soler, C.; Alquezar-Baeta, C.; Santolaria, P. Toward an integrative and predictive sperm quality analysis in Bos taurus. *Anim. Reprod. Sci.* **2017**, *181*, 108–114. [CrossRef]
9. Ruiz-Díaz, S.; Grande-Pérez, S.; Arce-López, S.; Tamargo, C.; Hidalgo, C.O.; Pérez-Cerezales, S. Changes in the cellular distribution of tyrosine phosphorylation and its relationship with the acrosomal exocytosis and plasma membrane integrity during in vitro capacitation of frozen/thawed bull spermatozoa. *Int. J. Mol. Sci.* **2020**, *21*, 2725. [CrossRef]
10. Vicente-Fiel, S.; Palacín, I.; Santolaria, J.P.; Fantova, E.; Quintin-Casorran, F.; Sevilla-Mur, E.; Yaniz, J.L. In vitro assessment of sperm quality from rams of high and low field fertility. *Anim. Reprod. Sci.* **2014**, *146*, 15–20. [CrossRef]
11. Yániz, J.L.; Palacín, I.; Vicente-Fiel, S.; Gosálvez, J.; López-Fernández, C.; Santolaria, P. Comparison of membrane-permeant fluorescent probes for sperm viability assessment in the ram. *Reprod. Domest. Anim.* **2013**, *48*, 598–603. [CrossRef]
12. Yaniz, J.; Alquezar-Baeta, C.; Yague-Martinez, J.; Alastruey-Benede, J.; Palacin, I.; Boryshpolets, S.; Kholodnyy, V.; Gadelha, H.; Perez-Pe, R. Expanding the Limits of Computer-Assisted Sperm Analysis through the Development of Open Software. *Biology* **2020**, *9*, 207. [CrossRef]
13. Parrish, J.J.; Susko-Parrish, J.; Winer, M.A.; First, N.L. Capacitation of bovine sperm by heparin. *Biol. Reprod.* **1988**, *38*, 1171–1180. [CrossRef]
14. Soler, C.; García, A.; Silvestre, M.; Sancho, M. The Trumorph℗® system: The new universal technique for the observation and analysis of the morphology of living sperm. *Anim. Reprod. Sci.* **2015**, *158*, 1–10. [CrossRef] [PubMed]
15. Bernecic, N.C.; Donnellan, E.; O'Callaghan, E.; Kupisiewicz, K.; O'Meara, C.; Weldon, K.; Lonergan, P.; Kenny, D.A.; Fair, S. Comprehensive functional analysis reveals that acrosome integrity and viability are key variables distinguishing artificial insemination bulls of varying fertility. *J. Dairy Sci.* **2021**, *104*, 11226–11241. [CrossRef] [PubMed]
16. Rahman, M.S.; Kwon, W.S.; Pang, M.G. Prediction of male fertility using capacitation-associated proteins in spermatozoa. *Mol. Reprod. Dev.* **2017**, *84*, 749–759. [CrossRef]
17. Kjaestad, H.; Ropstad, E.; Berg, K.A. Evaluation of Spermatological Parameters Used to Predict the Fertility of Frozen Bull Semen. *Acta Vet. Scand.* **1993**, *34*, 299–303. [CrossRef]
18. Buffone, M.G.; Hirohashi, N.; Gerton, G.L. Unresolved questions concerning mammalian sperm acrosomal exocytosis. *Biol. Reprod.* **2014**, *90*, 112. [CrossRef]
19. Georgadaki, K.; Khoury, N.; Spandidos, D.A.; Zoumpourlis, V. The molecular basis of fertilization (Review). *Int. J. Mol. Med.* **2016**, *38*, 979–986. [CrossRef] [PubMed]
20. Harrison, R.A.P.; Vickers, S.E. Use of fluorescent-probes to assess membrane integrity in mammalian spermatozoa. *J. Reprod. Fertil.* **1990**, *88*, 343–352. [CrossRef]
21. Brito, L.F.C.; Barth, A.D.; Bilodeau-Goeseels, S.; Panich, P.L.; Kastelic, J.P. Comparison of methods to evaluate the plasmalemma of bovine sperm and their relationship with in vitro fertilization rate. *Theriogenology* **2003**, *60*, 1539–1551. [CrossRef]
22. Kuisma, P.; Andersson, M.; Koskinen, E.; Katila, T. Fertility of frozen-thawed stallion semen cannot be predicted by the currently used laboratory methods. *Acta Vet. Scand.* **2006**, *48*, 14. [CrossRef] [PubMed]
23. Peña, A.I.; Quintela, L.A.; Herradon, P.G. Viability assessment of dog spermatozoa using flow cytometry. *Theriogenology* **1998**, *50*, 1211–1220. [CrossRef]
24. Ricci, G.; Perticarari, S.; Fragonas, E.; Giolo, E.; Canova, S.; Pozzobon, C.; Guaschino, S.; Presani, G. Apoptosis in human sperm: Its correlation with semen quality and the presence of leukocytes. *Hum. Reprod.* **2002**, *17*, 2665–2672. [CrossRef] [PubMed]
25. Garner, D.L.; Pinkel, D.; Johnson, L.A.; Pace, M.M. Assessment of spermatozoal function using dual fluorescent staining and flow cytometric analyses. *Biol. Reprod.* **1986**, *34*, 127–138. [CrossRef] [PubMed]
26. Tateno, H.; Krapf, D.; Hino, T.; Sánchez-Cárdenas, C.; Darszon, A.; Yanagimachi, R.; Visconti, P.E. Ca^{2+} ionophore A23187 can make mouse spermatozoa capable of fertilizing in vitro without activation of cAMP-dependent phosphorylation pathways. *Proc. Natl. Acad. Sci. USA* **2013**, *110*, 18543–18548. [CrossRef]

biology

Article

Relationship between Fertility Traits and Kinematics in Clusters of Boar Ejaculates

Vinicio Barquero [1], Eduardo R. S. Roldan [2], Carles Soler [3], Bernardo Vargas-Leitón [4], Francisco Sevilla [1], Marlen Camacho [1] and Anthony Valverde [1,*]

1 Animal Reproduction Laboratory, San Carlos Campus, Costa Rica Institute of Technology, School of Agronomy, Alajuela 223-21002, Costa Rica; vinicio1196@gmail.com (V.B.); fsevillabenavides@gmail.com (F.S.); mcamacho@itcr.ac.cr (M.C.)
2 Department of Biodiversity and Evolutionary Biology, Museo Nacional de Ciencias Naturales (CSIC), 28006 Madrid, Spain; roldane@mncn.csic.es
3 Department of Cellular Biology, Functional Biology and Physical Anthropology, University of Valencia, Campus Burjassot, C/Dr Moliner, 50, 46100 Burjassot, Spain; carles.soler@uv.es
4 Population Medicine Research Program, Veterinary Medicine School, National University of Costa Rica, Heredia 40104, Costa Rica; bernardo.vargas.leiton@una.cr
* Correspondence: anvalverde@itcr.ac.cr; Tel.: +506-2401-3223

Simple Summary: Swine reproduction efficiency is determined by the fertility potential of the sow and sperm quality. The objective of this study is to compare boar sperm motility and kinematic features to evaluate their relationships with reproductive success after artificial insemination (AI). In this study, the movement patterns of boar ejaculates were analyzed by a computer-assisted semen analysis (CASA)-Mot system, and the kinematic values of ejaculate clusters were assessed. The semen of the Pietrain boars showed more linear trajectory of the spermatozoa, while curvilinear velocity and oscillatory movement characterized the semen of the Duroc × Pietrain boars. The offspring of sows inseminated with Pietrain boars showed significantly lower number of stillbirths. In addition, ejaculate grouping into clusters did not have a predictive capacity on litter size variables. Nevertheless, the kinematic variables of the ejaculate may have a predictive, albeit reduced, capacity regarding litter size variables. The results of this study therefore open up possibilities for future assessments of fertility.

Citation: Barquero, V.; Roldan, E.R.S.; Soler, C.; Vargas-Leitón, B.; Sevilla, F.; Camacho, M.; Valverde, A. Relationship between Fertility Traits and Kinematics in Clusters of Boar Ejaculates. *Biology* **2021**, *10*, 595. https://doi.org/10.3390/biology10070595

Academic Editors: Pilar Santolaria Blasco, Rosaura Pérez-Pé and Tamara Leahy

Received: 31 May 2021
Accepted: 25 June 2021
Published: 28 June 2021

Abstract: The aim was to determine the relationship between kinematic parameters of boar spermatozoa and fertility rates of sow, as well as to assess the effect of sperm clusters on the fertility capacity of the ejaculate. Semen samples were collected from 11 sexually mature boars. Samples were analyzed by an ISAS®v1 CASA-Mot system for eight kinematic parameters. Ejaculate clusters were characterized using multivariate procedures, such as principal factors (PFs) analysis and clustering methods (the k-means model). Four different ejaculate clusters were identified from two kinematic PFs which involved linear trajectory and velocity. There were differences ($p < 0.05$) between the sperm kinematic variables by sire line. There was no statistical difference ($p > 0.05$) between dam lines and ejaculate clusters in fertility variables. The discriminant ability of the different kinematics of sperm variables to predict litter size fertility was analyzed using receiver operating characteristics (ROC) curve analysis. Curvilinear velocity (VCL), average path velocity (VAP), amplitude of lateral head displacement (ALH), and beat-cross frequency (BCF) showed significant, albeit limited, predictive capacity for litter size fertility variables (range: 0.55–0.58 area under curve, AUC). The kinematic analysis of the ejaculates in clusters did not have a predictive capacity for litter size variables.

Keywords: sperm; cluster; sows; motility; CASA-Mot; artificial insemination

1. Introduction

Several countries have implemented frequent artificial insemination (AI) in pigs [1–3], which has been associated with increased litter size [4–6]. AI increases the rate of genetic progress because it is designed to capture the advantages of heterosis; it increases the feed efficiency, growth, and litter size while also reducing problems related to boars [7]. One of the most important advantages of AI is that it requires no relocation of the sire, and a single ejaculate can be used to inseminate 10 to 20 sows [8]. The semen sample should be diluted and preserved at 17 °C [9] in order to extend useful sperm life, durability, and quality [10–12].

The sperm quality needs to be evaluated to understand the fertility and genetic value of the boar. Because of this, boars are housed in an artificial insemination station to optimize management, health, and fertility [8,13]. Computer-assisted semen analysis (CASA) has been used to evaluate the motility and kinematic parameters of the ejaculate, and more than 70% total motility must be achieved for the ejaculate to be considered for insemination of sows [14]. CASA systems have a pre-determined set-up to assess some species [15,16], boars included, but configurations must be validated to assure the precision and accuracy of the results.

Motility evaluation is based on kinematic patterns [16,17]. Sperm motion is obtained by capturing consecutive frames of a posterior structuration of the trajectory [17–19]. CASA-Mot systems provide accuracy in the results of motility and kinematic parameters, compared with the subjective methods used in the past [15]. However, this analysis provides large volumes of data [20] that must be evaluated using multivariate statistics [21] to reduce the dimensionality of the kinematic variables. Furthermore, for accuracy in the assessment, some factors must be considered, such as software [22], capture fields [23], recording time [24], counting chambers [25,26], and frame rate [27–29]. Therefore, to reduce the variance of the results, a 20 μm depth for counting chambers in boars has been recommended because it provides adequate space to promote the movement of the spermatozoa [23,30].

Several authors have indicated that the ejaculate is composed of a heterogeneous population of spermatozoa according to their kinematic variables [31–34]. Sperm subpopulations have been described in many species [20,26,29,35–39]; nevertheless, the biological meaning in terms of the physiological functions associated with fertility is still being studied [40]. Even today, there are multiple clustering approaches that can be used to estimate sperm subpopulations in the ejaculate, and different procedures are analyzed to determine the statistical and biological relevance of these subpopulations [41,42]. This paper characterized ejaculates into clusters for description. This approach was based on the idea that we do not know which spermatozoa fertilized the sow, and which sperm subpopulation contained that cell.

The principal parameters to determine sow fertility are litter size (LS), piglets born alive [8,43], stillbirth or mummified piglets [8], and farrowing or conception rate [44–47]. Several studies have shown that sow fertility has a positive relationship with some kinematic parameters of boar sperm, such as curvilinear velocity (VCL), straight line velocity (VSL), and beat cross frequency (BCF) [8,43,48–50]. The aim of the present study was to determine the relationship between kinematic parameters of boar spermatozoa and fertility rates of multiparous sows, using a commercial CASA-Mot system, as well as to determine the effect of sperm clusters on the fertility capacity of the ejaculate.

2. Materials and Methods

2.1. Animals

The experiment was conducted at a commercial swine farm (Agropecuaria Los Sagitarios S.A., Alajuela, Costa Rica) in 2019–2020 in the northwest of Costa Rica (Río Cuarto, 10°20′32″ N, 84°12′55″ W, Alajuela, Costa Rica, Central America), following the laws and regulations for experiments on live animals in Costa Rica. This study was performed following ethical principles, and with the approval of the Committee of Centro de Investigación y Desarrollo de la Agricultura Sostenible para el Trópico Húmedo at the Costa Rica Institute of Technology (CIDASTH-ITCR), according to Section 08/2020, article 1.0, DAGSC-100-2020.

Eleven sexually mature and healthy boars from two commercial terminal sire lines (SL: Duroc × Pietrain (n = 8) and Pietrain boars (n = 3)), 23.3 ± 8.5 months of age at the beginning of the experiment and with known fertility, were used as semen donors in this study. For the study, breeding boars were housed individually in well-ventilated pens with average temperature of 25.60 ± 2.94 °C during the time of the experiment. Ejaculates were collected in rainy season. Females came from four crossbred dam genetic lines (DL: York (Y), Landrace (L) and Pietrain (P); with the following crossing schemes YLP-50 ($\frac{1}{4}$ Y × $\frac{1}{4}$ L × $\frac{1}{2}$ P), YLP-75 ($^1/_8$ Y × $^1/_8$ L × $^3/_4$ P), YLP-87.5 ($^1/_{16}$ Y × $^1/_{16}$ L × $^7/_8$ P); and Y-L-50 ($\frac{1}{2}$ Y × $\frac{1}{2}$ L)). All the females were bred at the farm and they came from within maternal crossing schemes such as the continuous 3-generation cross between YLP hybrid sows and P boars. Mean sow parity for all dam genetic lines was 4.10 ± 2.76. The animals were fed with the standard breeder mixture (made on the farm), containing maize, soybean meal, mineral mixture, and common salt, to fulfill their nutrient requirements [51]. Concentrate was provided to pregnant sows—2.5 kg in the first 2/3 of gestation and 3 kg in the final third—and males consumed 2.5 kg per day and were provided with water ad libitum.

2.2. Fertility Trial

A total of 816 triple artificial inseminations performed with homospermic ejaculates were evaluated in 272 sows. These AIs were conducted randomly with 40 ejaculates from 11 males. Each genetic line of females was randomly inseminated with each genetic line of males. The ejaculates were used within 3 days of collection, and only those used to inseminate at least three females were evaluated. The mean number of inseminated sows per boar was 24.7 ± 10.1 females. All crossbred dam lines were inseminated with seminal doses from each sire line. Fertility rate was measured as females pregnant/total number of females inseminated (%). Total piglets born per litter (TPB), piglets born alive (PBA), stillbirth (SB), number of mummies (MP), and litter weight (LW; kg) were used as fertility variables, and those parameters were measured at the farrowing time.

2.3. Collection and Examination of Semen

Semen samples were collected in the morning, once per week, using the "gloved-hand" technique [52] and immediately placed in a water bath at 37 °C in the farm laboratory. In all cases, the sperm-rich fractions were collected and diluted 1:1 (vol:vol) by one-step with a commercial extender (Zoosperm ND5; Import-Vet, Barcelona, Spain). Insemination doses contained 3.7 ± 1.3 × 10^9 spermatozoa. From each boar, 3.64 ± 0.81 ejaculates were obtained. Samples from each ejaculate were evaluated for motility, and only ejaculates with at least 70% motile spermatozoa and 85% morphologically normal spermatozoa were used. The concentration was measured with Spermacue (Minitube, GmbH, Tiefenbach, Germany) following established protocols [24]. Samples were stored at 17 °C and were then transported to the laboratory in the same refrigerated conditions (17 °C) used for commercial distribution. A volume of one milliliter (1 mL) of mixed samples was placed in an Eppendorf® tube (Sigma-Aldrich, St. Louis, MO, USA) and maintained at 37 °C for 30 min before use.

2.4. Assessment of Sperm Variables

For the analysis of motility, ISAS® D4C20 disposable counting chambers (Proiser R+D., Paterna Spain) were used after being pre-warmed at 37 °C. After thorough mixing of the diluted semen samples, a volume of 2.7 μL was distributed along the counting chamber race by capillarity to fill it completely. Analyses were conducted using the CASA-Mot system ISAS®v1 (Integrated Semen Analysis System, Proiser R+D, Paterna, Spain) fitted with a video-camera (Proiser 782M, Proiser R+D), with 25 frames acquired per field at a frame rate of 50 Hz and final resolution of 768 × 576 pixels. The camera was attached to a microscope UB203 (UOP/Proiser R+D) with a 1× eyepiece and a 10× negative phase contrast objective (AN 0.25), and an integrated heated stage maintained at a constant temperature of 37.0 ± 0.5 °C. The CASA settings used were a particle area between 10 and

80 μm^2 and connectivity of 11 μm. The percentage of total motile cells and progressive motility (%) corresponded to spermatozoa swimming forward quickly in a straight line. The following parameters defined progressive motility: straightness (STR, straightness index) ≥45% and average path velocity (VAP) ≥25 μm·s^{-1}, defined as the average velocity over the smoothed cell path. A single technician carried out the assessments of sperm morphology. Sperm were classified as having normal or abnormal morphologic features following WHO strict criteria [53]. A total of 200 sperm were analyzed per slide; 100 sperm from each of two different locations on the slide were assessed. If the difference between the percentage of normal sperm in the two areas was 5% or less, then the mean value was calculated.

2.5. Computerized Kinematics Analysis

The CASA analyses were performed in seven microscope fields on a total of at least 600 cells per sample. The CASA-Mot variables assessed in this study included straight-line velocity (VSL, μm·s^{-1}), corresponding to the straight line from the beginning to the end of the track; curvilinear velocity (VCL, μm·s^{-1}), measured over the actual point-to-point track followed by the cell; average path velocity (VAP, μm·s^{-1}), the average velocity over the smoothed cell path; amplitude of lateral head displacement (ALH, μm), defined as the maximum of the measured width of the head oscillation as the sperm swims; beat-cross frequency (BCF, Hz), defined as the frequency with which the actual track crosses the smoothed track in either direction; motility (%), defined as the percentage of total motile cells; and progressive motility (%), corresponding to spermatozoa swimming rapidly forward in a straight line. Three progression ratios, expressed as percentages, were calculated from the velocity measurements described above: linearity of forward progression (LIN = VSL/VCL·100), straightness (STR = VSL/VAP·100), and wobble (WOB = VAP/VCL·100).

2.6. Statistical Analysis

The data obtained from the evaluations of all ejaculates and fertility were analyzed by descriptive statistics. Distribution properties for all variables were also explored using histograms and probability plots.

2.6.1. Multivariate Procedures

A subset of data was created with the means per ejaculate of all eight kinematic variables. Multivariate procedures were performed to identify ejaculate clusters from this subset of sperm kinematic data. This approach was founded on the fact that we do not know which spermatozoa fertilized the sow, and which sperm subpopulation contained that cell. All the values for the kinematic variables were standardized to avoid any scale effect. A principal factor analysis (PFA) was performed on these data to derive a small number of linear combinations that still retained as much information as possible from the original variables. Prior communalities for this analysis were estimated from the maximum absolute correlation coefficient between each variable and any other. The number of principal factors (PF) to be extracted was determined from the Kaiser criterion, namely by selecting only those with an eigenvalue >1. The KMO (Kaiser-Meyer-Olkin) statistic was also obtained [21] as a measure of dataset adequacy for factor extraction. As a rotation method, the varimax method with Kaiser normalization was used [54]. Correlations between factors and original kinematic variables were explored to better understand the meaning of the factors extracted.

Further, an analysis was conducted to classify the ejaculates into a reduced number of clusters, based on scores obtained from factor analysis. This was accomplished in two phases, combining hierarchical and non-hierarchical clustering procedures. First, factor scores for all ejaculates were clustered hierarchically using the Ward Minimum Variance method [55]. From this analysis, an optimal number of clusters was determined based on criteria such as the Cubic Clustering Criterion (CCC), Pseudo-T, Pseudo-F, and partial R^2.

Second, the optimal number of clusters obtained in the previous analysis was used as the target number of clusters in a non-hierarchical K-means cluster analysis [56].

ANOVA was further applied to evaluate statistical differences between clusters for all kinematic variables. The threshold for significance was defined as $p < 0.05$. Further, pairwise comparison between cluster means were performed by the Tukey-Kramer test. Results were presented as mean ± standard deviation of the mean. All data were analyzed using the SAS 9.4 [57] statistical program.

2.6.2. GLMM Model on Sow Fertility Parameters

Sow fertility variables were analyzed using the Generalized Linear Mixed Models (GLMM). The response variables were litter weight, litter size, piglets born alive, stillbirth, and number of mummies. A normal distribution with an identity link function was assumed for litter weight, while a Poisson distribution with a log identity link function was assumed for all other response variables. Ejaculate clusters, obtained from multivariate analysis, was considered as the main fixed independent factor in the model. Other fixed factors with potential effects on sow fertility were also added to the model, such as dam line, sire line, dam × sire line interaction, nested boar within sire line, month of farrowing, pregnancy length, different parities, and length of period between previous and present ejaculate. A random residual effect was also added to the model to account for correlations between different ejaculates obtained from the same boar. GLMM analysis was performed with the SAS 9.4 [57] statistical program.

2.6.3. ROC Analysis

The diagnostic test with a dichotomous outcome (positive/negative fertility test results) of the different kinematic semen variables to predict litter size fertility was analyzed using receiver operating characteristic (ROC) curve analysis. The approach of diagnostic test evaluation uses sensitivity and specificity as measures of accuracy of the test, in comparison with standard status (farrowing). The sensitivity (true positive rate) and specificity (true negative rate) of each kinematic variable vary across the different thresholds, and the sensitivity was inversely related with specificity. The plot of sensitivity versus 1-Specifity is called receiver operating characteristic (ROC) curve and the area under the curve (AUC). AUC varies from 0.5 (test with no discriminatory ability) to 1 (perfect discriminatory ability). An ROC was also used to calculate the elective breaking point (cut-off value) for each kinematic sperm variable. The analysis may also be used to determine the optimal cut-off value (optimal decision threshold).

3. Results

3.1. *Descriptive Analysis of Semen Evaluation*

Sperm concentration, volume of semen, and total spermatozoa in the ejaculate were 374.23 ± 129.24 × 10^6/mL, 231.98 ± 63.08 mL and 82.04 ± 23.73 × 10^9, respectively. The sperm concentration (million/mL) was 378.63 ± 134.98 in the Duroc × Pietrain crossbred and 361.00 ± 112.35 in the Pietrain. Total motility (%) of boar samples was 77.36 ± 11.17, with an overall range of 35.05–93.69%. The progressive motility of sperm (%) was 63.76 ± 11.96. Average total motility (%) for Duroc × Pietrain and Pietrain boars was 81.28 ± 7.76 and 65.61 ± 11.73, respectively ($p < 0.05$). The progressive motility (%) was 67.00 ± 10.05 (Duroc × Pietrain) and 54.04 ± 12.19 (Pietrain) ($p < 0.05$).

3.2. *Analysis of the Ejaculate Cluster Structure*

Principal factors analysis indicated a KMO statistic of 0.56, and final communality estimates were above 0.85 for all kinematic variables, except BCF (0.05). According to Kaiser criterion, two significant PF can be extracted from these data, both accounting for 98% of the total variance. The first PF, defined as linear trajectory (PF1), was responsible for 53% of the variance and was mainly associated with the kinematic variables LIN, STR, WOB, and VSL, with the largest correlation being for LIN (0.99). The second PF, defined as velocity (PF2), was strongly associated with the variables VCL, VAP, VSL, and ALH, with the largest correlation being for VCL (0.98) (Table 1). This factor also indicated that ejaculates whose sperm presented a linear trajectory had a relatively greater effect on the total variance than the ejaculates where sperm velocity was faster (Figure 1).

Table 1. Correlations between boar sperm kinematic variables (unrotated solution) and latent factors (PF1, PF2) *.

Variable	PF1	PF2
LIN	0.99	
STR	0.92	
WOB	0.89	
VSL	0.75	0.60
VCL		0.98
ALH		0.91
VAP		0.86
BCF		
Var Exp (%)	**53.1**	**44.9**

Var Exp: variance explained in each PF. Total variance explained: 98.0%. * Expresses the more important variables in each PF. Only eigenvectors >0.6 are presented. VCL: curvilinear velocity; VSL: straight-line velocity; VAP: average path velocity; LIN: linearity of forward progression; STR: straightness; WOB: wobble; ALH: amplitude of lateral head displacement; BCF: beat-cross frequency. Bold is recommended for the appreciation of the total variance explained.

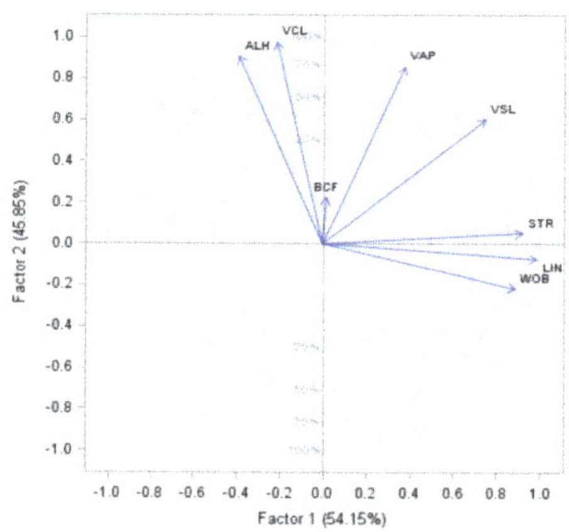

Figure 1. Distribution of factor loading of kinematic variables for boar spermatozoa on the plane conformed by two first principal factors (PFs with % variance explained). VCL: curvilinear velocity; VSL: straight-line velocity; VAP: average path velocity; LIN: linearity of forward progression; STR: straightness; WOB: wobble; ALH: amplitude of lateral head displacement; BCF: beat-cross frequency.

The optimal clustering level is obtained when local peaks of CCC, a high value of Pseudo F, and a low value of Pseudo T2 are combined together with a high value (Figure 2). This was approximately at group level 3; however, level 4 could be better by group location. The stability and accuracy of the grouping by cluster was contrasted by the coefficient of determination ($R^2 = 0.75$). The model was adjusted in the cluster procedures analysis with R^2 in each repetition for better validation (Figure 3).

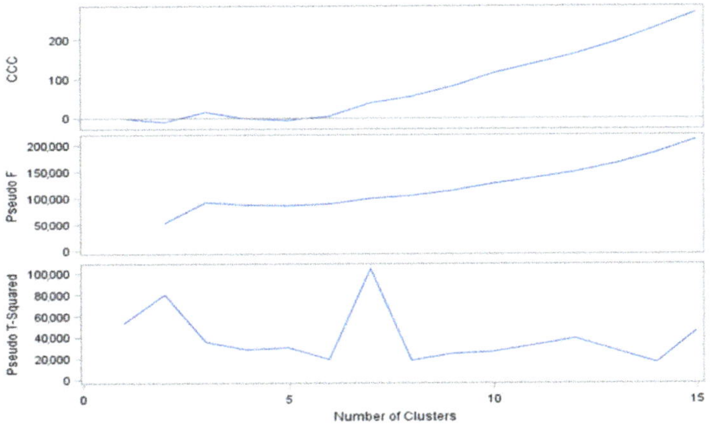

Figure 2. Optimal number of groups (ejaculate clusters) based on the statistical criteria of Cubic Clustering Criterion (CCC), Pseudo-T, Pseudo-F, and partial R^2.

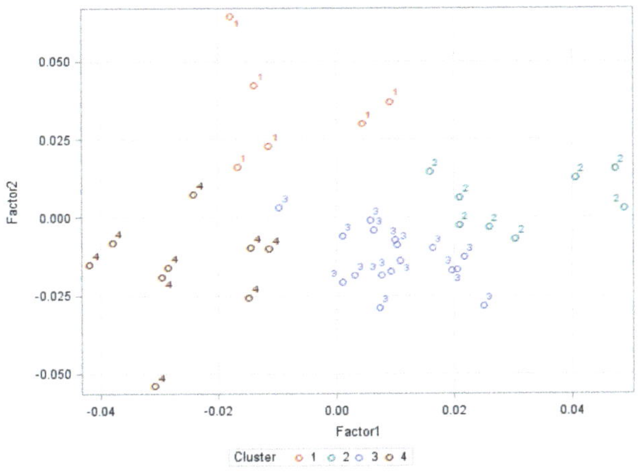

Figure 3. Distribution of ejaculates according to the principal factors value and prediction ellipse of the clusters (sperm populations).

Boar ejaculates were grouped into four clusters according to hierarchical Ward's minimum variance, followed by non-hierarchical k-means clustering procedures (Table 2). The kinematic parameters characterized the sperm movement in the ejaculate clusters (ECs). Cluster 1 (EC1) contained the sperm with the highest VCL and VAP (92.08 ± 5.12 μm·s^{-1}; 51.79 ± 3.28 μm·s^{-1} respectively). These ejaculates present a spermatozoa with highest BCF and ALH (8.93 ± 0.59 Hz; 3.46 ± 0.23 μm respectively). Cluster 2 (EC2) included sperm characterized by high VSL (42.91 ± 2.76 μm·s^{-1}), and the highest values of LIN and STR (57.17 ± 3.59%; 82.34 ± 1.50% respectively). Cluster 3 (EC3) contained the ejaculates whose spermatozoa had a high oscillatory movement, indicated by WOB and BCF (63.72 ± 2.86%; 8.73 ± 0.52 Hz respectively), and an intermediate value of STR (75.72 ± 5.58%). Cluster 4 (EC4) exhibited ejaculates whose sperm were less linear and progressive, as indicated by the lowest VSL and VAP (24.00 ± 4.68 μm·s^{-1}; 37.21 ± 5.56 μm·s^{-1} respectively), together with the lowest values of LIN and STR (37.26 ± 5.37%; 65.48 ± 6.66% respectively).

Table 2. Kinematic variables (mean ± SD) of the four ejaculate clusters (ECs).

Variable	EC1	EC2	EC3	EC4
VCL	92.08 ± 5.12 [a]	75.19 ± 4.54 [b]	65.48 ± 6.07 [c]	68.57 ± 9.98 [b,c]
VSL	38.09 ± 6.52 [a]	42.91 ± 2.76 [a]	31.69 ± 3.48 [b]	24.00 ± 4.68 [c]
VAP	51.79 ± 3.28 [a]	50.16 ± 3.43 [a]	41.30 ± 3.43 [b]	37.21 ± 5.56 [b]
LIN	42.23 ± 6.68 [c]	57.17 ± 3.59 [a]	49.53 ± 4.49 [b]	37.26 ± 5.37 [c]
STR	71.44 ± 7.51 [b,c]	82.34 ± 1.50 [a]	75.72 ± 5.58 [b]	65.48 ± 6.66 [c]
WOB	57.49 ± 4.54 [b]	67.35 ± 3.61 [a]	63.72 ± 2.86 [a]	55.45 ± 3.75 [b]
ALH	3.46 ± 0.23 [a]	2.73 ± 0.17 [b]	2.47 ± 0.20 [b]	2.70 ± 0.35 [b]
BCF	8.93 ± 0.59	8.36 ± 0.43	8.73 ± 0.52	8.44 ± 0.87

EC1: rapid, progressive and undulatory; EC2: medium velocity and linear progressive; EC3: slow, progressive and undulatory; EC4: slow velocity, nonlinear and non-progressive. Number of ejaculates = 40. VCL: curvilinear velocity ($\mu m \cdot s^{-1}$); VSL: straight line velocity ($\mu m \cdot s^{-1}$); VAP: average path velocity ($\mu m \cdot s^{-1}$); LIN: linearity of forward progression (%); STR: straightness (%); WOB: wobble (%); ALH: amplitude of lateral head displacement (μm); BCF: beat-cross frequency (Hz). SD: standard deviation. [a–c] Different letters indicate differences between ejaculate clusters. $p < 0.05$.

3.3. Relationship between Kinematics Cluster and Fertility

The mean fertility rate was 69.60 ± 21.67%. There were no differences between sire lines for this variable. There were differences ($p < 0.05$) between the sperm kinematic variables by sire line. The kinematic variables of semen of the Pietrain boars showed more linear trajectories of the spermatozoa, whereas semen from Duroc × Pietrain boars were characterized by curvilinear velocity and oscillatory movement of the sperm (Table 3). Fertility results based on the mean values of each cluster indicate differences ($p < 0.05$) between ejaculate clusters for total piglets born per litter (TPB), stillbirth (SB), and litter weight (LW; kg). There were no significant differences between ejaculate clusters ($p > 0.05$) for piglets born alive (PBA) and number of mummies (MP). However, some trends were observed. EC3 had higher PBA (10.25 ± 1.32), but higher MP (0.32 ± 0.17); moreover, LW was also higher (18.83 ± 1.71 kg). EC2 was characterized by intermedium fertility rates in all categories, exhibiting a higher value in PBA (9.55 ± 1.28) than EC1 and EC4. Finally, EC4 had the lightest litter (11.47 ± 2.03 kg); because of this, EC4 presents the lowest TPB (7.64 ± 1.13), and also reported the lowest SB (0.12 ± 0.07) and intermedium MP (0.19 ± 0.14). The fertility variables characterized according to ECs indicate that EC3 showed the highest litter size, while EC4 presented the lowest value of total piglets born per litter. EC2 and EC3 presented higher values of stillbirth than EC1 and EC4 (Table 4). The fertility variables characterized according to sire genetic line did not show differences ($p > 0.05$), except for significantly fewer stillbirths in Pietrain boars (Table 5).

Table 3. Kinematic variables (mean ± SEM) of boar ejaculates by sire genetic line.

Variable	Sire Line	
	Pietrain	Duroc × Pietrain
VCL	71.10 ± 0.18 [a]	76.15 ± 0.06 [b]
VSL	39.32 ± 0.11 [a]	32.25 ± 0.04 [b]
VAP	47.51 ± 0.11 [a]	43.70 ± 0.04 [b]
LIN	55.34 ± 0.14 [a]	43.68 ± 0.05 [b]
STR	80.03 ± 0.12 [a]	72.02 ± 0.04 [b]
WOB	67.58 ± 0.08 [a]	58.57 ± 0.03 [b]
ALH	2.61 ± 0.01 [a]	2.93 ± 0.02 [b]
BCF	8.16 ± 0.01 [a]	8.63 ± 0.02 [b]

Number of ejaculates = 40. VCL: curvilinear velocity ($\mu m \cdot s^{-1}$); VSL: straight line velocity ($\mu m \cdot s^{-1}$); VAP: average path velocity ($\mu m \cdot s^{-1}$); LIN: linearity of forward progression (%); STR: straightness (%); WOB: wobble (%); ALH: amplitude of lateral head displacement (μm); BCF: beat-cross frequency (Hz). SEM: standard error of the mean. [a–b] Different letters indicate differences between sire lines. $p < 0.05$.

Table 4. Fertility variables in pigs (mean ± SEM) by cluster of boar ejaculates.

Cluster of Ejaculates	Total Born per Litter	Piglets Born Alive	Stillbirth	Number of Mummies	Litter Weight at Birth (kg)
EC1	9.22 ± 1.21 [a,b]	8.91 ± 1.25 [a]	0.17 ± 0.09 [a]	0.08 ± 0.07 [a]	15.00 ± 1.88 [a,b]
EC2	10.37 ± 1.28 [a,b]	9.55 ± 1.28 [a]	1.33 ± 0.50 [b]	0.18 ± 0.09 [a]	15.72 ± 1.78 [a,b]
EC3	11.50 ± 1.37 [a]	10.25 ± 1.32 [a]	1.50 ± 0.61 [b]	0.32 ± 0.17 [a]	18.83 ± 1.71 [a]
EC4	7.64 ± 1.13 [b]	7.22 ± 1.14 [a]	0.12 ± 0.07 [a]	0.19 ± 0.14 [a]	11.47 ± 2.03 [b]

SEM: standard error of the mean. [a–b] Different letters indicate differences between clusters. $p < 0.05$.

Table 5. Porcine fertility variables (mean ± SEM.) by sire genetic line (percentage variation with respect to Pietrain in brackets).

Sire Line	Total Born per Litter	Piglets Born Alive	Stillbirth	Number of Mummies	Litter Weight at Birth (kg)
Pietrain	8.93 ± 0.93	8.74 ± 0.97	0.28 ± 0.11 [b]	0.13 ± 0.09	13.88 ± 1.45
Duroc × Pietrain	10.27 ± 0.65 (115.01%)	9.08 ± 0.61 (103.89%)	0.74 ± 0.16 [a] (264.29%)	0.24 ± 0.07 (184.61%)	15.58 ± 0.90 (112.2%)

SEM: standard error of the mean. [a–b] Different letters indicate differences between sire lines. $p < 0.05$.

3.4. Fertility Variables by Dam and Sire Genetic Line

Fertility parameters did not show significant differences ($p > 0.05$) between dam lines. The hybrid (Y-L-50) presents the highest value in TPB (10.44 ± 0.96), PBA (9.45 ± 0.92), and SB (0.65 ± 0.21). YLP-87.5 showed the lowest MP (0.13 ± 0.08), followed by Y-L-50 (0.15 ± 0.11). The heaviest litter was for YLP-75 (15.53 ± 1.32 kg), which represents 103.22% of the value of Y-L-50 (Figure 4). The Pietrain sire line had fewer SBs than Duroc × Pietrain (0.74 ± 0.16), but all the other parameters showed no differences ($p > 0.05$). Pietrain had lower values of MP (0.13 ± 0.09), but Duroc × Pietrain presented a higher value of TPB (10.27 ± 0.65) and PBA (9.08 ± 0.61) (Figure 4).

3.5. Predictive Capacity of Fertility

The sperm kinematic variables with significant results in the ROC curve analysis are presented in Table 6. VCL, VAP, ALH, and BCF showed significant, albeit limited, predictive capacity for litter size fertility variables (range: 0.55–0.58 AUC). Cut-off values, with their sensitivities and specificities, are also presented in Table 6. The best cut-off points to identify ejaculates with low fertility potential in relation to number of mummies were 70.70 $\mu m \cdot s^{-1}$ VCL, 42.40 $\mu m \cdot s^{-1}$ VAP, and 2.59 μm ALH. Similarly, ejaculate clusters showed limited predictive capacity for litter size variables (data not shown).

Table 6. Cut-off values of kinematic sperm variables significantly related to litter size fertility, calculated from receiver operating characteristic (ROC) curves.

Variable	Cut-Off Value	Sensitivity (%)	Specificity (%)	Area ROC	*p*-Value
		Total Born per Litter			
VCL	71.30	56.38	52.58	0.56	0.06
ALH	2.64	50.00	53.33	0.57	0.04
BCF	8.56	55.32	51.55	0.55	0.12
		Piglets Born Alive			
VCL	71.50	53.19	55.67	0.56	0.06
ALH	2.64	53.57	50.47	0.58	0.03
		Number of Mummies			
VCL	70.70	56.38	52.58	0.57	0.08
VAP	42.40	54.35	50.51	0.57	0.09
ALH	2.59	50.00	53.33	0.57	0.07

VCL: curvilinear velocity ($\mu m \cdot s^{-1}$); VAP: average path velocity ($\mu m \cdot s^{-1}$); ALH: amplitude of lateral head displacement (μm); BCF: beat-cross frequency (Hz).

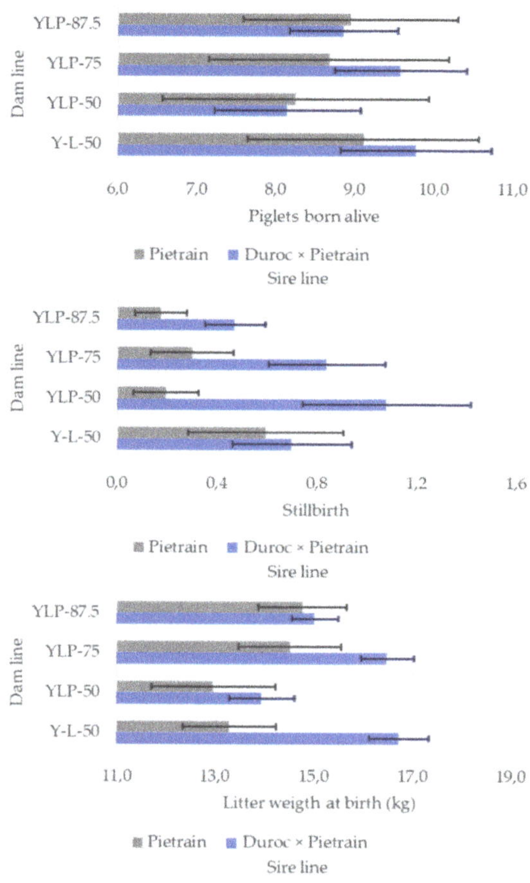

Figure 4. Dam genetic line effect on the values of piglets born alive, stillbirth, and litter weight at birth (kg). Y: York, L: Landrace, P: Pietrain, YLP-50 = ($\frac{1}{4}$ Y × $\frac{1}{4}$ L × $\frac{1}{2}$ P), YLP-75 = ($^1/_8$ Y × $^1/_8$ L × $^3/_4$ P), YLP-87.5 = ($^1/_{16}$ Y × $^1/_{16}$ L × $^7/_8$ P), Y-L-50: $^1/_2$ Y × $^1/_2$ L. $p < 0.05$.

4. Discussion

The number of sows inseminated by a single ejaculate depends on the concentration and quality of the semen [46]. This quality is measured by microscopic analysis of the samples, which allows maximization of the number of doses per boar [23]. The selection criteria of the genetic line or breed must be taken in the context of the farm's purpose [58] and the semen quality [59], because the fertility of the sows is related to semen characteristics, such as kinematic variables [8,43,48–50], morphometric values [60], DNA fragmentation [61], concentration [62], and viability [63]. Our results indicate a mean concentration of 3.7×10^9 spermatozoa per AI dose. Our multivariate analysis found four ejaculate clusters, where EC1 was characterized by rapid and progressive spermatozoa; this cluster had only ejaculates from Duroc × Pietrain boars. Moreover, EC4 was mainly ejaculates provided by Duroc × Pietrain boars, and was characterized by slow and non-progressive sperm. The major proportion of Pietrain ejaculates was in EC3; this cluster had ejaculates with slow and progressive spermatozoa. In EC2, both sire lines were better distributed, and the ejaculates were characterized by medium velocity and linear progressively motile spermatozoa.

Some authors have found that sperm parameters, such as motile spermatozoa, do not predict litter size [64,65]. Other studies have found similar results with respect to the effect of motile spermatozoa on fertility of litter size, indicating that when total motile spermatozoa is more than 60% and there is a concentration of 3×10^9 spermatozoa per AI dose, no relationship with fertility parameters is found [66,67]. This study showed that Duroc × Pietrain boars have better patterns in motile and progressively motile spermatozoa than Pietrain boars. These data could indicate that Duroc × Pietrain boars have the best fertility data, and the results of TPB showed that Duroc × Pietrain had the highest value, and that the value was relevant. Several authors have asserted that the kinematic values are related to fertility data of the sows, such as litter size [8,43,48], pregnancy rate [49,50], and fertility index [50]. The most related variables of kinematics are VSL [8,48], VCL [8,43,50], and BCF [8,49,50]. Our results showed that EC1 had the highest values of VCL and BCF, and an intermediary value of VSL. Multivariate procedures showed that the sire line had no effect ($p > 0.05$) on the fertility data. Non-relevance of the difference between boar racial groups could be due to the fact that the clusters are very close in Euclidean distance with respect to the centroids, and some ejaculates were assigned to a specific group (cluster) but maintained some motile spermatozoa and/or kinetic patterns similar to ejaculates from another cluster. The main obstacle to estimating the fertility of the boar is that it is necessary for a large number of sows to be inseminated over a long period of time, requiring an extended time period for the study, and during this time the boar fertility could change [65]. In our study, we determined that the coefficient of variation in motile spermatozoa ranged from 1.48% to 24.26%, while the percentage of progressively motile spermatozoa ranged from 5.41% to 37.69%. Differences between boars could be due to individual variability, breed, and age [8,49,68]. The ejaculates in the present study showed within and between boar differences. Furthermore, we characterize the assessed ejaculates on the basis of variables such as sperm velocity and progressiveness, being able to describe both rapid and progressive ejaculates as well as slow and non-progressive ejaculates. In our study, we characterized ejaculates into clusters for description. Other studies have described subpopulations of spermatozoa [26,29,33,34,36–39,69–71]. However, we believed it more appropriate to regard them as ejaculates because we do not know which spermatozoa fertilized the sow, and which sperm subpopulation contained that cell. Fertility is multifactorial, with semen having an effect only in the final result; on the other hand, the sow has many more influencing factors. Among the main factors that influence sow fertility are weather [72], housing conditions [73], nutritional status [72], duration of gestation [74], endocrine activity [72], and sow lifetime productivity [75]. In the present study, four maternal crosses were considered because maternal ability can influence phenotypic performance in species with numerous litters, such as pigs [76–78]. In the case of boars, two racial groups were used because of the selection objective in terminal crossings [79–81]. The results indicated that the sows of line Y-L-50 inseminated with Duroc × Pietrain boars presented better litter size yields; this is explained by the genetic potential provided by heterosis and complementarity [82]. There was no difference ($p > 0.05$) between dam lines in the fertility variables, even though those values showed highly variable percentages compared to Y-L-50. However, we suggest that there is biological and economic importance because of the genetic improvement achieved by selecting sows and boars to increase the number of live born piglets and the survival proportion [83]. The prenatal survival of pigs is of great importance [84], and this is linked to the quality of the oocyte [72], which is influenced by the nutrition of the sow [75]. To explain the relevance of the differences between ejaculate clusters, it is necessary to analyze the data; YLP-87.5 has the lowest values of SB and MP, therefore the value of PBA is intermediate even when it has a low TPB. On the other hand, Y-L-50 is the dam line with the highest TPB value, but it is also the one with the most SB, and therefore its PBA quantity is diminished. This study determined the probability of relevance from the Bayesian marginal posterior distribution to confront these findings with the significance $p < 0.05$ from each cluster of boar ejaculates, and in some cases indicated that differences between cluster could be considered irrelevant. Other studies carried out in other species, such as cattle, have described results that indicate that the differences between sperm subpopulations are not relevant [42]. On the other hand, the fertility variables showed differences between EC2 to EC4, which could be explained by the values of the sperm velocity variables and the velocity relationships. In this sense, EC2 presented the highest values of VSL, VAP, LIN, STR, and WOB, while EC4 was the cluster with the lowest values for the same variables. These results indicate the need to continue studying the cluster structure of pig ejaculates, and how these intervene in the fertility functionality of females.

5. Conclusions

We have shown that kinematic analysis of boar ejaculates reveals kinematically separate populations. There were differences between the sperm kinematic variables by sire line. However, there was no overall significant difference between dam lines assessed by multivariate procedures. The fertility variables characterized according to the sire genetic line did not show differences, except for significantly fewer stillbirths in Pietrain boars. Sperm kinematic variables may have a predictive capacity for litter size variables, albeit a limited one. Nevertheless, the analysis of the ejaculates into clusters did not have a predictive capacity for litter size variables.

Author Contributions: Conceptualization, A.V., E.R.S.R. and C.S.; methodology, V.B. and A.V.; software, V.B.; validation, V.B. and F.S.; formal analysis, M.C. and B.V.-L.; investigation, V.B., F.S., and A.V.; resources, A.V.; data curation, M.C. and B.V.-L.; writing—original draft preparation, V.B. and A.V.; writing—review and editing, V.B., A.V. and E.R.S.R.; visualization, A.V., M.C., E.R.S.R., and C.S.; supervision, A.V.; project administration, A.V.; funding acquisition, A.V. All authors have read and agreed to the published version of the manuscript.

Funding: This research was funded by Fundación para el Fomento y Promoción de la Investigación y Transferencia de Tecnología Agropecuaria de Costa Rica (FITTACORI), grant number F23-19 and the Costa Rica Institute of Technology (Vice-Chancellor's office of Research and Extension; VIE (Vicerrectoría de investigación y Extensión; Project-VIE-5402-2151-1015). The funders had no role in the design of the study; in the collection, analyses, or interpretation of data; in the writing of the manuscript, or in the decision to publish the results.

Institutional Review Board Statement: The study was conducted according to the approval of the ethical principles of the Committee of Centro de Investigación y Desarrollo de la Agricultura Sostenible para el Trópico Húmedo at the Costa Rica Institute of Technology (CIDASTH-ITCR) according to Section 08/2020, article 1.0, DAGSC-100-2020.

Informed Consent Statement: Not applicable.

Data Availability Statement: The data presented in this study are available within the article.

Acknowledgments: The authors thank the Costa Rica Institute of Technology (ITCR) and the Fundación para el Fomento y Promoción de la Investigación y Transferencia de Tecnología Agropecuaria de Costa Rica (FITTACORI) for financing this study. The authors are grateful to the staff of the Agropecuaria Los Sagitarios S.A. farm for supplying the boar ejaculates.

Conflicts of Interest: The authors declare no conflict of interest.

References

1. Waberski, D.; Petrunkina, A.M.; Töpfer-Petersen, E. Can external quality control improve pig AI efficiency? *Theriogenology* **2008**, *70*, 1346–1351. [CrossRef]
2. Maes, D.; López Rodríguez, A.; Rijsselaere, T.; Vyt, P.; Van Soom, A. Artificial Insemination in Pigs. In *Artificial Insemination in Farm Animals*; Manafi, E., Ed.; In Tech: Rijeka, Croatia, 2011; pp. 79–94.
3. Bonet, S.; Casas, I.; Holt, W.; Yeste, M. Boar reproduction: Fundamentals and new biotechnological trends. In *Boar Reproduction*; Springer-Verlag: Berlin/Heidelberg, Germany, 2013; p. 632. ISBN 9783642436260.
4. Myromslien, F.D.; Tremoen, N.H.; Andersen-Ranberg, I.; Fransplass, R.; Stenseth, E.B.; Zeremichael, T.T.; van Son, M.; Grindflek, E.; Gaustad, A.H. Sperm DNA integrity in Landrace and Duroc boar semen and its relationship to litter size. *Reprod. Domest. Anim.* **2019**, *54*, 160–166. [CrossRef]
5. Irgang, R.; Fávero, J.A.; Kennedy, B.W. Genetic parameters for litter size of different parities in Duroc, Landrace, and large white sows. *J. Anim. Sci.* **1994**, *72*, 2237–2246. [CrossRef] [PubMed]

6. Kemp, B.; Soede, N.M. Consequences of variation in interval from insemination to ovulation on fertilization in pigs. *J. Reprod. Fertil. Suppl.* **1997**, *52*, 79–89. [CrossRef]
7. Knox, R. Artificial insemination in pigs today. *Theriogenology* **2016**, *85*, 83–93. [CrossRef] [PubMed]
8. Broekhuijse, M.L.W.J.; Šoštarić, E.; Feitsma, H.; Gadella, B.M. Application of computer-assisted semen analysis to explain variations in pig fertility. *J. Anim. Sci.* **2012**, *90*, 779–789. [CrossRef]
9. Barranco, I.; Padilla, L.; Tvarijonaviciute, A.; Parrilla, I.; Martínez, E.A.; Rodriguez-Martinez, H.; Yeste, M.; Roca, J. Levels of activity of superoxide dismutase in seminal plasma do not predict fertility of pig AI-semen doses. *Theriogenology* **2019**, *140*, 18–24. [CrossRef] [PubMed]
10. Johnson, L.A.; Weitze, K.F.; Fiser, P.; Maxwell, W.M.C. Storage of boar semen. *Anim. Reprod. Sci.* **2000**, *62*, 143–172. [CrossRef]
11. Morrell, J.M. Artificial Insemination: Current and Future Trends. In *Artificial Insemination in Farm Animals*; Manafi, M., Ed.; INTECH: London, UK, 2011; ISBN 978-953-307-312-5.
12. Roca, J.; Parrilla, I.; Bolarin, A.; Martinez, E.A.; Rodriguez-Martinez, H. Will AI in pigs become more efficient? *Theriogenology* **2016**, *86*, 187–193. [CrossRef]
13. Lopez-Rodriguez, A.; Soom, A.V.; Arsenakis, I.; Maes, D. Boar management and semen handling factors affect the quality of boar extended semen. *Porc. Health Manag.* **2017**, *3*, 15. [CrossRef] [PubMed]
14. Flowers, W.L. Selection for boar fertility and semen quality–the way ahead. *Soc. Reprod. Fertil. Suppl.* **2009**, *66*, 67–78. [PubMed]
15. Amann, R.; Waberski, D. Computer-assisted sperm analysis (CASA): Capabilities and potential developments. *Theriogenology* **2014**, *81*, 5–17. [CrossRef] [PubMed]
16. Lu, J.C.; Huang, Y.F.; Lü, N.Q. Computer-aided sperm analysis: Past, present and future. *Andrologia* **2014**, *46*, 329–338. [CrossRef] [PubMed]
17. Valverde, A.; Madrigal-Valverde, M. Sistemas de análisis computadorizado de semen en la reproducción animal. *Agron. Mesoam.* **2018**, *29*, 449. [CrossRef]
18. Katz, D.F.; Dott, H.M. Methods of measuring swimming speed of spermatozoa. *J. Reprod. Fertil.* **1975**, *45*, 263–272. [CrossRef]
19. Mortimer, S.T. CASA-Practical aspects. *J. Androl.* **2000**, *21*, 515–524.
20. Valverde, A.; Madrigal-Valverde, M.; Castro-Morales, O.; Gadea-Rivas, A.; Johnston, S.; Soler, C. Kinematic and head morphometric characterisation of spermatozoa from the Brown Caiman (Caiman crocodilus fuscus). *Anim. Reprod. Sci.* **2019**, *207*, 9–20. [CrossRef] [PubMed]
21. Spencer, N.H. *Essentials of Multivariate Mata Mnalysis*; Chapman and Hall/CRC: New York, NY, USA, 2013; ISBN 9781466584792.
22. Amann, R.; Katz, D.F. Andrology Lab Corner: Reflections on CASA After 25 Years. *J. Androl.* **2004**, *25*, 317–325. [CrossRef]
23. Valverde, A.; Madrigal-Valverde, M. Evaluación de cámaras de recuento sobre parámetros espermáticos de verracos analizados con un sistema CASA-Mot. *Agron. Mesoam.* **2019**, *30*, 447–458. [CrossRef]
24. Valverde, A.; Madrigal-Valverde, M.; Lotz, J.; Bompart, D.; Soler, C. Effect of video capture time on sperm kinematic parameters in breeding boars. *Livest. Sci.* **2019**, *220*, 52–56. [CrossRef]
25. Bompart, D.; Vázquez, R.F.; Gómez, R.; Valverde, A.; Roldán, E.R.S.; García-Molina, A.; Soler, C. Combined effects of type and depth of counting chamber, and rate of image frame capture, on bull sperm motility and kinematics. *Anim. Reprod. Sci.* **2019**, *209*, 106169. [CrossRef]
26. Caldeira, C.; Hernández-Ibáñez, S.; Valverde, A.; Martin, P.; Herranz-Jusdado, J.G.; Gallego, V.; Asturiano, J.F.; Dzyuba, B.; Pšenička, M.; Soler, C. Standardization of sperm motility analysis by using CASA-Mot for Atlantic salmon (Salmo salar), European eel (Anguilla anguilla) and Siberian sturgeon (Acipenser baerii). *Aquaculture* **2019**, *502*, 223–231. [CrossRef]
27. Broekhuijse, M.L.W.J.; Šoštarić, E.; Feitsma, H.; Gadella, B.M. Additional value of computer assisted semen analysis (CASA) compared to conventional motility assessments in pig artificial insemination. *Theriogenology* **2011**, *76*, 1473–1486. [CrossRef]
28. Saravia, F.; Núñez-Martínez, I.; Morán, J.; Soler, C.; Muriel, A.; Rodríguez-Martínez, H.; Peña, F. Differences in boar sperm head shape and dimensions recorded by computer-assisted sperm morphometry are not related to chromatin integrity. *Theriogenology* **2007**, *68*, 196–203. [CrossRef]
29. Valverde, A.; Madrigal, M.; Caldeira, C.; Bompart, D.; de Murga, J.N.; Arnau, S.; Soler, C. Effect of frame rate capture frequency on sperm kinematic parameters and subpopulation structure definition in boars, analysed with a CASA-Mot system. *Reprod. Domest. Anim.* **2019**, *54*, 167–175. [CrossRef] [PubMed]
30. Castellini, C.; Dal Bosco, A.; Ruggeri, S.; Collodel, G. What is the best frame rate for evaluation of sperm motility in different species by computer-assisted sperm analysis? *Fertil. Steril.* **2011**, *96*, 24–27. [CrossRef]
31. Hirai, M.; Boersma, A.; Hoeflich, A.; Wolf, E.; Foll, J.; Aumüller, T.R.; Braun, J. Objectively measured sperm motility and sperm head morphometry in boars (Sus scrofa): relation to fertility and seminal plasma growth factors. *J. Androl.* **2001**, *22*, 104–110. [PubMed]
32. Thurston, L.; Watson, P.; Mileham, A.; Holt, W. Morphologically distinct sperm subpopulations defined by Fourier shape descriptors in fresh ejaculates correlate with variation in boar semen quality following cryopreservation. *J. Androl.* **2001**, *22*, 382–394.
33. Flores, E.; Taberner, E.; Rivera, M.M.; Peña, A.; Rigau, T.; Miró, J.; Rodríguez-Gil, J.E. Effects of freezing/thawing on motile sperm subpopulations of boar and donkey ejaculates. *Theriogenology* **2008**, *70*, 936–945. [CrossRef]

34. Ramió, L.; Rivera, M.M.; Ramírez, A.; Concha, I.I.; Peña, A.; Rigau, T.; Rodríguez-Gil, J.E. Dynamics of motile-sperm subpopulation structure in boar ejaculates subjected to "in vitro" capacitation and further "in vitro" acrosome reaction. *Theriogenology* **2008**, *69*, 501–512. [CrossRef]
35. Soler, C.; Contell, J.; Bori, L.; Sancho, M.; García-Molina, A.; Valverde, A.; Segarvall, J. Sperm kinematic, head morphometric and kinetic-morphometric subpopulations in the blue fox (Alopex lagopus). *Asian J. Androl.* **2017**, *19*, 154–159. [CrossRef]
36. Gallego, V.; Vílchez, M.C.; Peñaranda, D.S.; Pérez, L.; Herráez, M.P.; Asturiano, J.F.; Martínez-Pastor, F. Subpopulation pattern of eel spermatozoa is affected by post-activation time, hormonal treatment and the thermal regimen. *Reprod. Fertil. Dev.* **2015**, *27*, 529–543. [CrossRef]
37. Valverde, A.; Arenán, H.; Sancho, M.; Contell, J.; Yániz, J.; Fernández, A.; Soler, C. Morphometry and subpopulation structure of Holstein bull spermatozoa: Variations in ejaculates and cryopreservation straws. *Asian J. Androl.* **2016**, *18*, 851–857. [CrossRef]
38. Yániz, J.; Palacín, I.; Caycho, K.; Soler, C.; Silvestre, M.; Santolaria, P. Determining the relationship between bull sperm kinematic subpopulations and fluorescence groups using an integrated sperm quality analysis technique. *Reprod. Fertil. Dev.* **2018**, *30*, 919–923. [CrossRef] [PubMed]
39. Vásquez, F.; Soler, C.; Camps, P.; Valverde, A.; García-Molina, A. Spermiogram and sperm head morphometry assessed by multivariate cluster analysis results during adolescence (12-18 years) and the effect of varicocele. *Asian J. Androl.* **2016**, *18*, 824–830. [CrossRef] [PubMed]
40. Gomendio, M.; Roldan, E.R.S. Implications of diversity in sperm size and function for sperm competition and fertility. *Int. J. Dev. Biol.* **2008**, *52*, 439–447. [CrossRef]
41. Ibanescu, I.; Siuda, M.; Bollwein, H. Motile sperm subpopulations in bull semen using different clustering approaches – Associations with flow cytometric sperm characteristics and fertility. *Anim. Reprod. Sci.* **2020**, *215*, 106329. [CrossRef]
42. Víquez, L.; Barquero, V.; Soler, C.; Roldan, E.R.S.; Valverde, A. Kinematic Sub-Populations in Bull Spermatozoa: A Comparison of Classical and Bayesian Approaches. *Biology* **2020**, *9*, 138. [CrossRef]
43. Tremoen, N.H.; Gaustad, A.H.; Andersen-Ranberg, I.; van Son, M.; Zeremichael, T.T.; Frydenlund, K.; Grindflek, E.; Våge, D.I.; Myromslien, F.D. Relationship between sperm motility characteristics and ATP concentrations, and association with fertility in two different pig breeds. *Anim. Reprod. Sci.* **2018**, *193*, 226–234. [CrossRef]
44. Yamaguchi, S.; Funahashi, H. Effect of the addition of beta-mercaptoethanol to a thawing solution supplemented with caffeine on the function of frozen-thawed boar sperm and on the fertility of sows after artificial insemination. *Theriogenology* **2012**, *77*, 926–932. [CrossRef] [PubMed]
45. Fair, S.; Romero-Aguirregomezcorta, J. Implications of boar sperm kinematics and rheotaxis for fertility after preservation. *Theriogenology* **2019**, *137*, 15–22. [CrossRef] [PubMed]
46. Knox, R.; Esparza-Harris, K.C.; Johnston, M.E.; Webel, S.K. Effect of numbers of sperm and timing of a single, post-cervical insemination on the fertility of weaned sows treated with OvuGel®. *Theriogenology* **2017**, *92*, 197–203. [CrossRef]
47. De Rensis, F.; Kirkwood, R.N. Control of estrus and ovulation: Fertility to timed insemination of gilts and sows. *Theriogenology* **2016**, *86*, 1460–1466. [CrossRef]
48. Holt, C.; Holt, W.V.; Moore, H.D.M.; Reed, H.C.B.; Curnock, R.M. Objectivily measured boar sperm motility parameters correlate with the outcomes of on-farm inseminations: Results of two fertility trials. *J. Androl.* **1997**, *18*, 312–323. [CrossRef] [PubMed]
49. Schulze, M.; Ruediger, K.; Mueller, K.; Jung, M.; Well, C.; Reissmann, M. Development of an in vitro index to characterize fertilizing capacity of boar ejaculates. *Anim. Reprod. Sci.* **2013**, *140*, 70–76. [CrossRef]
50. Winters, R.A.; Hamilton, D.N.; Bhatnagar, A.S.; Fitzgerald, R.; Bovin, N.; Miller, D.J. Porcine sperm binding to oviduct cells and glycans as supplements to traditional laboratory semen analysis. *J. Anim. Sci.* **2018**, *96*, 5265–5275. [CrossRef]
51. National Research Council. *Nutrient Requirements of Swine*; National Academies Press: Cambridge, MA, USA, 2012.
52. Hancock, J.; Hovell, G. The collection of boar semen. *Vet. Rec.* **1959**, *71*, 664–665.
53. World Health Organization. *WHO Laboratory Manual for the Examination and Processing of Human Semen*; World Health Organization: Geneva, Switzerland, 2010; ISBN 9789241547789.
54. Kaiser, H.F. The varimax criterion for analytic rotation in factor analysis. *Psychometrika* **1958**, *23*, 187–200. [CrossRef]
55. Murtagh, F.; Legendre, P. Ward's Hierarchical Agglomerative Clustering Method: Which Algorithms Implement Ward's Criterion? *J. Classif.* **2014**, *31*, 274–295. [CrossRef]
56. Kaufman, L.; Rousseuw, P.J. Finding Groups in Data: An Introduction to Cluster Analysis. *Biometrics* **1991**, *47*, 788. [CrossRef]
57. Statistical Analysis Systems, SAS. *The SAS system for Windows, Release 9.4*; Statistical Analysis Systems Institute: Cary, NC, USA, 2013; p. 556.
58. Merks, J.; Ducro-Steverink, D.; Feitsma, H. Management and genetic factors affecting fertility in sows. *Reprod. Domest. Anim.* **2000**, *35*, 261–266. [CrossRef]
59. Morrell, J.M. Effect of colloid centrifugation on boar sperm quality during storage and function in in vitro fertilization. *Theriogenology* **2019**, *137*, 122–126. [CrossRef]
60. García-Herreros, M.; Aparicio, I.M.; Barón, F.J.; García-Marín, L.J.; Gil, M.C. Standardization of sample preparation, staining and sampling methods for automated sperm head morphometry analysis of boar spermatozoa. *Int. J. Androl.* **2006**, *29*, 553–563. [CrossRef]
61. Soler, C.; Cooper, T.; Valverde, A.; Yániz, J. Afterword to Sperm morphometrics today and tomorrow special issue in Asian Journal of Andrology. *Asian J. Androl.* **2016**, *18*, 895–897. [CrossRef] [PubMed]

62. Amann, R. Seminal Can Sample the Fertility Be Predicted Potential of a Accurately? *J. Androl.* **1989**, *10*, 89–98. [CrossRef]
63. Sutkeviciene, N.; Riskeviciene, V.; Januskauskas, A.; Zilinskas, H.; Andersson, M. Assessment of sperm quality traits in relation to fertility in boar semen. *Acta Vet. Scand.* **2009**, *51*, 53. [CrossRef]
64. Tsakmakidis, I.A.; Lymberopoulos, A.G.; Khalifa, T.A. Relationship between sperm quality traits and field-fertility of porcine semen. *J. Vet. Sci.* **2010**, *11*, 151–154. [CrossRef] [PubMed]
65. Gadea, J. Sperm factors related to in vitro and in vivo porcine fertility. *Theriogenology* **2005**, *63*, 431–444. [CrossRef] [PubMed]
66. Flowers, W.L. Management of boars for efficient semen production. *J. Reprod. Fertil. Suppl.* **1997**, *52*, 67–78.
67. Xu, X.; Pommier, S.; Arbov, T.; Hatchings, B.; Sotto, W.; Foxcroft, G.R. In Vitro Maturation and Fertilization Techniques for Assessment of Semen Quality and Boar Fertility. *J. Anim. Sci.* **1998**, *76*, 3079–3089. [CrossRef]
68. Flowers, W.L. Genetic and phenotypic variation in reproductive traits of AI boars. *Theriogenology* **2008**, *70*, 1297–1303. [CrossRef]
69. Ramón, M.; Jiménez-Rabadán, P.; García-Álvarez, O.; Maroto-Morales, A.; Soler, A.; Fernández-Santos, M.; Pérez-Guzmán, M.; Garde, J. Understanding Sperm Heterogeneity: Biological and Practical Implications. *Reprod. Domest. Anim.* **2014**, *49*, 30–36. [CrossRef] [PubMed]
70. Yániz, J.; Capistrós, S.; Vicente-Fiel, S.; Hidalgo, C.; Santolaria, P. A comparative study of the morphometry of sperm head components in cattle, sheep, and pigs with a computer-assisted fluorescence method. *Asian J. Androl.* **2016**, *18*, 840–843. [CrossRef]
71. Jodar, M.; Soler-Ventura, A.; Oliva, R. Semen proteomics and male infertility. *J. Proteomics* **2017**, *162*, 125–134. [CrossRef]
72. De Rensis, F.; Ziecik, A.J.; Kirkwood, R.N. Seasonal infertility in gilts and sows: Aetiology, clinical implications and treatments. *Theriogenology* **2017**, *96*, 111–117. [CrossRef] [PubMed]
73. Jansen, J.; Kirkwood, R.N.; Zanella, A.J.; Tempelman, R.J. Influence of gestation housing on sow behavior and fertility. *J. Swine Heal. Prod.* **2007**, *15*, 132–136.
74. Oliviero, C.; Kothe, S.; Heinonen, M.; Valros, A.; Peltoniemi, O. Prolonged duration of farrowing is associated with subsequent decreased fertility in sows. *Theriogenology* **2013**, *79*, 1095–1099. [CrossRef] [PubMed]
75. Patterson, J.; Foxcroft, G. Gilt management for fertility and longevity. *Animals* **2019**, *9*, 434. [CrossRef] [PubMed]
76. Canario, L.; Bidanel, J.P.; Rydhmer, L. Genetic trends in maternal and neonatal behaviors and their association with perinatal survival in french large white swine. *Front. Genet.* **2014**, *5*, 410. [CrossRef]
77. Yang, K.; Jeon, J.; Kwon, K.; Choi, H.; Kim, J.; Lee, J. Effect of different parities on reproductive performance, birth intervals, and tail behavior in sows. *J. Anim. Sci. Technol.* **2019**, *61*, 147–153. [CrossRef]
78. Wientjes, J.G.M.; Soede, N.M.; Knol, E.F.; van den Brand, H.; Kemp, B. Piglet birth weight and litter uniformity: Effects of weaning-to-pregnancy interval and body condition changes in sows of different parities and crossbred lines. *J. Anim. Sci.* **2013**, *91*, 2099–2107. [CrossRef]
79. Safranski, T.J. Genetic selection of boars. *Theriogenology* **2008**, *70*, 1310–1316. [CrossRef]
80. Chang, H.-L.; Lai, Y.-Y.; Wu, M.-C.; Sasaki, O. Genetic correlations between male reproductive traits and growth traits in growth performance tested Duroc, Landrace and Yorkshire breed boars. *Anim. Sci. J.* **2017**, *88*, 1258–1268. [CrossRef] [PubMed]
81. Do, D.N.; Strathe, A.B.; Jensen, J.; Mark, T.; Kadarmideen, H.N. Genetic parameters for different measures of feed efficiency and related traits in boars of three pig breeds1. *J. Anim. Sci.* **2013**, *91*, 4069–4079. [CrossRef] [PubMed]
82. Cassady, J.P.; Young, L.D.; Leymaster, K.A. Heterosis and recombination effects on pig reproductive traits. *J. Anim. Sci.* **2002**, *80*, 2303–2315. [CrossRef]
83. Knol, E.F.; Leenhouwers, J.I.; Van der Lende, T. Genetic aspects of piglet survival. *Livest. Prod. Sci.* **2002**, *78*, 47–55. [CrossRef]
84. Peltoniemi, O.A.T.; Björkman, S.; Oliviero, C. Parturition effects on reproductive health in the gilt and sow. *Reprod. Domest. Anim.* **2016**, *51*, 36–47. [CrossRef]

MDPI
St. Alban-Anlage 66
4052 Basel
Switzerland
Tel. +41 61 683 77 34
Fax +41 61 302 89 18
www.mdpi.com

Biology Editorial Office
E-mail: biology@mdpi.com
www.mdpi.com/journal/biology

www.ingramcontent.com/pod-product-compliance
Lightning Source LLC
Chambersburg PA
CBHW061519210926
51850CB00008B/1967
9783036584379